THE THEORY OF GROUPS

BY

HANS J. ZASSENHAUS

SECOND EDITION

CHELSEA PUBLISHING COMPANY
NEW YORK

Reprinted at New York, N. Y., 1974

Printed on 'long-life' acid-free paper

Library of Congress Catalog Card 56-13058

Library of Congress Classification QA171.Z312 1958

Dewey Decimal Classification 512.86

International Standard Book Number 0-8284-0053-9

Printed in the United States of America

PREFACE

Investigations published within the last fifteen years have greatly deepened our knowledge of groups and have given wide scope to group-theoretic methods. As a result, what were isolated and separate insights before, now begin to fit into a unified, if not yet final, pattern. I have set myself the task of making this pattern apparent to the reader, and of showing him, as well, in the group-theoretic methods, a useful tool for the solution of mathematical and physical problems.

It was a course by E. Artin, given in Hamburg during the Winter Semester of 1933 and the Spring Semester of 1934, which started me on an intensive study of group theory. In this course, the problems of the theory of finite groups were transformed into problems of general mathematical interest. While any question concerning a single object [e.g., finite group] may be answered in a finite number of steps, it is the goal of research to divide the infinity of objects under investigation into classes of types with similar structure.

The idea of O. Hölder for solving this problem was later made a general principle of investigation in algebra by E. Nöther. We are referring to the consistent application of the concept of homomorphic mapping. With such mappings one views the objects, so to speak, through the wrong end of a telescope. These mappings, applied to finite groups, give rise to the concepts of normal subgroup and of factor group. Repeated application of the process of diminution yields the composition series, whose factor groups are the finite simple groups. These are, accordingly, the bricks of which every finite group is built. How to build is indicated—in principle at least—by Schreier's extension theory. The Jordan-Hölder-Schreier theorem tells us that the type and the number of bricks is independent of the diminution process. The determination of all finite simple groups is still the main unsolved problem.

After an exposition of the fundamental concepts of group theory in Chapter I, the program calls for a detailed investigation of the concept of homomorphic mapping, which is carried out in Chapter II. Next, Chapter III takes up the question of how groups are put together from their simple components. According to a conjecture of Artin, insight into the nature of simple groups must depend on further research on p-groups. The elements of the theory of p-groups are expounded in

Chapter IV. Finally, Chapter V describes a method by which solvable factor groups may be split off from a finite group.

For the concepts and methods presented in Chapter II, particularly those in § 7, one may also consult v. d. Waerden, *Moderne Algebra* I (Berlin 1937). [English translation: *Modern Algebra,* New York, 1949]. The first part of Chapter III follows a paper by Fitting, while the proof of the basis theorem for abelian groups, and Schreier's extension theory, are developed on the basis of a course by Artin. The presentation of the theory of p-groups makes use of a paper by P. Hall. The section on monomial representations and transfers into a subgroup has also been worked out on the basis of a course by Artin. In addition one should consult the bibliography at the end of the book.

Many of the proofs in the text are shorter and—I hope—more transparent than the usual, older, ones. The proof of the Jordan-Hölder-Schreier theorem, as well as the proofs in Chapter IV, §§ 1 and 6, owe their final form to suggestions of E. Witt.

I am grateful to Messrs. Brandt, Fitting, Koethe, Magnus, Speiser, Threlfall and v. d. Waerden for their valuable suggestions in reading the manuscript. I also wish to thank Messrs. Hannink and Koluschnin for their help.

The group-theoretic concepts taken up in this book are developed from the beginning. The knowledge required for the examples and applications corresponds to the contents of, say, the book by Schreier and Sperner, *Analytische Geometrie und Algebra,* Part I (Leipzig, 1935) [English translation: *Introduction to Modern Algebra and Matrix* Theory, Chelsea Publishing Company, New York, 1951]. A historical introduction to group theory may be found in the book by Speiser, *Theorie der Gruppen von endlicher Ordnung* (Berlin, 1937).

I would suggest to the beginner that he familiarize himself first with Chapters I and II, Chapter III, §§ 1, 3, 4, 6, and 7, and Chapter IV, §§ 1 and 3, and also with the corresponding exercises. Then the program outlined in this preface will become clear to him.

Hans Zassenhaus

PREFACE TO THE SECOND EDITION

The revision made in this work consists almost entirely of additions to the material of the first edition. In particular, there have been added in Chapters I and II some remarks and exercises concerning semi-groups, and in Chapter II an introduction to the theory of lattices. The remainder of the new material is to be found in Appendixes A-G, each of which is closely related to one of the chapters of the book.

Since the appearance of the first edition of this work, lattice theory has been developed, by the combined efforts of O. Ore, A. Kurosh, J. von Neumann, G. Birkhoff, and others, into an independent discipline of modern mathematics. As a consequence, attention has been drawn to certain aspects of abstract group theory, in particular, by H. Wielandt's work on the lattice formed by the subnormal subgroups of a finite group. An account of the connections between lattice theory and group theory, which I consider promising for further investigation, has accordingly been added (see Appendix B).

Appendix C is an introduction to the theory of products and groups with generators and defining relations; the latter is one of the most powerful tools for the construction of groups, which is the main theme of Chapter III.

Many advanced exercises have been supplied illustrating both lattice-theoretical ideas and the extension of group-theoretical concepts to multiplicative domains (see Appendixes A, D, E, and F).

I am grateful to Professor C. Williams-Ayoub, Professor D. G. Higman, Professor B. Noonan, and to the editor for their valuable suggestions in reading the manuscript.

Finally, I wish to express my appreciation of the encouragement and assistance given me both by the Summer Research Institute of the Canadian Mathematical Congress at Kingston and by the Institute for Advanced Study at Princeton.

HANS ZASSENHAUS

McGILL UNIVERSITY
July, 1956

CONTENTS

I. ELEMENTS OF GROUP THEORY

§ 1. The Axioms of Group Theory

DEFINITION: A *group* is a set in which an operation called multiplication is defined under which there corresponds to each ordered pair x, y of elements of the set a unique third element z of the set. z is called the *product* of the factors x and y, written $z = xy$. For this multiplication we have

I. *The associative law:* $a(bc) = (ab)c$.

II. *The existence of a left identity* e *with the property* $ea = a$ *for all elements* a *of the group.*

III. *The solvability of the equation* $xa = e$ *for all elements* a *of the group.*

The associative law states that a product of three factors is determined solely by the order of its factors, its value being independent of the insertion of parentheses.

We assert: A product of arbitrarily many factors is determined solely by the order of its factors.

In order to prove this, let n be a number greater than three and assume the statement true for products of fewer than n factors. We write, for every $m < n$, a product of m factors $a_1, a_2, \ldots, a_m$ —in that order—as $P = a_1 \cdot a_2 \cdot \ldots \cdot a_m$ and have thus designated, unambiguously, an element of the group.

Now let P be a product of the n factors $a_1, a_2, \ldots, a_n$. After all of the parentheses have been removed except the last two pairs, P can be decomposed into two factors

$$P_1 = a_1 \cdot a_2 \cdot \ldots \cdot a_m$$

and

$$P_2 = a_{m+1} \cdot \ldots \cdot a_n,$$

with $0 < m < n$. We shall show that P is equal to the particular product $a_1 \cdot (a_2 \cdot \ldots \cdot a_n)$ and so we may assume $m > 1$. Then

$$\begin{aligned} P = P_1 P_2 &= (a_1 \cdot \ldots \cdot a_m)(a_{m+1} \cdot \ldots \cdot a_n) \\ &= (a_1(a_2 \cdot \ldots \cdot a_m))(a_{m+1} \cdot \ldots \cdot a_n) \\ &= a_1((a_2 \cdot \ldots \cdot a_m)(a_{m+1} \cdot \ldots \cdot a_n)) \\ &= a_1(a_2 \cdot \ldots \cdot a_n) . \end{aligned}$$

A non-empty system of elements in which multiplication is defined and is associative is called a *semi-group.*

For example the natural numbers form a semi-group under ordinary multiplication or addition as the operation.

The rational integers (positive, negative, and zero) form an additive group and a multiplicative semi-group. The rational numbers different from zero form a multiplicative group. All rational numbers form an additive group.

We assert that in a group every left unit e is also a right unit, (i.e., $ae = a$ holds for all group elements a.) In order to prove this, we solve $xa = e$ and $yx = e$. Then

$$(yx)a = ea = a$$
$$= y(xa) = ye = y(ee) = (ye)e = ae.$$

Similarly, $ye = y$, hence $y = a,\ ax = xa = e.$

We call one of the solutions of the equation $xa = e$ the *inverse element* of a and denote it by a^{-1}. Thus

$$aa^{-1} = a^{-1}a = e.$$

If $xa = b$, then on right multiplication by a^{-1}, it follows that

$$ba^{-1} = (xa)a^{-1} = x(aa^{-1}) = xe = x.$$

Conversely $ba^{-1} \cdot a = b \cdot a^{-1}a = be = b$. Thus the equation $xa = b$ has one and only one solution, $x = ba^{-1}$. Similarly it follows that the equation $ay = b$ has one and only one solution, $y = a^{-1}b$.

Multiplication in a group has a unique inverse.

The element e is called the identity or unit element of the group. It is uniquely determined as the solution of either of the equations $ax = a$ or $ya = a$. Similarly the inverse a^{-1} of the element a is uniquely determined as the solution of the equation $xa = e$ or $ay = e$.

The product of n equal factors a is denoted by a^n. Furthermore, if we set $a^0 = e,\ a^1 = a$ and $a^{-n} = (a^{-1})^n$ then the two power rules

$$a^n \cdot a^m = a^{n+m},$$
$$(a^n)^m = a^{nm},$$

are valid for arbitrary integral exponents n, m, as can be shown by induction.

Axioms II. and III. are not symmetric; they can be replaced by the two symmetric axioms:

II. a. *A group is non-empty.*

III. a. *Multiplication has an inverse, i.e., the equations*

$$xa = b$$

and

$$ay = b$$

are solvable for all pairs of elements a, b *of the group.*

Obviously II. a. is an immediate consequence of II., and III. a. follows from I.-III. If, conversely, I., II. a., III. a. are assumed, then we can find an element a in the given set and solve the equations $ea = a,\ ay = b$. From this it follows that

$$eb = e(ay) = (ea)y = ay = b$$

for all elements b.

Thus II. is valid. III. is a consequence of III. a.

A group which consists of a finite number of elements is called a *finite group*. The number of its elements is called its *order*. The order of an infinite group is defined to be zero.

In every group the *cancellation laws* hold:

III. b. $ax = ay$ *implies* $x = y$.
III. c. $xa = ya$ *implies* $x = y$.

THEOREM 1: *A finite semi-group in which the cancellation laws hold is a group.*

In order to prove this, let $a_1, a_2, \ldots, a_n$ be the finite number of elements and let a be a particular element. From III. b. it follows that the n elements $aa_1, aa_2, \ldots, aa_n$ are all distinct and so $ay = b$ is solvable for every pair a, b in the semi-group. The solvability of $xa = b$ follows similarly from the other cancellation law.

An abstract group is completely known if each of its elements is represented by a symbol and the product of any two symbols in any given order is exhibited.

The multiplication rule is given conveniently by a square table, in which the products in a row have the same left factor and the products in a column have the same right factor.

The multiplication tables of groups having at most three elements are the following:

Z_1

	e
e	e

Z_2

	e	a
e	e	a
a	a	e

Z_3

	e	a	b
e	e	a	b
a	a	b	e
b	b	e	a

The different multiplication tables of a group can be transformed into one another by row interchanges and column interchanges.

The existence of unique inverses is equivalent to the fact that each group element occurs exactly once in every row and column.

In order to exhibit[1] the associative law we agree to put the unit element of the group in the upper left corner of the square table. If we call the row starting with a , the a -row, and the column headed by b the b -column, then we find the product of a by b at the intersection of the a -row and b -column. The initial elements of each row and column may thus be omitted.

A table, constructed as above, is called *normal,* if in addition every element of the main diagonal is the identity element of the group. For example, the normal multiplication tables for groups of four and five elements are as follows:

Z_4

e	a	b	c
c	e	a	b
b	c	e	a
a	b	c	e

D_4

e	a	b	c
a	e	c	b
b	c	e	a
c	b	a	e

Z_5

e	a	b	c	d
d	e	a	b	c
c	d	e	a	b
b	c	d	e	a
a	b	c	d	e

The element a_{ik} at the intersection of i-th row and the k -th column is $a_{i1}a_{k1}^{-1}$, so that the rectangle rule

$$a_{ik}a_{kl} = a_{il}$$

holds. This may be seen from the following section of the table:

$$\begin{array}{ccc} a & \cdots\cdots & ab \\ \vdots & & \vdots \\ e & \cdots\cdots & b \end{array}$$

The rectangle rule is equivalent to the associative law.

The problem of abstract group theory is to examine all multiplication tables in which Axioms I.-III. are satisfied.

§ 2. Permutation Groups

For finite groups, the problem stated at the close of the last section can be solved by trial. For example, it can easily be established that the

[1] Brandt, Über eine Verallgemeinerung des Gruppenbegriffs, *Math. Ann.* 96 (1927) p. 365.

only multiplication tables for groups whose order is at most five are those which we have given previously. We can see, however, even from these first examples, that the direct verification of the associative law is time-consuming.

We must look about for more serviceable realizations of abstract groups. Naturally we require that the multiplication table be determined easily from the realization. An example of a domain in which arbitrary abstract groups can be realized is the group of permutations of a set of objects.

We denote single-valued mappings of a given set $\mathfrak{M}$ onto itself by lower case Greek letters, and elements of the set itself by lower case Roman letters. Let πx be the image of x under the mapping π. Any two single-valued mappings π, ϱ can be combined into a third single-valued mapping $\pi\varrho$ according to the rule $(\pi\varrho)x = \pi(\varrho x)$. The associative law is valid for this relation, since

$$(\pi(\varrho\sigma))x = \pi((\varrho\sigma)x) = \pi(\varrho(\sigma x)) = (\pi\varrho)(\sigma x) = ((\pi\varrho)\sigma)x.$$

The identity mapping $\underline{1}$, defined by $\underline{1}\, x = x$, is the unit element of this multiplicative set of mappings.

The single-valued mappings of a set onto itself form a semi-group with unit element.

A one-to-one mapping of a given set onto itself is called a permutation.

A permutation is a single-valued mapping π, for which $\pi x = a$ is solvable for every a and for which $\pi x = \pi y$ implies $x = y$. Therefore $\pi x = a$ is uniquely solvable for every a, and the solution is designated by $\pi^{-1}a$. $\pi(\pi^{-1}a) = a$, for every a. Therefore $\pi\pi^{-1} = \underline{1}$. Similarly $\pi^{-1}(\pi a) = a$, and therefore $\pi^{-1}\pi = \underline{1}$. Conversely, if the single-valued mapping π has an inverse mapping π^{-1}, defined by $\pi\pi^{-1} = \pi^{-1}\pi = \underline{1}$, then π is a permutation, since the equation $\pi x = a$ has the solution $\pi^{-1}a$ and $\pi x = \pi y$ implies $\pi^{-1}\pi x = \pi^{-1}\pi y$ and therefore $x = y$.

The inverse of the permutation π is the permutation π^{-1}; and if π, ϱ are two permutations, then the two-sided (i.e., right and left) inverse of $\pi\varrho$ is $\varrho^{-1}\pi^{-1}$. We conclude that the totality of permutations of the objects of a set form a group.

In order to see at a glance the effect of a single-valued mapping π we write it

$$\begin{pmatrix} x, & y, & \dots \\ \pi x, & \pi y, & \dots \end{pmatrix} \qquad (\textit{functional notation}).$$

Here $x, y, \ldots$ run through the elements of the given set in any order. Under every element is placed its image element. A shorter functional notation for π is $\binom{x}{\pi x}$. Multiplication is indicated by

$$\binom{x}{\varrho x}\binom{x}{\pi x} = \binom{x}{\varrho(\pi x)}. \quad *$$

π is a permutation, then, if and only if every element of $\mathfrak{M}$ occurs exactly once in the lower row of the parenthesis symbol indicated above. $\pi = \begin{pmatrix} x, & y, & \ldots \\ \pi x, & \pi y, & \ldots \end{pmatrix} = \binom{x}{\pi x}$, which indicates the mapping $x \to \pi(x)$. Then

$$\pi^{-1} = \begin{pmatrix} \pi x, & \pi y, & \ldots \\ x, & y, & \ldots \end{pmatrix} = \binom{\pi x}{x}.$$

Groups whose elements are permutations of a given set and are also multiplied like the permutations are called *permutation groups.*

THEOREM 2: *Every group can be represented as a permutation group* (Cayley).

Proof: We take the permuted objects to be the elements of the group. The mapping $\pi_a = \binom{x}{ax}$ is a permutation since $ax = b$ has a unique solution x. From the associative law it follows that the corresponding permutations multiply like the group elements. Since $\pi_a e = a$, the correspondence $\pi_a \longleftrightarrow a$ is one-to-one. The parenthesis notation for π_a is derived from the multiplication table of the group by writing the a-row under the e-row. This permutation group is called the (left) regular permutation group of the given abstract group.

The group of all permutations of a finite set of n *things is denoted by* $\mathfrak{S}_n$ *and is called the symmetrical group on* n *objects.* The permuted objects may be numbered from 1 to n, and we may think not of the permuted objects themselves but merely of their numbers. The latter are permuted just as the objects to which they correspond. Every permutation may be written uniquely as $\begin{pmatrix} 1, & 2, & \ldots, & n \\ i_1, & i_2, & \ldots, & i_n \end{pmatrix}$, where $i_1, i_2, \ldots, i_n$ run through the n integers $1, 2, \ldots, n$ in a definite order. We shall refer to these n consecutive integers hereafter as the *ciphers* of the permutation. Since there are $n!$ permutations of n elements, $\mathfrak{S}_n$ has the order $n!$.

---

* *Editor's note:* Since these permutations are mappings applied as operators from the *left* it follows that in the product $\varrho\pi$ the permutation π is followed by the permutation ϱ. This is contrary to the order used by such authors as Burnside, Speiser, and Dubreil.

The permutations of n letters can be written still more simply in cycle notation.

A permutation π is called d-cycle if π permutes cyclically a certain set of d letters $i_1, i_2, \ldots, i_d$:

$$\pi i_m = i_{m+1}, \ \pi i_d = i_1 \ (m = 1, 2, \ldots, d-1)$$

and if π leaves every other letter fixed. For example

$\begin{pmatrix} 1 & 2 & 3 & 4 \\ 2 & 1 & 3 & 4 \end{pmatrix}$ is a 2-cycle (transposition) and $\begin{pmatrix} 1 & 2 & 3 & 4 \\ 2 & 3 & 1 & 4 \end{pmatrix}$ is a 3-cycle.

We may then denote the d-cycle by $(i_1, i_2, \ldots, i_d)$. However the same d-cycle has d different cycle notations, one for each different initial symbol.

Every permutation of n letters can be written as the product of disjoint cycles (*i.e., cycles having no letter in common*).

For example $\begin{pmatrix} 1 & 2 & 3 & 4 & 5 \\ 5 & 3 & 4 & 2 & 1 \end{pmatrix} = (15)(234)$. This decomposition is naturally unique up to the order of factors, as regards the set of elements in any cycle.

In order to prove the above, let π be a permutation of n letters $1, 2, \ldots, n$. Among the $n+1$ letters $1, \pi 1, \ldots, \pi^n 1$ certainly two are equal. Let $\pi^i 1 = \pi^k 1$ with $i > k \geqq 0$ be the first equation of this sort. If $k > 0$, then we could conclude that $\pi^{i-1} 1 = \pi^{k-1} 1$. Therefore $k = 0$ and $z_1 = (1, \pi 1, \ldots, \pi^{i-1} 1)$ is an i-cycle. Now we construct a cycle z_2 containing a letter not occurring in z_1. Continue this process. z_2 must be disjoint from z_1 and since finally all the letters are used, π is a product of disjoint cycles.

(1) represents uniquely the identical permutation $\underline{1}$. If the 1-cycles are deleted from the set of other permutations in $\mathfrak{S}_n$, then the cycle notation remains unambiguous, e.g.,

$$\begin{pmatrix} 1 & 2 & 3 & 4 & 5 \\ 2 & 1 & 4 & 3 & 5 \end{pmatrix} = (12)(34)(5) = (12)(34).$$

Multiplication of permutations in cycle notation can easily be carried out. E.g., to calculate $(123)(45)(234)$, proceed as follows: The cycle farthest to the right containing 1 indicates $1 \to 2$. The cycle farthest to the right containing 2 indicates $2 \to 3$, the one farthest to the right containing 3, but to the left of the one just used, gives $3 \to 1$. Hence (12) is one cycle of the product. Continuing to work from right to left,* $3 \to 4 \to 5$, $5 \to 4$, $4 \to 2 \to 3$, giving (354). Hence

* *Editor's Note:* In Burnside, Speiser, et al., the procedure would be to start from the left and work to the right.

$$(123)\,(45)\,(234) = (12)\,(354).$$

The simplest non-identical permutations are the transpositions.

Every cycle of n *letters is a product of* $n-1$ *transpositions*

$(12), (23), \ldots, (n-1, n)$,

i.e. *every interchange of* n *letters can be arrived at by interchange of neighboring letters.*

This follows from

$$(1) \qquad (i, i+k) = (i+k-1, i+k)\ldots(i+1, i+2)$$
$$(i, i+1)\,(i+1, i+2)\ldots(i+k-1, i+k)$$

and

$$(2) \qquad (i_1, i_2, \ldots, i_m) = (i_1, i_2)\,(i_2, i_3)\ \ldots\ (i_{m-1}, i_m).$$

DEFINITION: *A permutation of* $n > 1$ *letters is called even or odd according to whether the number*

$$\varepsilon_\pi = \prod_{i<k} \frac{\pi k - \pi i}{k - i} \qquad \text{(here } \prod \text{ indicates ordinary product.)}$$

is equal to $+1$ *or* -1.[1]

If π and ϱ are two permutations in $\mathfrak{S}_n$, then

$$\varepsilon_{\varrho\pi} = \prod_{i<k} \frac{\varrho\pi k - \varrho\pi i}{k-i} = \prod_{i<k} \frac{\varrho\pi k - \varrho\pi i}{\varrho k - \varrho i} \cdot \prod_{i<k} \frac{\varrho k - \varrho i}{k-i} = \varepsilon_\varrho \cdot \varepsilon_\pi .$$

Thus all the even permutations form a group. It is called the alternating permutation group of n *letters and is denoted by* $\mathfrak{A}_n$. The transposition $(j, j+1)$ is an odd permutation as can immediately be seen. A permutation is even or odd according to whether it is the product of an even or odd number of transpositions.

From (1) and (2) it follows that an m-cycle is even or odd according to whether m is odd or even. An arbitrary permutation is even or odd according to whether the number of cycles with an even number of members in its decomposition is even or odd.

To every even permutation π there corresponds an odd permutation $(12)\,\pi$, and this correspondence is one-to-one, i.e., there are as many even as odd permutations.

The alternating permutation group on n letters thus has order $\frac{1}{2}n!$.

---

[1] For each ordered pair of digits $i < k$ just one of the two differences $k-i$, $i-k$ appears in the numerator. The occurrence of $i-k$ is called an *inversion* in π. ε_π is the product of as many factors -1 as there are inversions, and π is even or odd according to the number of inversions in π. The number of inversions and hence the value of ε_π will not change if a permutation ϱ is applied to each digit in numerator and denominator.

§ 3. Investigation of Axioms

If the e-row is made equal to the e-column by means of appropriate row and column interchanges in the multiplication tables of § 1, then for these special cases the tables are symmetric about the main diagonal. In a group whose order is less than 6 the equation $ab = ba$ is valid.

We call a group *abelian* (or *commutative*) if the commutative law

IV. $$ab = ba \quad \text{holds.}$$

In an abelian group a product of n factors is uniquely determined by its factors, irrespective of order and insertion of parentheses.

We must show that $a_1 \cdot a_2 \dots \cdot a_n = a_{i_1} \cdot a_{i_2} \dots \cdot a_{i_n}$, where $\begin{pmatrix} 1, & 2, & \dots, & n \\ i_1, & i_2, & \dots, & i_n \end{pmatrix}$ is a permutation. Since every interchange of n factors can be effected by the interchange of neighboring factors, we merely have to prove that

$$a_1 \cdot a_2 \dots \cdot a_i \cdot a_{i+1} \dots \cdot a_n = a_1 \cdot a_2 \dots \cdot a_{i+1} \cdot a_i \dots \cdot a_n$$

This follows from the associative and commutative laws.

In general, groups are non-commutative, e.g., $\mathfrak{S}_3$ has a multiplication table which is not symmetric:

		e	a	b	c	d	f
	e	e	a	b	c	d	f
(123)	$= a$	a	b	e	d	f	c
(132)	$= b$	b	e	a	f	c	d
(12)	$= c$	c	f	d	e	b	a
(13)	$= d$	d	c	f	a	e	b
(23)	$= f$	f	d	c	b	a	e

The independence of axiom IV. from the group axioms I.-III. is shown by the above example. Similarly we show that the axioms I.-III. are independent of one another.

1. III. does not follow from I., II. and the solvability of $ax = e$, e.g.,

	e	e'
e	e	e'
e'	e	e'

2. There are multiplicative domains in which II., III.a., IV. are valid but I. is not, e.g.,

	e	a	b	c	d	f
e	e	a	b	c	d	f
a	a	b	c	d	f	e
b	b	c	e	f	a	d
c	c	d	f	e	b	a
d	d	f	a	b	e	c
f	f	e	d	a	c	b

we have

$$(ab)b = f$$
$$a(bb) = a.$$

§ 4. Subgroups

DEFINITION: A subset $\mathfrak{U}$ of a given group $\mathfrak{G}$ is called a *subgroup* if the elements of $\mathfrak{U}$ form a group with the multiplication defined for $\mathfrak{G}$.

$\mathfrak{G}$ and e are trivial subgroups of $\mathfrak{G}$. A subgroup different from $\mathfrak{G}$ is called a *proper subgroup*. A subgroup different from $\mathfrak{G}$ and e is called a *non-trivial subgroup*. A proper subgroup $\mathfrak{U}$ is called a *largest* (*maximal*) *subgroup* if there is no subgroup of $\mathfrak{G}$ containing $\mathfrak{U}$ and different from $\mathfrak{U}$ and $\mathfrak{G}$. The subgroup $\mathfrak{U}$ is called a *smallest* (*minimal*) *subgroup* if e is the largest proper subgroup of $\mathfrak{U}$.

DEFINITION: *Two elements a and b are called right congruent under $\mathfrak{U}$ if $a = bU$ where $U \in \mathfrak{U}$.*[1] *Thus two elements are called right congruent if they differ by a factor on the right which is in $\mathfrak{U}$.* We denote the right congruence of a to b by $a \equiv b(\mathfrak{U}r)$. This symbol, $\equiv$, has the following three properties:

1. $a \equiv a$ (since $a = ae,\ e \in \mathfrak{U}$);
2. $a \equiv b$ implies $b \equiv a$ ($a = bU$ implies $b = aU^{-1}$);
3. $a \equiv b,\ b \equiv c$ implies $a \equiv c$ ($a = bU_1,\ b = cU_2$ implies $a = cU_2U_1$ where $U_2U_1 \in \mathfrak{U}$).

A right congruence may be multiplied on the right by a factor from $\mathfrak{U}$ and by any factor on the left. Thus from $a \equiv b(\mathfrak{U}r)$ it follows that $xa \equiv xb(\mathfrak{U}r)$ and conversely. Also either side of a right congruence may be multiplied on the right by an element of $\mathfrak{U}$.

All the elements congruent to an element a form the left coset* belonging to a. Every element of the group belongs to one and only one left coset. Since the mapping $U \to aU$ is one-one, there are as many elements in each left coset as there are in $\mathfrak{U}$. The number of different left cosets is called the *index* of $\mathfrak{U}$ in $\mathfrak{G}$, and is denoted by $\mathfrak{G}:\mathfrak{U}$.

---

[1] $U \in \mathfrak{U}$ is read: The element U belongs to the set $\mathfrak{U}$.

* The terms *residue class, coset* and *remainder class* are synonymous. (*Ed.*)

DEFINITION: *A system of elements which contains exactly one element from each left coset is called a system of right representatives.*

To each representative system of the cosets of $\mathfrak{G}$ there corresponds a mapping $G \to \bar{G}$, which maps each element G of $\mathfrak{G}$ onto its representative $\bar{G}$. A representative function of the left cosets of $\mathfrak{G}$ is characterized as a single-valued function $G \to \bar{G}$ defined on $\mathfrak{G}$ with the three properties

1. $\bar{\bar{G}} = \bar{G}$
2. $\bar{G}^{-1}G \in \mathfrak{U}$
3. $\overline{GU} = \bar{G}$ for all U belonging to $\mathfrak{U}$.

Furthermore the rule $\overline{HG} = \overline{H\bar{G}}$ is valid. Such a mapping will be called a *right representative function* of $\mathfrak{G}$ with respect to $\mathfrak{U}$, written $\mathfrak{G}(\mathfrak{U}r)$.

Let $\{a_i\}$ be a system of right representatives of $\mathfrak{G}$ with respect to $\mathfrak{U}$ and $\{b_k\}$ a system of right representatives of $\mathfrak{U}$ with respect to the subgroup $\mathfrak{U}$ of $\mathfrak{V}$. We will show that $\{a_i b_k\}$ is then a system of right representatives of $\mathfrak{G}$ with respect to $\mathfrak{V}$:

From $$a_i b_k \equiv a_l b_m (\mathfrak{V}r)$$
it follows that $$a_i b_k \equiv a_l b_m (\mathfrak{U}r),$$
whence $a_i \equiv a_l (\mathfrak{U}r)$. Hence $i = l$.
Therefore $a_i b_k \equiv a_i b_m (\mathfrak{V}r)$,
whence $b_k \equiv b_m (\mathfrak{V}r)$. Hence $k = m$.

If a belongs to $\mathfrak{G}$, then $a = a_i U$ has a solution $U \in \mathfrak{U}$, and $U = b_k \cdot V$ has a solution $V \in \mathfrak{V}$. Hence $a \equiv a_i b_k (\mathfrak{V}r)$, Q.E.D. We therefore have:

If $G \to \bar{G}$ is a representative function $\mathfrak{G}(\mathfrak{U}r)$ and $U \to \underline{U}$ a representative function $\mathfrak{U}(\mathfrak{V}r)$, then $G \to \bar{\bar{G}} = \bar{G}\, \underline{\bar{G}^{-1}G}$ is a representative function

$$\mathfrak{G}(\mathfrak{V}r).$$

We see, then, that the formula $\mathfrak{G} : \mathfrak{V} = (\mathfrak{G} : \mathfrak{U})(\mathfrak{U} : \mathfrak{V})$ holds for indices. If $\mathfrak{G}$ is finite, then $\mathfrak{G} : e = \mathfrak{G} : 1$ is the order of $\mathfrak{G}$, and so the following relation holds:

$$\mathfrak{G} : 1 = (\mathfrak{G} : \mathfrak{U})(\mathfrak{U} : 1).$$

We state this relation in the form—

$$\textit{Number of cosets } \mathfrak{G}(\mathfrak{U}r) = \frac{\textit{Number of elements in group}}{\textit{Number of elements in subgroup}}$$

(i.e., the order of any subgroup divides the order of the group.)

We call two elements a and b *left congruent* with respect to $\mathfrak{U}$ if $a = Ub$ with $U \in \mathfrak{U}$, and we write $a \equiv b(\mathfrak{U}l)$. The three rules mentioned above are also valid for the left congruence. Cancellation and multiplication on the right of a left congruence preserves the congruence. Either side of a left congruence may be altered on the left only by an element in $\mathfrak{U}$. The definitions of right coset and system of left representatives are analagous to those of left coset and system of right representatives.

A left residue (representative) function is characterized by the three properties:

1. $\overline{\overline{G}} = \overline{G}$
2. $G\overline{G}^{-1} \in \mathfrak{U}$
3. $\overline{UG} = G$

for all $U \in \mathfrak{U}$.

From the right congruence $a \equiv b(\mathfrak{U}r)$ follows the left congruence $a^{-1} \equiv b^{-1}(\mathfrak{U}l)$ and conversely. Therefore if $\{a_i\}$ is a system of left representatives, then $\{a_i^{-1}\}$ is a system of right representatives.

A group has just as many right residue classes as left residue classes with respect to a subgroup. Moreover,

THEOREM 3: *If the index of a group with respect to a subgroup is finite, then the right and left cosets have a common system of representatives.*

If $\mathfrak{U}$ is finite, then r right residue classes contain at most r left residue classes. The same is true if only $\mathfrak{G} : \mathfrak{U}$ is finite, as follows from a remark on p. 41.

We shall prove the more general theorem:

THEOREM 4: *If a set $\mathfrak{M}$ is subdivided into n disjoint classes in two ways and if any r classes of the first subdivision contain at most r classes of the second subdivision, then the two subdivisions have a common system of representatives.*

The first to prove Theorem 4 (in the language of graph theory) was D. König (*Uber Graphen und ihre Anwendungen auf Determinantentheorie und Mengenlehre,* Math. Ann., vol. 77 (1916), pp. 453-465). Frobenius claimed the theorem for matrix theory whereas, van der Waerden, O. Sperner, P. Hall, and W. Maak claimed it for set theory.

Eventually it was treated by H. Weyl, Halmos, and P. and H. Vaugham as the solution of a marriage problem (Amer. J. Math. vol. 72 (1950), pp. 214-215).

Proof: Let $\mathfrak{M} = \Sigma\mathfrak{A}_i = \Sigma\mathfrak{B}_i$ be the two decompositions. The incidence matrix $A = (a_{ik})$, where $a_{ik} = 0$ if $\mathfrak{A}_i$ and $\mathfrak{B}_k$ are distinct, $a_{ik} = 1$ otherwise, is normal in the sense that A is a quadratic matrix with its coefficients equal to 1 or 0, so that for every submatrix consisting entirely of 0's (zero-submatrix) the total number of rows and columns does not exceed the degree of A. We have to prove that a normal $(n \times n)$-matrix $A = (a_{ik})$ can be rearranged (by application of a suitable row permutation as well as a suitable column permutation) so that $a_{11} = a_{22} = \ldots = a_{nn} = 1$.

This is clear for $n = 1$. Apply induction on n. If $n > 1$ we wish to show that A can be rearranged so that for some r between 1 and $n-1$ both the top left $(r \times r)$-minor and the bottom right $(n-r) \times (n-r)$-minor are normal. Then, by the induction hypothesis, A can be rearranged so that $a_{11} = \ldots = a_{rr} = 1$; moreover $a_{r+1,r+1} = \ldots = a_{nn} = 1$.

Indeed, if there is a $r \times (n-r)$-zero-submatrix then A can be rearranged in such a way that $a_{ik} = 0$ if $1 \leqq i \leqq r < k \leqq n$. Now, if the top left $(r \times r)$-minor were not normal, then, after further rearrangement of A, we would have $a_{ik} = 0$ for $1 \leqq i \leqq s \leqq k \leqq n$ and some s between 1 and $r-1$, contradicting the normality of A. Hence the top left $(r \times r)$-minor of A is normal. Similarly it follows that the bottom $(n-r) \times (n-r)$-minor of A is normal. If, however, there is no $r \times (n-r)$-zero-submatrix of A, then every $(n-1) \times (n-1)$-submatrix is normal, and we simply rearrange A so that $a_{11} = 1$.

A Remark on Congruence Relations

A congruence relation R is defined in a set if for two elements a, b of

a is congruent to b: $a \equiv b$ or $R(a, b)$

a is non-congruent to b: $a \equiv b$ or $\sim R(a, b)$.[1]

A *normal congruence* satisfies the following three requirements:

1. (Reflexitivity) Every element is congruent to itself.

2. (Symmetry) The sides of a congruence may be interchanged: $a \equiv b$ implies $b \equiv a$.

[1] $\sim$ means *not*.

3. (Transitivity) $a \equiv b$, $b \equiv c$ implies $a \equiv c$.

For example, the ordinary equality relation in the set is a normal congruence relation.

Exercise: To a normal congruence relation corresponds a decomposition of the given set into disjoint classes in accordance with the rule:

Exactly those elements of the set which are congruent to a are put into the class $\mathfrak{K}_a$. Two classes are regarded as equal if they are the same subset of the given set. Two classes having any element in common are equal.

Exercise: If the set $\mathfrak{M}$ has a decomposition $\mathfrak{M} = \sum_{i=1}^{\omega} \mathfrak{K}_i$ into disjoint non-empty subsets $\mathfrak{K}_i$, then this decomposition is the class decomposition which corresponds to the following normal congruence relation:

a is congruent to b if a and b lie in the same subset of the decomposition.

a is not congruent to b if a and b do not lie in same subset of the decomposition.

A subset $\mathfrak{S}$ of a given set $\mathfrak{M}$ is called a *residue system* relative to a normal congruence relation, if $\mathfrak{S}$ contains exactly one element from each class, this element being called the *representative* of the class.

We obtain the residue system by choosing an element from each class and forming the subset $\mathfrak{S}$ of $\mathfrak{M}$ consisting of precisely these chosen elements. Then there corresponds to every residue system $\mathfrak{S}$ a *representative function* which associates an $\bar{a}$ of $\mathfrak{S}$ to every element a in $\mathfrak{M}$ according to the rule: $\bar{a}$ is the element of $\mathfrak{S}$ congruent to a.

Exercise: A single-valued function on a given set $\mathfrak{M}$, which maps a on $\bar{a}$, is a representative function if and only if $\bar{\bar{a}} = \bar{a}$.

Here, given the representative function, the congruence relation is defined by the rule:

$$a \text{ is congruent to } b, \text{ if } \bar{a} = \bar{b};$$
$$a \text{ is non-congruent to } b, \text{ if } \bar{a} \neq \bar{b}.$$

Exercise: If the left cancellation rule: $ab \equiv ac$ implies $b \equiv c$, holds for a normal congruence relation in a group $\mathfrak{G}$, then the relation is a right congruence with respect to the subgroup $\mathfrak{U}$ which consists of all elements congruent to e.

If the right cancellation rule: $ba \equiv ca$ implies $b \equiv c$, holds, then the normal congruence is a left congruence with respect to the subgroup $\mathfrak{U}$ which consists of all the elements congruent to e.

§ 5. Cyclic Groups

A group is called *cyclic* if it can be generated by one of its elements through multiplication and the taking of inverses (i.e., the group consists precisely of the set of all powers of the element a, positive, negative and zero).

The group $\mathfrak{G}$ generated by a is denoted by (a). Every element of $\mathfrak{G}$ is a power of a.

We wish to determine the subgroups $\mathfrak{U}$ of a cyclic group $\mathfrak{G}$. If $\mathfrak{U}$ is different from (e), then $\mathfrak{U}$ contains a power of a with an exponent different from zero. Since, if a^m lies in $\mathfrak{U}$, a^{-m} does also, we can assume that for some $m > 0$, a^m lies in $\mathfrak{U}$. Let d be the smallest of these natural numbers m. Then $e, a, a^2, \ldots, a^{d-1}$ must be mutually non-congruent with respect to $\mathfrak{U}$; therefore $\mathfrak{G}:\mathfrak{U} \geqq d$. Every rational integer m can be put in the form $m = qd + r$ where the quotient q is a rational integer and the remainder r is a non-negative integer less than d. The element a^m in $\mathfrak{G}$ has the form $a^r \cdot (a^d)^q$, therefore $a^m \equiv a^r$ and $\mathfrak{G}:\mathfrak{U} \leqq d$. From the two inequalities it follows that $\mathfrak{G}:\mathfrak{U} = d$ and that $e, a, a^2, \ldots, a^{d-1}$ is a system of representatives of $\mathfrak{G}$ with respect to $\mathfrak{U}$. $\mathfrak{U}$ consists of all powers of a^d.

Every subgroup of a cyclic group is cyclic. The index of a subgroup different from e is finite, and for every divisor $d > 0$ of $\mathfrak{G}:1$, there is only the one subgroup (a^d) of index d.

We shall see later that this last property characterizes the cyclic groups.

If $\mathfrak{G}$ has an order n different from zero, then two of the powers $a^0 = e, a, \ldots, a^n$ are equal. From $a^r = a^s$ it follows that $a^{r-s} = e$; thus a power of a with positive exponent lies in the subgroup e. Since $\mathfrak{G} : e = n$, we have $(a^n) = e$ and $\mathfrak{G}$ consists of the n elements $e, a, a^2, \ldots, a^{n-1}$.

n is the smallest positive number for which $a^n = e$. If $a^x = e$ then x is divisible by n.

DEFINITION: In an arbitrary group, the order of the cyclic subgroup generated by the element a is called the *order of the element* a. The order of an element is therefore either zero or the smallest positive number for which $a^n = e$.

The order of a group element is a divisor of the order of the group.

For a finite group of order N we have as a consequence the analog of the Fermat theorem, for groups:

$$a^N = e.$$

How can the order of a permutation in $\mathfrak{S}_n$ be read off from its decomposition into cycles?

If $\pi = (i_1, i_2, \ldots, i_d)$ is a d-cycle, then $\pi^d = \underline{1}$. If $0 < y < d$, then $\pi^y i_1 = i_{1+y} \neq i_1$, , therefore $\pi^y \neq \underline{1}$. The order of a d-cycle is d. Now let $\pi = z_1 \cdot z_2 \cdot \ldots \cdot z_k$ be the cycle representation of π, where z_i is a d_i-cycle. If $\pi^d = \underline{1}$ then $z_i^d = \underline{1}$; thus d_i is a divisor of d. The least common multiple d' of all the d_i is a divisor of d. Since conversely $z_i^{d'} = \underline{1}$ and therefore $\pi^{d'} = \underline{1}$, we have:

The order of a permutation of n letters is equal to the least common multiple of the orders of cycles in its cycle representation.

§ 6. Finite Rotation Groups

As an example of the meaning of the previous concepts, let us examine the finite rotation groups.

The rotations of cartesian three-dimensional space about the fixed point O have the following properties:

1. Every rotation about O permutes the points of the unit sphere $\mathfrak{K}$ with center O and is uniquely determined by its effect on the points of the surface of the unit sphere.

2. Two rotations carried out consecutively produce a rotation.

If σ and τ are two rotations about O, then $\sigma\tau$ is the rotation which transforms the point P into the point $\sigma(\tau P)$.

3. A rotation either leaves all points on $\mathfrak{K}$ fixed or it leaves exactly two points fixed.

In the latter case, the two fixed points are called the *poles* of the rotation. The rotation which leaves all points fixed, is denoted by $\underline{1}$.

4. A rotation angle φ_σ is associated with every rotation σ. φ_σ is uniquely determined to within addition of an integral multiple of 2π. Two rotations σ, τ with two common fixed points satisfy $\varphi_{\sigma\tau} \equiv \varphi_\sigma + \varphi_\tau (2\pi)$.

We wish to know which multiplication tables represent finite multiplicative domains $\mathfrak{G}$ of rotations.

Since $\mathfrak{G}$ consists of a finite number of permutations, $\mathfrak{G}$ is a finite group. The unit element of $\mathfrak{G}$ is $\underline{1}$. Let the number N of rotations in $\mathfrak{G}$ be greater than 1.

We say two points are *conjugate* under $\mathfrak{G}$ if there is a rotation in $\mathfrak{G}$ which sends one of the two points into the other. The finite number of poles of rotations in $\mathfrak{G}$ fall into classes of conjugate poles; let us call them $\mathfrak{P}_1, \mathfrak{P}_2, \ldots, \mathfrak{P}_H$.

All the rotations in $\mathfrak{G}$ which have the same pole P, together with $\underline{1}$, form a subgroup $\mathfrak{g}$. We call $\mathfrak{g}$ the subgroup belonging to P.

The p poles conjugate to P have the form $\sigma_1 P, \sigma_2 P, \ldots, \sigma_p P$ with $\sigma_i \in \mathfrak{G}$. If $\sigma \in \mathfrak{G}$ then $\sigma P = \sigma_i P$ is solvable, i.e.,

$$\sigma_i^{-1}.\ \sigma P = \underline{1} P = P,\ \sigma_i^{-1}\sigma \in \mathfrak{g},\ \sigma \equiv \sigma_i(\mathfrak{g}r).$$

Thus all the rotations in $\mathfrak{G}$ fall into p left cosets with respect to $\mathfrak{g}$ and these are determined by their effect on P. $\mathfrak{g}$ is one of these complexes and contains $n = N/p$ elements. P is called n-tuple pole in $\mathfrak{G}$.

If $\mathfrak{g}$ belongs to P, we determine the group belonging to τP.

If σ_1 is such that $\sigma_1 \tau P = \tau P$, it follows that $\tau^{-1}\sigma_1\tau P = P$; therefore $\sigma = \tau^{-1}\sigma_1\tau \in \mathfrak{g}$ and $\sigma_1 = \tau\sigma\tau^{-1}$. If $\sigma P = P$, then $\tau\sigma\tau^{-1}(\tau P) = \tau P$. Therefore the group $\tau\mathfrak{g}\tau^{-1}$ belongs to τP. In that account we also say: $\tau\sigma\tau^{-1}$ is *conjugate* to σ.

We determine the number of poles of rotations in $\mathfrak{G}$. There are exactly $2(N-1)$, since there are precisely $N-1$ non-identity rotations in $\mathfrak{G}$. On the other hand there are exactly $n_i = N/p_i$ rotations in $\mathfrak{G}$ which leave a pole of the i-th class fixed. Hence a totality of $p_i(n_i - 1)$ non-identical rotations leave *some* pole of the i-th class fixed.

Therefore $$2N - 2 = \sum_1^H p_i(n_i - 1),$$

i.e., $$2(1 - 1/N) = \sum_1^H (1 - 1/n_i).$$

From the further conditions $N \geqq n_i \geqq 2$ it follows that $2 \leqq H \leqq 3$. Furthermore

I. if $H = 2$, $n_1 = n_2 = N$ arbitrary > 1.

II. if $H = 3$, $2 = n_1 \leqq n_2 \leqq n_3$, $n_2 \leqq 3$:

$n_1 = n_2 = 2,\ n_3 = N/2,$

$n_1 = 2,\ n_2 = n_3 = 3,\ N = 12,$

$n_1 = 2,\ n_2 = 3,\ n_3 = 4,\ N = 24,$

$n_1 = 2,\ n_2 = 3,\ n_3 = 5,\ N = 60.$

I. $H=2$: All rotations $\neq \underline{1}$ have the same poles. Let φ_σ be the smallest of all the positive rotation angles corresponding to rotations in $\mathfrak{G}$. If τ is any rotation in $\mathfrak{G}$ there exists a rational integer m such that $m\varphi_\sigma \leqq \varphi_\tau < (m+1)\varphi_\sigma$. Since $\varphi_{\sigma^{-1}} = -\varphi_\sigma$, $\varphi_{\sigma^m} = m\varphi_\sigma$, we have $0 \leqq \varphi_{\sigma^{-m}\tau} < \varphi_\sigma$. Therefore $\varphi_{\sigma^{-m}\tau} = 0$ and $\tau = \sigma^m$ because of property 4 of rotations.

$\mathfrak{G}$ is a cyclic group of order N generated by σ, where $\varphi_\sigma = 2\pi/N$, and is designated by Z_N.

II.a. $H = 3$, $n_1 = n_2 = 2$, $n_3 = N/2$.

$\mathfrak{P}_3$ consists of two poles P, Q, and therefore a rotation in $\mathfrak{G}$ either will leave both poles fixed or will interchange the poles; thus $\mathfrak{g}_P = \mathfrak{g}_Q$. $\mathfrak{G}$ decomposes into $\mathfrak{g}$ and $\tau\mathfrak{g}$; the square of any rotation σ in $\mathfrak{G}$, which does not lie in $\mathfrak{g}$, is $\underline{1}$, since it leaves the fixed points of σ fixed, and also leaves P and Q fixed.

From $\tau^2 = (\tau\sigma)^2 = \underline{1}$ for all σ in $\mathfrak{g}$, it follows that $\tau\sigma = \sigma^{-1}\tau$. Therefore

$$\tau\sigma_1 \cdot \tau\sigma_2 = \sigma_1^{-1}\tau\tau\sigma_2 = \sigma_1^{-1}\sigma_2.$$

Since $\mathfrak{g}$ is cyclic, by **I.**, the multiplication table of $\mathfrak{G}$ is uniquely determined by its order. The table on page 9 shows $\mathfrak{G}$ for $N=6$. $\mathfrak{G}$ is called a dihedral group and is denoted by $\boldsymbol{D_N}$.

II.b. $n_1 = 2,\ n_2 = n_3 = 3,\ N = 12.$

The eleven rotations $\neq \underline{1}$ permute the four triple poles of the second (and third) class in 3-cycles and double transpositions. Thus $\mathfrak{G}$ is the alternating permutation group on the four triple poles of one of the latter two classes. $\mathfrak{G}$ is called the tetrahedral group.

II.c. $n_1 = 2,\ n_2 = 3,\ n_3 = 4,\ N = 24.$

The eight triple poles fall into four pairs of poles, such that a rotation in $\mathfrak{G}$ either has both poles of a pair as fixed points or else has neither. A rotation σ which takes each of the four pole pairs into itself has the identity as its square. If $\sigma \neq \underline{1}$, then σ interchanges the two poles in each pair and since, for every τ in $\mathfrak{G}$, $\tau\sigma\tau^{-1}$ has the same property, $\sigma\tau\sigma\tau^{-1} = \underline{1}$. If, however, τ is a rotation of order 3, then $\sigma\tau$ interchanges the poles of τ and consequently $\sigma\tau\sigma\tau = \underline{1}$. But this would give $\tau^{-1} = \tau$, $\tau^2 = 1$, a contradiction, and so σ must be $\underline{1}$ and $\mathfrak{G}$ is the symmetric permutation group of its four pair of triple poles. $\mathfrak{G}$ is called the octahedral group.

II.d. $n_1 = 2,\ n_2 = 3,\ n_3 = 5,\ N = 60.$

The 30 poles fall into 15 pairs of double poles, such that a rotation in $\mathfrak{G}$ leaves neither or both of the poles of each pair fixed. Let (PQ) be one of these pairs and let σ be a rotation $\neq \underline{1}$ in $\mathfrak{G}$ with poles P, Q.

There exists in $\mathfrak{G}$ a rotation τ which maps P onto Q. $\tau\sigma\tau^{-1}$ leaves the point Q fixed; therefore since Q is a double pole,

$$\tau\sigma\tau^{-1} = \sigma,\ \tau\sigma = \sigma\tau,\ Q \neq \tau Q = \tau\sigma Q = \sigma\tau Q.$$

Since τQ is a pole of σ, it follows that $\tau Q = P$. If conversely ϱ is a rotation in $\mathfrak{G}$ which leaves (PQ) fixed, then either ϱ or $\varrho\tau^{-1}$ leaves each of the points P, Q fixed. Of the elements of $\mathfrak{G}$ only $\underline{1}, \sigma, \tau, \sigma\tau$ leave the pole-pair (PQ) fixed. These four rotations are exactly those rotations in $\mathfrak{G}$ which commute with σ. The square of a rotation in $\mathfrak{G}$ which leaves (PQ) fixed has more than two fixed points and therefore is $\underline{1}$. From this we conclude: If a pole-pair (PQ) remains fixed under a rotation in $\mathfrak{G}$, then the three rotations $\neq \underline{1}$, which leave (PQ) fixed, leave the pole-pairs of each of them fixed. In this way the 30 double poles fall into five sextuples of poles which are permuted by $\mathfrak{G}$. By their effect upon a sextuple, the 60 rotations of $\mathfrak{G}$ fall into five complexes, each consisting of twelve rotations. All the rotations which leave a sextuple fixed form a subgroup of order 12 which has double and triple poles only; therefore the subgroup is the tetrahedral group. This tetrahedral group is generated by its elements of order 3. If a rotation of order 3 leaves each of the five sextuples of poles fixed, then all the rotations in $\mathfrak{G}$ of order 3 have this property, since they are all conjugate to one another under $\mathfrak{G}$. Then all the rotations in the tetrahedral group which belong to a sextuple leave every sextuple fixed. A rotation of order 2 does not have this property. Therefore the 59 rotations $\neq \underline{1}$ in $\mathfrak{G}$ permute the five pole sextuples in either a 3-cycle, a 5-cycle, or a double transposition. $\mathfrak{G}$ is the alternating permutation group on its five sextuples of double poles. $\mathfrak{G}$ is called the icosahedral group.

The names of the last four types are related to the regular polyhedra whose vertices are poles of the third class. Geometrically it can be seen that $\mathfrak{G}$ consists of all the rotations of space which carry the corresponding regular polyhedron into itself. Conversely, from the existence of the regular polyhedra we can deduce the existence of the rotation groups named after them.

In cases b) – d), the poles of the second class are the vertices of the dual regular polyhedra: tetrahedron, hexahedron (cube), dodecahedron. If the poles of third (second) class are at the vertices of the regular polyhedron, then the poles of second (third) class lie on the lines from O to the midpoints of the faces.

The double poles of the first class lie on the lines from O to the midpoints of the edges. The five sextuples of double poles of the icosahedral group are similar to the five vertex sextuples of the five octahedra inscribed in the icosahedron.

§ 7. Calculus of Complexes

In order to know the structure of a given group $\mathfrak{G}$, we must investigate its subsets.

We call any subset of a semi-group $\mathfrak{G}$ a *complex*. Let the empty subset be denoted by 0.

The set-theoretic relations of complexes are expressed by means of the symbols $=, \subseteq, \subset, \cap, \cup, +, -$. The equality of two complexes is expressed by $\mathfrak{K}_1 = \mathfrak{K}_2$ i.e., $\mathfrak{K}_1$ and $\mathfrak{K}_2$ contain the same elements. The three well-known rules are valid for this equality relation.

$\mathfrak{K}_1 \subseteq \mathfrak{K}_2$ means that the complex $\mathfrak{K}_1$ is contained in the complex $\mathfrak{K}_2$, i.e., every element in $\mathfrak{K}_1$, lies in $\mathfrak{K}_2$. Equivalent to this is $\mathfrak{K}_2 \supseteq \mathfrak{K}_1$ i.e., $\mathfrak{K}_2$ contains $\mathfrak{K}_1$. We have the rules:

a) $\mathfrak{K} \subseteq \mathfrak{K}$

b) If $\mathfrak{K}_1 \subseteq \mathfrak{K}_2$ and $\mathfrak{K}_2 \subseteq \mathfrak{K}_3$ then $\mathfrak{K}_1 \subseteq \mathfrak{K}_3$.

The equality of two complexes $\mathfrak{K}_1, \mathfrak{K}_2$ is equivalent to: $\mathfrak{K}_1 \subseteq \mathfrak{K}_2$ and $\mathfrak{K}_2 \subseteq \mathfrak{K}_1$.

If the complex $\mathfrak{K}_1$ is a proper subset of $\mathfrak{K}_2$, we denote this condition by $\mathfrak{K}_1 \subset \mathfrak{K}_2$, i.e., $\mathfrak{K}_1$ lies in $\mathfrak{K}_2$ but there is an element in $\mathfrak{K}_2$ which is not in $\mathfrak{K}_1$. The following two rules are valid:

a) $\mathfrak{K} \not\subset \mathfrak{K}$.

b) $\mathfrak{K}_1 \subset \mathfrak{K}_2$, $\mathfrak{K}_2 \subset \mathfrak{K}_3$ imply $\mathfrak{K}_1 \subset \mathfrak{K}_3$.

The totality of all elements which lie simultaneously in n given complexes $\mathfrak{K}_1, \mathfrak{K}_2, \ldots, \mathfrak{K}_n$ is called the *intersection* of the $\mathfrak{K}_i$. It is denoted by $\mathfrak{K}_1 \cap \mathfrak{K}_2 \cap \mathfrak{K}_3 \cap \ldots \cap \mathfrak{K}_n$.

The following rules are valid for the intersection:

$$\mathfrak{K} \cap \mathfrak{K} = \mathfrak{K},$$

$$\mathfrak{K}_1 \cap \mathfrak{K}_2 = \mathfrak{K}_2 \cap \mathfrak{K}_1 \quad \text{(commutative law)}$$

$$(\mathfrak{K}_1 \cap \mathfrak{K}_2) \cap \mathfrak{K}_3 = \mathfrak{K}_1 \cap (\mathfrak{K}_2 \cap \mathfrak{K}_3) \qquad \text{(associative law)}$$

$$(\mathfrak{K}_1 \cap \mathfrak{K}_2 \cap \ldots \cap \mathfrak{K}_n) \cap (\mathfrak{K}_{n+1} \cap \ldots \cap \mathfrak{K}_{n+m}) = \mathfrak{K}_1 \cap \mathfrak{K}_2 \cap \ldots \cap \mathfrak{K}_{n+m}.$$

The inequality $\mathfrak{K}_1 \subseteq \mathfrak{K}_2$ is equivalent to $\mathfrak{K}_1 \cap \mathfrak{K}_2 = \mathfrak{K}_1$. $\mathfrak{K}_1 \subset \mathfrak{K}_2$ is equivalent to $\mathfrak{K}_1 \cap \mathfrak{K}_2 = \mathfrak{K}_1$ and $\mathfrak{K}_1 \neq \mathfrak{K}_2$.

If $\mathfrak{k}_i \subseteq \mathfrak{K}_i$, then $(\mathfrak{k}_1 \cap \mathfrak{k}_2) \subseteq (\mathfrak{K}_1 \cap \mathfrak{K}_2)$.

The totality of all elements that lie either in $\mathfrak{K}_1$ or in $\mathfrak{K}_2 \ldots$ or in $\mathfrak{K}_n$ is called the *sum* of $\mathfrak{K}_1, \mathfrak{K}_2, \ldots, \mathfrak{K}_n$. It is denoted by $\mathfrak{K}_1 \cup \mathfrak{K}_2 \cup \ldots \cup \mathfrak{K}_n$. The above four rules are valid if $\cap$ is replaced by $\cup$. The relation $\mathfrak{K}_1 \subseteq \mathfrak{K}_2$ is equivalent with $\mathfrak{K}_1 \cup \mathfrak{K}_2 = \mathfrak{K}_2$. $\mathfrak{K}_1 \subset \mathfrak{K}_2$ is equivalent with $\mathfrak{K}_1 \cup \mathfrak{K}_2 = \mathfrak{K}_2$ and $\mathfrak{K}_1 \neq \mathfrak{K}_2$. If $\mathfrak{k}_i \subseteq \mathfrak{K}_i$, then $\mathfrak{k}_1 \cup \mathfrak{k}_2 \subseteq \mathfrak{K}_1 \cup \mathfrak{K}_2$.
The relation between sum and intersection is distributive:

$$\mathfrak{K}_1 \cap (\mathfrak{K}_2 \cup \mathfrak{K}_3) = (\mathfrak{K}_1 \cap \mathfrak{K}_2) \cup (\mathfrak{K}_1 \cap \mathfrak{K}_3),$$

$$\mathfrak{K}_1 \cup (\mathfrak{K}_2 \cap \mathfrak{K}_3) = (\mathfrak{K}_1 \cup \mathfrak{K}_2) \cap (\mathfrak{K}_1 \cup \mathfrak{K}_3).$$

The sum of pairwise disjoint complexes $\mathfrak{K}_1, \mathfrak{K}_2, \ldots, \mathfrak{K}_n$ is denoted by $\mathfrak{K}_1 + \mathfrak{K}_2 + \cdots + \mathfrak{K}_n$.

The rules for the symbol $\cup$ remain valid if $+$ is substituted for $\cup$ everywhere and it is assumed that the sets on the left connected by the plus symbol are disjoint. The second distributive law is an exception.

If $\mathfrak{K}_1$ is contained in $\mathfrak{K}_2$, then the difference set, denoted by $\mathfrak{K}_2 - \mathfrak{K}_1$, consists of those elements of $\mathfrak{K}_2$which are not in $\mathfrak{K}_1$. It follows that: $\mathfrak{K}_2 = \mathfrak{K}_1 + (\mathfrak{K}_2 - \mathfrak{K}_1)$,and $\mathfrak{K}_2 - \mathfrak{K}_1$ is uniquely determined by this equation.

Beside these set-theoretic operations we also introduce the *product* of n complexes $\mathfrak{K}_1, \mathfrak{K}_2, \ldots, \mathfrak{K}_n$: $\mathfrak{K}_1 \cdot \mathfrak{K}_2 \cdot \ldots \cdot \mathfrak{K}_n$ is the set of all products $x_1 \cdot x_2 \cdot \ldots \cdot x_n$, where $x_i \in \mathfrak{K}_i$ and n is a positive integer. We have:

$$\mathfrak{K}_1(\mathfrak{K}_2\mathfrak{K}_3) - (\mathfrak{K}_1\mathfrak{K}_2)\mathfrak{K}_3 = \mathfrak{K}_1\mathfrak{K}_2\mathfrak{K}_3 \quad \text{(tho associative law)}.$$

The combination of product with sum or intersection satisfies

$$\mathfrak{K}_1(\mathfrak{K}_2 \cap \mathfrak{K}_3) \subseteq \mathfrak{K}_1\mathfrak{K}_2 \cap \mathfrak{K}_1\mathfrak{K}_3,$$
$$\mathfrak{K}_1(\mathfrak{K}_2 \cup \mathfrak{K}_3) = \mathfrak{K}_1\mathfrak{K}_2 \cup \mathfrak{K}_1\mathfrak{K}_3.$$

If $\mathfrak{k}_i \subseteq \mathfrak{K}_i$, then $\mathfrak{k}_1\mathfrak{k}_2 \subseteq \mathfrak{K}_1\mathfrak{K}_2$.

In a group $\mathfrak{G}$ we define the *inverse complex* of a non-empty complex as the complex consisting of all the inverses of the elements of $\mathfrak{K}$. It is denoted by $\mathfrak{K}^{-1}$.

$\mathfrak{K}\mathfrak{K}^{-1} \supseteq e$, but $\mathfrak{K}\mathfrak{K}^{-1} = e$ if and only if $\mathfrak{K}$ consists of exactly one element. Furthermore:

$$(\mathfrak{K}^{-1})^{-1} = \mathfrak{K}$$
$$(\mathfrak{K}_1 \cup \mathfrak{K}_2)^{-1} = \mathfrak{K}_1^{-1} \cup \mathfrak{K}_2^{-1},$$
$$(\mathfrak{K}_1 \cap \mathfrak{K}_2)^{-1} = \mathfrak{K}_1^{-1} \cap \mathfrak{K}_2^{-1},$$
$$(\mathfrak{K}_1 \cdot \mathfrak{K}_2)^{-1} = \mathfrak{K}_2^{-1} \cdot \mathfrak{K}_1^{-1},$$

and if $\mathfrak{k} \subseteq \mathfrak{K}$, then $\mathfrak{k}^{-1} \subseteq \mathfrak{K}^{-1}$.

Necessary and sufficient conditions that a complex $\mathfrak{U}$ be a subgroup, are:

$$\mathfrak{U} \neq 0,$$
$$\mathfrak{U}\mathfrak{U} \subseteq \mathfrak{U},$$
$$\mathfrak{U}^{-1} \subseteq \mathfrak{U}.$$

The latter two conditions can be replaced by

$$\mathfrak{U}\mathfrak{U}^{-1} \subseteq \mathfrak{U},$$

for then $e \subseteq \mathfrak{U}$, and so it follows that $\mathfrak{U}^{-1} \subseteq \mathfrak{U}$. Taking inverses in this inequality, we get $\mathfrak{U} \subseteq \mathfrak{U}^{-1}$; therefore $\mathfrak{U} = \mathfrak{U}^{-1}$, $\mathfrak{U}\mathfrak{U} \subseteq \mathfrak{U}$.

The intersection of two subgroups is itself a subgroup. The product of two subgroups is a subgroup when the two factors can be interchanged.

If a non-empty complex $\mathfrak{U}$ contains only a finite number of elements, then the condition $\mathfrak{U}\mathfrak{U} \subseteq \mathfrak{U}$ is necessary and sufficient for $\mathfrak{U}$ to be a subgroup, since the cancellation laws hold in $\mathfrak{U}$.

Let $\mathfrak{K}$ be a non-empty complex. Let $\mathfrak{K}_1 = \mathfrak{K} \cup \mathfrak{K}^{-1}$. Then $\mathfrak{K}_1^{-1} = \mathfrak{K}_1$ and $\mathfrak{K}_1 \cup \mathfrak{K}_1{}^2 \cup \mathfrak{K}_1{}^3 \ldots$ is a subgroup of $\mathfrak{G}$ which lies in every subgroup which contains $\mathfrak{K}$. The subgroup is called the subgroup generated by $\mathfrak{K}$ and is denoted by $\{\mathfrak{K}\}$. We set $\{0\} = e$.

Then the following rules hold:

$$\{\mathfrak{K}_1 \wedge \mathfrak{K}_2\} \subseteq \{\mathfrak{K}_1\} \wedge \{\mathfrak{K}_2\},$$

$$\{\mathfrak{K}_1 \cup \mathfrak{K}_2\} = \Big\{ \{\mathfrak{K}_1\} \cup \{\mathfrak{K}_2\} \Big\}.$$

If $\mathfrak{k} \subseteq \mathfrak{K}$ then $\{\mathfrak{k}\} \subseteq \{\mathfrak{K}\}$; furthermore $\{\mathfrak{K}^{-1}\} = \{\mathfrak{K}\}$.

The following useful rule of the calculus of subgroups can be proven.

If $\mathfrak{u} \subseteq \mathfrak{U}$ and $\mathfrak{v} \subseteq \mathfrak{V}$, then

$$\mathfrak{U} \wedge \mathfrak{u}\mathfrak{v} \wedge \mathfrak{V} = (\mathfrak{u} \wedge \mathfrak{V}) \cdot (\mathfrak{U} \wedge \mathfrak{v}).$$

PROOF: It is immediate that $\mathfrak{U} \wedge \mathfrak{u}\mathfrak{v} \wedge \mathfrak{V} \supset (\mathfrak{u} \wedge \mathfrak{V}) \cdot (\mathfrak{U} \wedge \mathfrak{v})$. Moreover, let $x \in \mathfrak{U} \wedge \mathfrak{u}\mathfrak{v} \wedge \mathfrak{V}$. Then x is of the form uv where $u \in \mathfrak{u}$, $v \in \mathfrak{v}$. Since $x \in \mathfrak{U}$, it follows that $v \in \mathfrak{U}$. Since $x \in \mathfrak{V}$, $u \in \mathfrak{V}$. Consequently x is in $(\mathfrak{u} \wedge \mathfrak{V}) \cdot (\mathfrak{U} \wedge \mathfrak{v})$, whence the rule follows.

If we set $\mathfrak{V} = \mathfrak{G}$ in the rule, we obtain:

If $\mathfrak{u} \subseteq \mathfrak{U}$ and $\mathfrak{v}$ is arbitrary then $\mathfrak{u} \cdot (\mathfrak{U} \wedge \mathfrak{v}) = \mathfrak{U} \wedge \mathfrak{u}\mathfrak{v}$.

We consider an ordered ascending chain of subgroups of a group, i.e, an ordered set of subgroups for which $\mathfrak{U} \subset \mathfrak{V}$ implies $\mathfrak{U} \subseteq \mathfrak{V}$.

The sum of all $\mathfrak{U}$ in this subgroup chain, which has an arbitrary cardinal number of members, is itself a subgroup which we denoted by $\mathfrak{V}$

If a complex $\mathfrak{K}$ has no element in common with any member of the chain, then the intersection of $\mathfrak{V}$ and $\mathfrak{K}$ is empty.

We prove the following existence theorem on maximal subgroups.

THEOREM 5: *If $\mathfrak{K}$ is an arbitrary complex in $\mathfrak{G}$ and $\mathfrak{U}$ is a subgroup disjoint from $\mathfrak{K}$, then among the subgroups which contain $\mathfrak{U}$ and are disjoint from $\mathfrak{K}$, there exists a maximal one $\mathfrak{V}$. Thus $\mathfrak{V}$ is defined as a subgroup of $\mathfrak{G}$ such that:*

1. $\mathfrak{U} \subseteq \mathfrak{V}$,
2. $\mathfrak{V} \wedge \mathfrak{K} = 0$.
3. $\{\mathfrak{V}, x\} \wedge \mathfrak{K} = 0$ implies $x \in \mathfrak{V}$.

We consider the elements of $\mathfrak{G}$ as well ordered: $e < \nu_2 < \nu_3 \ldots$. We define an ascending chain of subgroups $\mathfrak{U}_e \subseteq \mathfrak{U}_{\nu_2} \ldots$ by means of transfinite induction: $\mathfrak{U}_e = \mathfrak{U}$. Assume that the subgroup $\mathfrak{U}_\nu$ has already been defined for all $\nu < \omega$ and that it has been shown that $\mathfrak{U}_\nu \subseteq \mathfrak{U}_\mu$ for $\nu \leqq \mu < \omega$ and that $\mathfrak{U}_\nu \cap \mathfrak{K} = 0$. Then let $\mathfrak{U}_\omega$ be the union Σ_ω of all $\mathfrak{U}_\nu$ for $\nu < \omega$ if $\{\Sigma_\omega, \omega\} \cap \mathfrak{K} \neq 0$, but let $\mathfrak{U}_\omega = \{\Sigma_\omega, \omega\}$ if $\{\Sigma_\omega, \omega\} \cap \mathfrak{K} = 0$. Since Σ_ω is a subgroup, $\mathfrak{U}_\omega$ is also a subgroup and $\mathfrak{U}_\nu \subseteq \mathfrak{U}_\omega$ for $\nu \leqq \omega$. Furthermore $\Sigma_\omega \cap \mathfrak{K} = 0$ by the construction of Σ_ω, therefore $\mathfrak{U}_\omega \cap \mathfrak{K} = 0$.

The union $\mathfrak{V}$ of all the $\mathfrak{U}_\omega$ is the maximal subgroup the existence of which was to be proven.

§ 8. The Concept of Normal Subgroup

What condition must a subgroup $\mathfrak{U}$ of a group $\mathfrak{G}$ satisfy in order that left congruency shall be equivalent to right congruency?

From $au \equiv a(\mathfrak{U}r)$ where $u \in \mathfrak{U}$ it should follow that

$$au \equiv a(\mathfrak{U}l),$$

and therefore

$$aua^{-1} \equiv e(\mathfrak{U}l).$$

If, conversely,

$$aua^{-1} \equiv e(\mathfrak{U}l),$$

then

$$au \equiv a$$

for both left and right congruency.

We come upon the *normality condition*

$$a\mathfrak{U}a^{-1} \subseteq \mathfrak{U}.$$

We arrive at this same condition if we ask when congruences can be multiplied. Then it should follow from $a \equiv a(\mathfrak{U}l)$ and $u \equiv e(\mathfrak{U}l)$ that: $au \equiv a(\mathfrak{U}l)$ and this implies $a\mathfrak{U}a^{-1} \subseteq \mathfrak{U}$. If conversely $x\mathfrak{U}x^{-1} \subseteq \mathfrak{U}$ for all x in $\mathfrak{G}$, then we can drop the l, r-symbols from the congruences and it follows from

$$a \equiv b$$
$$c \equiv d,$$

that

$$ac \equiv bc$$
$$bc \equiv bd,$$

and therefore

$$ac \equiv bd.$$

DEFINITION: *A subgroup $\mathfrak{N}$ of $\mathfrak{G}$ for which $x\mathfrak{N}x^{-1} \subseteq \mathfrak{N}$ holds for all x in $\mathfrak{G}$ is called a* ***normal subgroup.***

Left congruency is equivalent to right congruency if both are with respect to the same normal subgroup. Congruences with respect to a normal subgroup may be multiplied together.

From $x\mathfrak{N}x^{-1} \subseteq \mathfrak{N}$ and $x^{-1}\mathfrak{N}x \subseteq \mathfrak{N}$, it follows that

$$\mathfrak{N} = xx^{-1}\mathfrak{N}xx^{-1} \subseteq x\mathfrak{N}x^{-1},$$

and therefore $x\mathfrak{N}x^{-1} = \mathfrak{N}$,

$$x\mathfrak{N} = \mathfrak{N}x.$$

A normal subgroup commutes with every complex.

If conversely a subgroup commutes with every complex, then it is a normal subgroup, since $x\mathfrak{U} = \mathfrak{U}x$ implies $x\mathfrak{U}x^{-1} = \mathfrak{U}$.

The product of a normal subgroup and a subgroup $\mathfrak{U}$ *is a subgroup.*

DEFINITION: *A group with no non-trivial normal subgroups is said to be* ***simple.*** *Any other group is called* ***composite.***

A group without a non-trivial subgroup is simple. Moreover,

THEOREM 6: *A group with no non-trivial subgroups is cyclic of prime order, or consists of merely the unit element e.*

Proof: *If* $\mathfrak{G} \neq e$, then there is an element $a \neq e$ in $\mathfrak{G}$. By hypothesis $\mathfrak{G} = (a)$. If $\mathfrak{G}$ were infinite then $(a) \neq (a^2) \neq e$, and consequently $\mathfrak{G}$ is finite. If p is a prime dividing $\mathfrak{G}:1$ then $(a) \neq (a^p)$ and therefore $a^p = e$ and $\mathfrak{G}:1 = p$.

The converse was seen earlier.

A congruence relation in a multiplicative domain is said to be *multiplicative* if $a \equiv b$, $c \equiv d$ implies $ac \equiv bd$.

Example: In the multiplicative group of positive real numbers the relation:

a is congruent to b, if $a \geqq b$,
a is not congruent to b, if $a < b$,

is a multiplicative congruence relation.

Exercise: A multiplicative normal congruence relation in a group is the congruence relation of the group of elements with respect to the normal subgroup consisting of all the elements congruent to e.

§ 9. Normalizer, Class Equation

The following investigation shows the meaning of the concepts of subgroup and of right congruence.

Let $\mathfrak{P}$ be a group of permutations of the objects of a given set $\mathfrak{M}$.

DEFINITION: *Two objects in the set $\mathfrak{M}$ are said to be conjugate under the permutation group $\mathfrak{P}$ ($\mathfrak{P}$-conjugate) if there is a permutation in $\mathfrak{P}$ which maps one of the two objects onto the other.*

The relation "a is conjugate to b" fulfills our three requirements:

1. a is $\mathfrak{P}$-conjugate to itself since the identity permutation in $\mathfrak{P}$ maps a on itself.

2. If a is $\mathfrak{P}$-conjugate to b, then there is a permutation in $\mathfrak{P}$ which maps a onto b. The permutation which is the inverse of the latter lies likewise in $\mathfrak{P}$ and maps b onto a. Thus b is conjugate to a.

3. If $b = \pi a$, $c = \varrho b$, then $\varrho\pi$ as well as π, ϱ lies in $\mathfrak{P}$, and $\varrho\pi a = \varrho b = c$; therefore a is $\mathfrak{P}$-conjugate to c.

Under the action of a permutation group a set splits into disjoint classes of $\mathfrak{P}$-conjugate elements.

We call a class of $\mathfrak{P}$-conjugate objects of a set $\mathfrak{M}$ a *system of transitivity* for the permutation group $\mathfrak{P}$. The system of transitivity in which a lies consists of all πa with $\pi \in \mathfrak{P}$.

How many objects lie in a system of transitivity? The answer is given by THEOREM 7: *All the permutations of a permutation group $\mathfrak{P}$ which leave an object a of the permuted set $\mathfrak{M}$ fixed, form the subgroup $\mathfrak{P}_a$ of $\mathfrak{P}$ belonging to a. All the objects $\mathfrak{P}$-conjugate to a can be found as images of a, each once, under the permutations of a right representative system of $\mathfrak{P}$ with respect to $\mathfrak{P}_a$. Therefore the number of objects which are $\mathfrak{P}$-conjugate to a is equal to the index of $\mathfrak{P}_a$ in $\mathfrak{P}$.*

Proof: Let $\mathfrak{P}_a$ be the set of permutations in $\mathfrak{P}$ which leave a fixed. $\underline{1}$ belongs to $\mathfrak{P}_a$. If π belongs to $\mathfrak{P}_a$, then π^{-1} is in $\mathfrak{P}$ and $\pi^{-1}a = \pi^{-1}(\pi a) = a$, and therefore π^{-1} also belongs to $\mathfrak{P}_a$. If ϱ and π are in $\mathfrak{P}_a$ then $\varrho\pi$ is in $\mathfrak{P}$ and $\varrho\pi a = \varrho(\pi a) = \varrho a = a$; therefore $\varrho\pi$ is also in $\mathfrak{P}_a$.

$\mathfrak{P}_a$ is a subgroup of $\mathfrak{P}$.

If the permutations ϱ and π in $\mathfrak{P}$ have the same effect on a then they are right congruent with respect to $\mathfrak{P}_a$ since $\pi a = \varrho a$ implies $\varrho^{-1}\pi a = a$, $\varrho^{-1}\pi \in \mathfrak{P}_a$, $\varrho \equiv \pi(\mathfrak{P}_a r)$, and conversely. If, then, $\pi \to \bar{\pi}$ is a right representative function of $\mathfrak{P}$ with respect to $\mathfrak{P}_a$, then every conjugate πa of a is equal to $\bar{\pi} a$ and $\bar{\pi} a = \bar{\varrho} a$ implies $\bar{\bar{\pi}} = \bar{\bar{\varrho}} = \bar{\pi} = \bar{\varrho}$, as was to be shown.

DEFINITION: *We say that two subsets of the set $\mathfrak{M}$ are* **conjugate under the permutation group** $\mathfrak{P}$ *if there is a permutation in $\mathfrak{P}$ which maps one subset onto the other.*

Since $\mathfrak{P}$ also permutes the subsets of $\mathfrak{M}$, the above statements remain valid if "object" is replaced by "subset".

However, we denote by $\mathfrak{P}_{(\mathfrak{m})}$ the subgroup of permutations in $\mathfrak{P}$ which map a subset $\mathfrak{m}$ of $\mathfrak{M}$ onto itself, whereas $\mathfrak{P}_{\mathfrak{m}}$ will denote the subgroup of all permutations in $\mathfrak{P}$ which leave each element of $\mathfrak{m}$ fixed.

We now take the set of all elements of a group $\mathfrak{G}$ as an example of a permuted set.

For every element x in $\mathfrak{G}$, we define the "*x-transformation*" as the single-valued mapping $\binom{a}{x\,a\,x^{-1}}$. $x\,a\,x^{-1}$ is called the x-transform of a. The x-transformations of $\mathfrak{G}$ form a permutation group, since $\binom{a}{e\,a\,e^{-1}}=\binom{a}{a}$ is the identity permutation $\underline{1}$:

$$(1) \qquad \binom{a}{x\,a\,x^{-1}}\binom{a}{y\,a\,y^{-1}} = \binom{a}{x\,y\,a(x\,y)^{-1}}, \text{ and in particular}$$

$$(2) \qquad \binom{a}{x\,a\,x^{-1}}\cdot\binom{a}{x^{-1}\,a\,x} = \binom{a}{a} = \underline{1}.$$

The group of transformations of $\mathfrak{G}$ is denoted by $J_{\mathfrak{G}}$ or simply by J.

DEFINITION: Two complexes in $\mathfrak{G}$ are said to be *conjugate* (*under* $\mathfrak{G}$) if one complex is the transform of the other: $\mathfrak{K}_2 = x\,\mathfrak{K}_1 x^{-1}$, or equivalently, $x\,\mathfrak{K}_1 = \mathfrak{K}_2 x$.

From equations (1), (2) we immediately see that in $\mathfrak{G}$, all elements x whose corresponding transformations lie in a given subgroup of J form a subgroup of $\mathfrak{G}$. We can therefore define, in accordance with Theorem 7:

The *normalizer* $N_{\mathfrak{K}}$ of the complex $\mathfrak{K}$ is the subgroup consisting of all elements x of $\mathfrak{G}$ which transform $\mathfrak{K}$ into itself: $x\,\mathfrak{K}\,x^{-1} = \mathfrak{K}$, or equivalently $x\,\mathfrak{K} = \mathfrak{K}\,x$.

If $x_1, x_2, \ldots$ is a representative system of $\mathfrak{G}$ with respect to $N_{\mathfrak{K}}$, then $x_1 \mathfrak{K}\,x_1^{-1}$, $x_2 \mathfrak{K}\,x_2^{-1}, \ldots$ are the complexes conjugate to $\mathfrak{K}$, each occurring exactly once; and conversely. Thus

The number of complexes conjugate to a given complex is equal to the index of its normalizer.

The group $\mathfrak{G}$ falls into classes of conjugate elements relative to the transformations in J, giving the direct decomposition $\mathfrak{G} = \mathfrak{C}_1 + \mathfrak{C}_2 + \cdots$. The number of classes of conjugate elements of a group is called the *class-number* of the group. The direct decomposition

$$\mathfrak{G} = \mathfrak{C}_1 + \mathfrak{C}_2 + \cdots + \mathfrak{C}_r$$

of the group $\mathfrak{G}$ into classes of conjugate elements corresponds to the equation

$$(3) \qquad \mathfrak{G}:1 = h_1 + h_2 + \ldots + h_r \qquad (\textit{class equation})$$

where h_i is the number of elements in $\mathfrak{C}_i$.

DEFINITION: All the elements of a group $\mathfrak{G}$ which transform each element of $\mathfrak{G}$ into itself, i.e., those which commute with every element of $\mathfrak{G}$, form a subgroup called the *center* $\mathfrak{z}(\mathfrak{G})$ of $\mathfrak{G}$. $\mathfrak{z}$ is obviously the intersection of all the normalizers of elements in $\mathfrak{G}$.

It follows from the definition that the center is an abelian normal subgroup. The center is just that domain of all elements which are transformed into themselves by every element in $\mathfrak{G}$. Therefore we may write the class-equation as follows:

(4) $\mathfrak{G}:1 = \mathfrak{z}:1 + \sum_{h_i > 1} h_i.$

It is important in the above to note that the summation is performed over some *group indices* different from 1.

The subgroups which are transformed into themselves by every element in $\mathfrak{G}$ are precisely the normal subgroups of $\mathfrak{G}$.

The *normalizer* of an arbitrary subgroup $\mathfrak{U}$ of $\mathfrak{G}$ is the (uniquely determined) maximal subgroup containing $\mathfrak{U}$ as normal subgroup.

We wish to determine the classes of conjugate elements in the symmetric and alternating permutation groups of n letters.

Let π and ϱ be two permutations in $\mathfrak{S}_n$; then $\varrho\pi\varrho^{-1}(\varrho x) = \varrho\pi x$, and therefore $\varrho\pi\varrho^{-1} = \begin{pmatrix} \varrho x \\ \varrho\pi x \end{pmatrix}$, i.e.: *The ϱ-transform of π originates from π by replacing the letter x by ϱx in the functional symbol for π. The same also holds for the cycle symbol.*

Two permutations are conjugate under $\mathfrak{S}_n$ if and only if they have cycle decompositions with like groupings.

Let π be a product of a_1 1-cycles, a_2 2-cycles, . . . , a_n n-cycles. Then the number of permutations that commute with π is just as large as the number of formally different ways that π can be written as a product of first a_1 1-cycles, then a_2 2-cycles, and finally a_n n-cycles, and this is $a_1!\,1^{a_1} \cdot a_2!\,2^{a_2} \ldots a_n!\,n^{a_n}$. Consequently the class $\mathfrak{C}_\pi$ of elements conjugate to π under $\mathfrak{S}_n$ contains $\dfrac{n!}{a_1!\,1^{a_1} \ldots a_n!\,n^{a_n}}$ permutations.

Now let $n > 1$, $\pi_1 = (12)\pi(12)^{-1}$. Every permutation in $\mathfrak{C}_\pi$ is conjugate either to π or π_1 under $\mathfrak{A}_n$, the alternating group. Therefore $\mathfrak{C}_\pi$ decomposes into two classes under $\mathfrak{A}_n$, each with an equal number of elements, or it does not decompose. The latter takes place if and only if π commutes with an odd permutation. This last is equivalent to the condition: There is an $a_{2i} > 0$ or an $a_{2i+1} > 1$.

§ 10. A Theorem of Frobenius

The following theorem is not yet fitted into a wider context in a satisfactory way.

THEOREM OF FROBENIUS: *The number of solutions of $x^n = c$, where c belongs to a fixed class $\mathfrak{C}$ of h elements conjugate under a finite group $\mathfrak{G}$ of order N, is divisible by the greatest common divisor of hn and N.*

Proof: The complex consisting of those elements in $\mathfrak{G}$ whose n-th powers lie in the complex $\mathfrak{K}$ is denoted by $\mathfrak{A}_{\mathfrak{K},n}$. Let $A_{\mathfrak{K},n}$ be the number of elements in $\mathfrak{A}_{\mathfrak{K},n}$. If $N=1$, then the theorem is true. Now let $N > 1$ and let the theorem be proven for groups whose order is less than N. If $n=1$, then $A_{\mathfrak{C},n} = h$. Therefore the statement is true. Now let $n > 1$ and let the statement be proven for all smaller n. (We are using induction twice.) Since the elements in $\mathfrak{C}$ are conjugate under $\mathfrak{G}$, $A_{\mathfrak{C},n} = h \cdot A_{c,n}$. $\mathfrak{A}_{c,n}$ lies in the normalizer N_c of c. If $h > 1$, then, applying the induction hypothesis to N_c, we find* that $\left(n, \frac{N}{h}\right) / A_{c,n}$, and therefore $(hn, N)/A_{\mathfrak{C},n}$.

Now let $h = 1$. If $n = n_1 n_2$, $(n_1, n_2) = 1$, $n_1, n_2 \neq 1$ and if $\mathfrak{D} = \mathfrak{A}_{\mathfrak{C},n_2}$ then $\mathfrak{A}_{\mathfrak{C},n} = \mathfrak{A}_{\mathfrak{D},n_1}$. By the induction hypothesis (n_1, N) is a divisor of $A_{\mathfrak{D},n_1}$, and therefore also a divisor of $A_{\mathfrak{C},n}$. Similarly it follows that (n_2, N) is a divisor of $A_{\mathfrak{C},n}$ and since n_1 is relatively prime to n_2, we have (n, N) as divisor of $A_{\mathfrak{C},n}$.

It can now be assumed that $n = p^a$ is the a-th power of a prime number p with $a > 0$. If p divides the order ϱ of c, then an element x in $\mathfrak{A}_{c,n}$ has the order $n \cdot \varrho$. Then exactly n elements of $\mathfrak{A}_{c,n}$ lie in (x), and all these n elements generate the same subgroup, namely (x). The number of elements in $\mathfrak{A}_{c,n}$ is consequently divisible by n.

Finally we may assume that n is relatively prime to the order of the center element c. All the elements of the center whose order is prime to n form a subgroup $\mathfrak{g}$ of $\mathfrak{G}$ of order g prime to n.[1] Since every element in $\mathfrak{g}$ is an n-th power[2], the equation $c_1 = c_2 x^n$ is solvable in $\mathfrak{g}$ for every pair of elements c_1, c_2, and since $\mathfrak{g}$ lies in the center of $\mathfrak{G}$, we have $A_{c_1,n} = A_{c_2,n}$. It now follows from the class equation that

$$N = \sum_{\mathfrak{C} \nsubseteq \mathfrak{g}} A_{\mathfrak{C},n} + g \cdot A_{c,n}.$$

In the above, N and all the $A_{\mathfrak{C},n}$ with $\mathfrak{C} \nsubseteq \mathfrak{g}$ are divisible by (n, N). Therefore $g \cdot A_{\mathfrak{C},n}$ is divisible by (n, N). Since $(g, n) = 1$, we have

$$(n, N)/A_{c,n}, \qquad \text{Q.E.D.}$$

* Since the index of N_c equals h. (*Ed.*)
[1] See Exercises 2, 3 at the end of the Chapter.
[2] See Exercise 3 at the end of the Chapter.

Exercises

1. The complex of all n-th powers of elements of a complex $\mathfrak{K}$ in a group is denoted by $\mathfrak{K}^{\underline{n}}$. The complex of all elements in $\mathfrak{K}$ whose n-th power is equal to e is denoted by $\mathfrak{K}_{\underline{n}}$.

We have

$$\mathfrak{K}^{\underline{1}} = \mathfrak{K}, \quad \mathfrak{K}^{\underline{-1}} = \mathfrak{K}^{-1}, \quad \mathfrak{K}^{\underline{nm}} = (\mathfrak{K}^{\underline{n}})^{\underline{m}}, \quad \mathfrak{K}_{\underline{n}} \cap \mathfrak{K}_{\underline{m}} = \mathfrak{K}_{\underline{(n,\,m)}}.$$

2. If a commutes with b then

$$(ab)^n = a^n b^n$$

and the order of ab is a divisor of the least common multiple of the orders of a and b.

3. If the rational integer n is relatively prime to the order of $\mathfrak{G}$ then $\mathfrak{G}^{(n)} = \mathfrak{G}$.

(Exercises 4-6 in Burnside.)

4. In a group $\mathfrak{G}$ if the equation

$$(ab)^n = a^n b^n$$

holds for every pair a, b of group elements, then $\mathfrak{G}^{\underline{n}}$ and $\mathfrak{G}_{\underline{n}}$ are subgroups of $\mathfrak{G}$.

Then, moreover, $\mathfrak{G} : \mathfrak{G}^{\underline{n}} = \mathfrak{G}_{\underline{n}} : 1$.

(Hint: The elements of $\mathfrak{G}$ whose n-th power is a fixed element of $\mathfrak{G}$ form a (right) coset of $\mathfrak{G}$ with respect to $\mathfrak{G}_{\underline{n}}$.)

$\mathfrak{G}^{\underline{n-1}}$ commutes elementwise with $\mathfrak{G}^{\underline{n}}$. (Young.)

5. If $\mathfrak{U}$ and $\mathfrak{V}$ are finite subgroups of the group $\mathfrak{G}$, then $\mathfrak{U}\mathfrak{V}$ contains exactly $\dfrac{(\mathfrak{U}:1)\,(\mathfrak{V}:1)}{(\mathfrak{U}\cap\mathfrak{V}:1)}$ elements.

6. If the index of the normal subgroup $\mathfrak{N}$ of a finite group $\mathfrak{G}$ is relatively prime to the order n of $\mathfrak{N}$, then $\mathfrak{N}$ contains every subgroup of $\mathfrak{G}$ whose order is a divisor of n. (Use Exercise 5.)

7. The alternating permutation group of $n > 2$ letters can be generated by $(123), (124), \ldots, (12n)$.

8. A well known puzzle requires that 15 numbered stones on a board divided into 16 squares be moved horizontally and vertically until we obtain the situation of Fig. 1, p. 30.

We may assume that in the initial position the lower right corner of the board is vacant, so that the initial position can be described uniquely, with the use of Fig. 1, by a permutation of the fifteen letters. It is to be shown that Fig. 1 is attainable precisely when the permutation for the initial position is even. (Generalization?)

9. If $\mathfrak{N}$ is a normal subgroup of the finite group $\mathfrak{G}$, then a normal multiplication table of $\mathfrak{G}$ can be constructed so that it is possible to divide the table into squares having the following properties:

1). Each square contains the same number of compartments. (The number of squares is $(\mathfrak{G}:\mathfrak{N})^2$.

2). The rows of each square are the same to within the order of elements.

3). The square in the upper left corner contains exactly the elements of $\mathfrak{N}$ (Example, Fig. 2).

What sort of elements are in a square?

Conversely if it is possible to divide a normal multiplication table of $\mathfrak{G}$ into squares, such that 1., 2. hold and e is in the upper left corner of a square, then it is to be shown that we have a division into squares with respect to a normal subgroup.

What divisions of $\mathfrak{G}$ occur if we omit the condition that the multiplication table be normal?

1	2	3	4
5	6	7	8
9	10	11	12
13	14	15	

Fig. 1.

e	a	b	c
a	e	c	b
b	c	e	a
c	b	a	e

Fig. 2

10. If $\mathfrak{F}$ is a set of complexes of a given group with the properties:
1). Every element in $\mathfrak{G}$ is in at least one of the complexes of $\mathfrak{F}$.
2). No complex in $\mathfrak{F}$ is a proper subset of any other complex of $\mathfrak{F}$.
3). The product of two complexes in $\mathfrak{F}$ is contained in a third complex of $\mathfrak{F}$, then $\mathfrak{F}$ is the set of cosets of $\mathfrak{G}$, with respect to a normal subgroup, and $\mathfrak{F}$ is a group.

Additional Exercises

11. Let $\mathfrak{U}$, $\mathfrak{V}$ be two subgroups of finite index in the group $\mathfrak{G}$. Denote by $\mathfrak{U}\mathfrak{V}\!:\!\mathfrak{U}$ the number of right cosets modulo $\mathfrak{U}$ in $\mathfrak{U}\mathfrak{V}$ and by $\mathfrak{U}\mathfrak{V}\!:\!\mathfrak{V}$ the number of left cosets modulo $\mathfrak{V}$ in $\mathfrak{U}\mathfrak{V}$. Show that

a) $\mathfrak{U}\mathfrak{V}\!:\!\mathfrak{U} = \mathfrak{V}\mathfrak{U}\!:\!\mathfrak{U} = \mathfrak{V}\!:\!(\mathfrak{U} \wedge \mathfrak{V})$.

b) $\mathfrak{G}\!:\!(\mathfrak{U} \wedge \mathfrak{V}) = (\mathfrak{G}\!:\!\mathfrak{U})\,(\mathfrak{U}\mathfrak{V}\!:\!\mathfrak{V}) = (\mathfrak{U}\mathfrak{V}\!:\!\mathfrak{U})\,(\mathfrak{G}\!:\!\mathfrak{V}) \leqq (\mathfrak{G}\!:\!\mathfrak{U})\,(\mathfrak{G}\!:\!\mathfrak{V})$.

c) If $\mathfrak{G}\!:\!\mathfrak{U}$ and $\mathfrak{G}\!:\!\mathfrak{V}$ are coprime then $\mathfrak{G}\!:\!(\mathfrak{U} \wedge \mathfrak{V}) = (\mathfrak{G}\!:\!\mathfrak{U})\,(\mathfrak{G}\!:\!\mathfrak{V})$.

12. Let $a_1, a_2, \ldots, a_n$ be n elements, not necessarily distinct, of a group of order n. Show that there exist integers p and q, $1 \leqq p \leqq q \leqq n$, such that $a_p a_{p+1} \ldots a_q = e$. (*Moser.*)

13. Let $\mathfrak{G}$ be a group and let $\mathfrak{K}$ be a complex consisting of the elements $a_1, a_2, \ldots, a_n$ of $\mathfrak{G}$ such that $\mathfrak{K}^2$ does not contain e. Consider the n^2 (not necessarily distinct) elements of $\mathfrak{G}$ of the form $a_i a_j$ and prove that at most $n(n-1)/2$ of these products are themselves in $\mathfrak{K}$. (*Hint:* One has $a_k a_j^{-1} = a_i$ as often as $a_i a_j = a_k$.)

Seek the best estimate for the number of elements in $\mathfrak{K}^2 - \mathfrak{K}$ depending only on n. (*Moser.*)

14. *Decomposition with respect to a double module.*

a) For any two subgroups $\mathfrak{U}$ and $\mathfrak{V}$ of a group $\mathfrak{G}$ a normal congruence relation is defined by the rule: $a \equiv b$ (mod $\mathfrak{U}$, $\mathfrak{V}$) if $b = uav$ with $u \in \mathfrak{U}$, $v \in \mathfrak{V}$.

b) For a given representative system $a_1, a_2, \ldots$ modulo $\mathfrak{U}$, $\mathfrak{V}$ there is the decomposition $\mathfrak{G} = \sum \mathfrak{U} a_i \mathfrak{V}$ of $\mathfrak{G}$ into residue classes with respect to the double module $\mathfrak{U}$, $\mathfrak{V}$. The residue class $\mathfrak{U} a_i \mathfrak{V}$ consists of $\mathfrak{V}\!:\!(\mathfrak{V} \wedge a_i^{-1}\mathfrak{U} a_i)$ right cosets of $\mathfrak{G}$ modulo $\mathfrak{U}$ or of $\mathfrak{U}\!:\!(\mathfrak{U} \wedge a_i \mathfrak{V} a_i^{-1})$ left cosets of $\mathfrak{G}$ modulo $\mathfrak{V}$.

c) Is it possible to interpret the left congruence modulo a given subgroup $\mathfrak{U}$ as a congruence with respect to a suitable double module? Is there a similar possibility for right congruence? What is the congruence with respect to a double module $\mathfrak{N}$, $\mathfrak{N}$ if $\mathfrak{N}$ is a normal subgroup of $\mathfrak{G}$?

15. A set $\mathfrak{G}$ is called a *groupoid* if for certain ordered pairs of elements a, b of $\mathfrak{G}$ the product ab is uniquely defined in $\mathfrak{G}$ in such a way that

I. $a(bc) = (ab)c$

in the sense that whenever one side of the equation can be formed in $\mathfrak{G}$, the other side can also be formed, and both sides are equal,

II. with any two elements a, b of $\mathfrak{G}$ there can be associated at least one element x such that both ax and xb are defined in $\mathfrak{G}$,

III. if a, b have a common left multiplier x such that both xa and xb are defined, then the equation $ay = b$ can be solved, and if c, d have a common right multiplier, then the equation $ua = b$ can be solved.

Trivial example: Let Σ be a system of sets with the same cardinality. The set $\mathfrak{G}(\Sigma)$ of all one-to-one correspondences between any two (not necessarily different) sets in Σ is a groupoid, if the one-to-one mapping π of S_1 onto S_2 and the one-to-one mapping ρ of S_3 onto S_4 are combined to form the mapping $\pi\rho$ if, and only if $S_4 = S_1$ where the product $\pi\rho$ is defined according to the rule $(\pi\rho)x = \pi(\rho x)$ for $x \in S_3$ which first occurred in § 2.

We may interpret a groupoid $\mathfrak{G}$ as a semi-group with special properties by doing the following:

(1) Introduce a new symbol n.

(2) Extend the given multiplication in $\mathfrak{G}$, enlarged by n, by defining $ab = n$ whenever ab is not defined in $\mathfrak{G}$ and $an = na = nn = n$ for any element a in $\mathfrak{G}$.

16. Using the definition of a groupoid $\mathfrak{G}$ as given in the preceding exercise prove that:

a) For each element a there exists a left unit e_a and a right unit ${}_ae$ such that $e_a a = a = a\,{}_ae$.

b) If a, b have a common left multiplier then a left unit of a is a left unit of b. Show that $e_a e_a$ is defined, and hence $e_a e_a = e_a$.

c) For each left unit e_a of a there is an inverse a^{-1} satisfying $aa^{-1} = e_a$. Deduce from the equations

$$e_a' = e_a e_a' = (aa^{-1})\, e_a' = a(a^{-1} e_a')$$
$$a^{-1} e_a' = a^{-1}$$

which are valid for any left unit e_a' of a that there is only one left unit of a. Similarly, show there is only one right unit of a.

d) $a^{-1}a = {}_ae$, $e_{a^{-1}} = {}_ae$, ${}_{a^{-1}}e = e_a$.

e) There is only one inverse of a.

f) $(a^{-1})^{-1} = a$.

g) (Uniqueness of division.) If $ay = b$, then $y = a^{-1}b$. If $uc = d$, then $u = dc^{-1}$.

h) $(ab)^{-1} = b^{-1}a^{-1}$.

i) Those elements having a given unit e as their left and right unit form a group $\mathfrak{G}_e$.

(*Exercises* 17-21 *inclusive extend the concepts of* § 4.)

17. Any congruence relation R generates a normal congruence relation R^* as follows: $R^*(a, b)$ is true if and only if there is a chain of elements $a = a_1, a_2, \dots, a_n = b$ such that for any two consecutive elements a_i, a_{i+1} at least one of the three statements

$$a_i = a_{i+1}, \quad R(a_i, a_{i+1}), \quad R(a_{i+1}, a_i)$$

is true. R^* is called the *ancestral relation* of R.

Example: The relation "b is child of a" generates the relation "c and d have a common ancestor."

18. Show that:

a) By *symmetrization* of the binary relation R on a set $\mathfrak{M}$ one obtains the symmetric relation R^s defined by: $a\,R^s\,b$ if and only if $a\,R\,b$ or $b\,R\,a$, such that every symmetric relation implying R also implies R^s;

b) By forming the *ancestral relation* of R one obtains the transitive relation R^a defined by: $a\,R^a\,b$ if and only if there is a finite chain: $a = a_0$, $a_0\,R\,a_1$, $a_1\,R\,a_2, \dots,$ $a_{n-1}\,R\,a_n$, $a_n = b$ $(n > 0)$ such that every transitive relation implying R also implies R^a (note that the ancestral relation of 'a is parent of b' is 'a is ancestor of b');

c) The ancestral relation of a symmetric relation is symmetric, but the symmetrization of a transitive relation need not be transitive;

d) By *normalization* of R one obtains the normal relation R^n defined by: $a R^n b$ if and only if there is a finite chain $a = a_0$, $a_{2i} R a_{2i+1}$ or $a_{2i} = a_{2i+1}, a_{2i+1}$ or $a_{2i+2} R a_{2i+2} = a_{2i+1}$ $(i = 0, 1, 2, \ldots, n-1;\ n > 0)$, $a_{2n} = b$, such that every normal relation implying R also implies R^n,

e) If R is reflexive and symmetric, then $R^n = R^a$.

19. A *multiplicative domain* is a set $\mathfrak{G}$ in which a multiplication is uniquely defined. For any subset $\mathfrak{K}$ of $\mathfrak{G}$ we define the right congruence modulo $\mathfrak{K}$ as the ancestral relation of the relation $a \equiv ak$ for $a \in \mathfrak{G}$, $k \in \mathfrak{K}$. Similarly, we define left congruence modulo $\mathfrak{K}$ as the ancestral relation generated from $a \equiv ka$ for $a \in \mathfrak{G}$, $k \in \mathfrak{K}$.

With these definitions we can introduce left cosets, right representative systems, right cosets, left representative systems modulo $\mathfrak{K}$.

Both left and right congruence modulo the empty set coincide with the equality relation in $\mathfrak{G}$.

If $\mathfrak{G}$ has a unit element e and if e is in $\mathfrak{K}$, then the right coset $\mathfrak{U}$ represented by e is closed under multiplication and coincides with the left coset represented by e modulo $\mathfrak{K}$. $\mathfrak{U}$ is the smallest subset of $\mathfrak{G}$ containing $\mathfrak{K}$ and closed under multiplication.

If $\mathfrak{G}$ is a semigroup then right (left) congruence modulo $\mathfrak{K}$ coincides with right (left) congruence modulo $\mathfrak{U}$.

Give examples in which the number of left cosets is different from the number of right cosets modulo $\mathfrak{K}$.

20. If $\mathfrak{U}$ is a subgroup of the semigroup $\mathfrak{G}$, then right (left) congruence modulo $\mathfrak{U}$, as defined in the preceding exercise, has the same meaning as in groups. Each coset contains as many elements as $\mathfrak{U}$.

21. If $\mathfrak{G}$ is a group, then right (left) congruence modulo a subset $\mathfrak{K}$ coincides with the same relation modulo the smallest subgroup of $\mathfrak{G}$ containing $\mathfrak{K}$.

(*Exercises* 22-28 *extend the concepts of* § 8.)

22. Every subgroup of an abelian semigroup is normal.

23. The ancestral relation of a multiplicative and reflexive congruence relation is a multiplicative normal congruence relation.

24. Let $\mathfrak{G}$ be a multiplicative domain with a subset $\mathfrak{K}$. The ancestral relation of the relation "$a \equiv b$ if and only if there is a factorization of both a and b with the same number of factors, and with the same distribution of brackets such that corresponding factors are either right congruent modulo $\mathfrak{K}$ or left congruent modulo $\mathfrak{K}$" is a normal multiplicative congruence relation $NM(\mathfrak{K})$.

If $\mathfrak{K}$ is empty, then $NM(\mathfrak{K})$ coincides with the equality relation.

It is always true that all elements of $\mathfrak{K}$ belong to the same residue class $\mathfrak{K}^*$ modulo $NM(\mathfrak{K})$. Show that $NM(\mathfrak{K})$ coincides with $NM(\mathfrak{K}^*)$ and that $\mathfrak{K}^*$ is closed under multiplication.

25. A *normal divisor* of the multiplicative domain $\mathfrak{G}$ may be defined as a non-empty subset $\mathfrak{N}$, closed under multiplication, such that a product in $\mathfrak{N}$ of several elements of $\mathfrak{G}$ remains in $\mathfrak{N}$ after any one of its factors has been multiplied from the left or the right by an element of $\mathfrak{N}$.

Show that any normal divisor $\mathfrak{N}$ of $\mathfrak{G}$ is a residue class (called the unit residue class) of the multiplicative normal congruence relation $NM(\mathfrak{N})$ generated by the rule $au \equiv ua \equiv a$ for all $a \in \mathfrak{G}$, $u \in \mathfrak{N}$, and, conversely, show that the subset $\mathfrak{K}^*$ constructed at the end of the preceding exercise is a normal divisor of $\mathfrak{G}$. (*E. Lyapin*).

26. If there is a unit residue class $\mathfrak{N}$, as defined in the preceding exercise, for a certain multiplicative normal congruence relation R, then $\mathfrak{N}$ is a normal divisor of $\mathfrak{G}$ and the other residue classes with respect to R are obtained by uniting some of the residue classes re $NM(\mathfrak{N})$, i.e. R is somewhat of a blurring of $NM(\mathfrak{N})$.

27. $\mathfrak{G}$ is a normal divisor of $\mathfrak{G}$. What is $NM(\mathfrak{G})$?

28. In a group, the notion of the normal divisor as defined in 25. coincides with the notion of normal subgroup.

II. THE CONCEPT OF HOMOMORPHY AND GROUPS WITH OPERATORS

§ 1. Homomorphisms

1. *The Concept of Homomorphy.*

Let $\mathfrak{G}$ and $\mathfrak{G}^*$ be sets in which a multiplication is uniquely defined (multiplicative domains).

DEFINITION: A single valued mapping of the elements in $\mathfrak{G}$ onto a certain subset of $\mathfrak{G}^*$ is called a *homomorphy,* if the product of two elements is mapped onto the product of the image elements.

Example: The mapping, defined on page 8, of $\mathfrak{S}_n$ into the group consisting of ± 1 is a homomorphy.

If the image of x is denoted by σx then σ must satisfy the functional equation:

$$\sigma(xy) = \sigma x \cdot \sigma y \ .$$

The homomorphy is said to be a *homomorphic mapping* or a *homomorphism* if every element of $\mathfrak{G}^*$ is an image element. $\mathfrak{G}^*$ is said to be homomorphic to $\mathfrak{G}$. We denote this by: $\mathfrak{G} \sim \mathfrak{G}^*$.

A homomorphy is a mapping *into* $\mathfrak{G}^*$ while a homomorphism is a mapping *onto* $\mathfrak{G}^*$.

Example: From Chapter I. § 6 we see that the mapping of the group of surface rotations of a regular tetrahedron into $\mathfrak{S}_4$ is a homomorphy, but onto $\mathfrak{A}_4$ the mapping is a homomorphism.

Under every homomorphy the set $\overline{\mathfrak{G}}$ of image elements is homomorphic to $\mathfrak{G}$.

The relation of homomorphy is transitive. If $x \to \sigma x$ is a homomorphy of $\mathfrak{G}$ into $\mathfrak{G}^*$ *and* $x^* \to \tau x^*$ a homomorphy of $\mathfrak{G}^*$ into $\mathfrak{G}^{**}$, then the product of τ by σ is defined by means of the equation:

$$\tau \sigma x = \tau(\sigma x) \ .$$

$\tau\sigma$ is a single-valued mapping of the elements of $\mathfrak{G}$ onto a certain subset of $\mathfrak{G}^{**}$.

Since

$$(\tau\sigma)(xy) = \tau(\sigma(xy)) = \tau(\sigma x \cdot \sigma y) = \tau(\sigma x) \cdot \tau(\sigma y) = \tau\sigma x \cdot \tau\sigma y,$$

$\tau\sigma$ is a homomorphy of $\mathfrak{G}$ into $\mathfrak{G}^{**}$. We have the following rule for calculation with homomorphies:

If $\tau\sigma$ *and* $\varrho\tau$ *are defined, then* $\varrho(\tau\sigma)$ *and* $(\varrho\tau)\sigma$ *are also defined and*

$$\varrho(\tau\sigma) = (\varrho\tau)\sigma = \varrho\tau\sigma.$$

The defining equation $(\sigma\tau)x = \sigma(\tau x)$ shows that $\sigma\tau x$ can be written instead of $(\sigma\tau)x$ without misunderstanding. *Thus the homomorphy relation is transitive.*

From $\mathfrak{G} \sim \mathfrak{G}^*$, $\mathfrak{G}^* \sim \mathfrak{G}^{**}$ it follows that $\mathfrak{G} \sim \mathfrak{G}^{**}$.

The homomorphy relation is reflexive. That is, the identity mapping $\underline{1}_{\mathfrak{G}}$ of $\mathfrak{G}$, defined by $\underline{1}_{\mathfrak{G}}x = x$, has $\mathfrak{G}$ as its set of images.

If we speak of the product of two homomorphies, then it will be assumed at the same time that it is definable in terms of the above relations.

In this sense

$$\sigma\underline{1}_{\mathfrak{G}^*} = \sigma \quad \text{and} \quad \underline{1}_{\mathfrak{G}}\sigma = \sigma.$$

The image of the complex $\mathfrak{K}$ in $\mathfrak{G}$ under the homomorphy σ is denoted by $\sigma\mathfrak{K}$. If $\mathfrak{K}$ is a multiplicative subdomain, then $\mathfrak{K}$ is mapped homomorphically onto $\sigma\mathfrak{K}$ by σ .

We say σ *induces* a homomorphy of $\mathfrak{K}$ into $\mathfrak{G}^*$.

The homomorphic image of a group $\mathfrak{U}$ *in* $\mathfrak{G}$ *is a group*:

If $\mathfrak{U}$ is a subgroup of $\mathfrak{G}$, then it follows from $x, y \in \mathfrak{U}$ that $xy \in \mathfrak{U}$; *therefore* $\sigma x \cdot \sigma y = \sigma(xy) \in \sigma\mathfrak{U}$; $(\sigma x \cdot \sigma y)\,\sigma z = \sigma x\,(\sigma y \cdot \sigma z) = \sigma(xyz)$.

Furthermore $ex = xe = x$, $\sigma e \cdot \sigma x = \sigma x \cdot \sigma e = \sigma x$; therefore σe is the unit element of $\sigma\mathfrak{U}$. We have $\sigma(xx^{-1}) = \sigma x \cdot \sigma(x^{-1}) = \sigma e$ and therefore $\sigma(x^{-1})$ is the inverse of σx . Therefore $\sigma\mathfrak{U}$ is a group.

If $\overline{\mathfrak{U}}$ *is a subgroup of the image domain* $\mathfrak{G}$, *then the set of all the elements of* $\mathfrak{G}$ *whose image is in* $\overline{\mathfrak{U}}$ *forms a subgroup* $\mathfrak{U}$ *of* $\mathfrak{G}$, *and*

$$\overline{\mathfrak{U}} = \sigma\mathfrak{U}.$$

Every element in $\overline{\mathfrak{U}}$ is of the form σx for x in $\mathfrak{U}$; if $x, y \in \mathfrak{U}$, then $\sigma(xy) = \sigma x \cdot \sigma y \in \overline{\mathfrak{U}}$, $xy \in \mathfrak{U}$, $\sigma(x^{-1}) = (\sigma x)^{-1} \in \overline{\mathfrak{U}}$, $x^{-1} \in \mathfrak{U}$.

2. *The Isomorphy Concept.*

If a group $\mathfrak{G}$ is mapped onto the group $\overline{\mathfrak{G}}$ homomorphically, then multiplication in $\overline{\mathfrak{G}}$ parallels that in $\mathfrak{G}$. However, we consider two groups as the same in abstract group theory only if their tables differ merely in notation,order of rows and columns: Homomorphic groups are not always equivalent in the abstract sense. If, for example, the group $\mathfrak{G}$ contains more than one element, then there is a homomorphism of $\mathfrak{G}$ which maps every element of $\mathfrak{G}$ onto the unit element, but the tables of $\mathfrak{G}$ and e have a different number of rows.

Those homomorphic mappings under which the table of the group is preserved are called *isomorphic mappings*. That the homomorphic mapping is *one-one* is necessary and sufficient for the latter.

DEFINITION: A homomorphy σ of the multiplicative domain $\mathfrak{G}$ into the multiplicative domain $\mathfrak{G}^*$ is called an *isomorphy,* if $\mathfrak{G}$ is mapped onto the set $\overline{\mathfrak{G}}$ of image elements in a one-one manner, i.e., if $\sigma x = \sigma y$ implies $x = y$.

An isomorphy which is a also homomorphic mapping is called an *isomorphic mapping* (*isomorphism*). $\mathfrak{G}$ is isomorphic to $\overline{\mathfrak{G}}$ under every isomorphy.

Example: The group of rotations of an equilateral triangle is isomorphic to $\mathfrak{S}_3$; the group of rotations of a regular tetrahedron is isomorphic to $\mathfrak{A}_4$ (§ 6 of the previous chapter).

The existence of an isomorphic mapping of $\mathfrak{G}$ onto $\overline{\mathfrak{G}}$ is denoted by $\mathfrak{G} \simeq \overline{\mathfrak{G}}$.

The three well-known rules hold for isomorphism:

1) The identity isomorphism maps $\mathfrak{G}$ onto itself.

2) If $\mathfrak{G} \simeq \overline{\mathfrak{G}}$, $\overline{\mathfrak{G}} \simeq \overline{\overline{\mathfrak{G}}}$, then $\mathfrak{G} \simeq \overline{\overline{\mathfrak{G}}}$, since $\sigma\tau x = \sigma\tau y$ implies $\tau x = \tau y$ and $x = y$.

3) If $\mathfrak{G} \simeq \overline{\mathfrak{G}}$, then every element y in $\overline{\mathfrak{G}}$ can be written uniquely in the form $y = \sigma x$. $\sigma^{-1} y = x$ now defines an isomorphic mapping of $\overline{\mathfrak{G}}$ onto $\mathfrak{G}$: Let $y_1 = \sigma x_1$, $y_2 = \sigma x_2$; then

$$\sigma^{-1}(y_1 y_2) = \sigma^{-1}(\sigma x_1 \cdot \sigma x_2) = \sigma^{-1}(\sigma(x_1 x_2)) = x_1 x_2 = \sigma^{-1} y_1 \cdot \sigma^{-1} y_2 .$$

Moreover, the mapping σ^{-1} is one-one, since σ is one-one. Therefore it follows from $\mathfrak{G} \simeq \overline{\mathfrak{G}}$ that $\overline{\mathfrak{G}} \simeq \mathfrak{G}$. Calculation with inverse mappings satisfies the following rules:

If the equation $\tau\sigma = 1_{\mathfrak{G}}$ is solvable for a homomorphy σ, then σ is an isomorphy, since $\sigma x = \sigma y$ implies $\tau\sigma x = \tau\sigma y$ and therefore $x = y$.

An isomorphic mapping can also be defined as a homomorphy for which $\tau\sigma = 1_{\mathfrak{G}}$ and $\sigma\varrho = 1_{\mathfrak{G}^*}$ are solvable. Then for all $y \in \mathfrak{G}^*: y = \sigma\varrho y$ and therefore $\mathfrak{G}^* = \overline{\mathfrak{G}}$. Moreover $\tau = \varrho = \sigma^{-1}$.

3. *Factor Group. Isomorphy Theorems.*

Under what circumstances is it possible to read off the multiplication table of a homomorphic image of a given group $\mathfrak{G}$ from the multiplication table of $\mathfrak{G}$ itself?

We have first the following theorem:

If $\mathfrak{N}$ is a normal subgroup of $\mathfrak{G}$, then there is a homomorphism σ of $\mathfrak{G}$ under which the set of elements of $\mathfrak{G}$ mapped onto σe is precisely $\mathfrak{N}$.

We set $$\sigma a = a\mathfrak{N},$$
Then $\sigma(ab) = ab\mathfrak{N} = a(b\mathfrak{N})\mathfrak{N} = a\mathfrak{N} \cdot b\mathfrak{N} = \sigma a \cdot \sigma b.$

From the above we realize that the set of cosets of $\mathfrak{G}$ with respect to a normal subgroup of $\mathfrak{G}$ form a group homomorphic to $\mathfrak{G}$. The group of residue classes of a group $\mathfrak{G}$ with respect to a normal subgroup $\mathfrak{N}$ is called a *factor group* and is denoted by $\mathfrak{G}/\mathfrak{N}$. The order of a factor group is equal to $\mathfrak{G}:\mathfrak{N}$. The unit element of the factor group is the normal subgroup $\mathfrak{N}$.

If, conversely, the left cosets, formed with respect to a subgroup, form a group under the usual complex multiplication, then the subgroup is a normal subgroup, as we saw previously.

FIRST ISOMORHISM THEOREM: *Under a homomorphism σ of a given group $\mathfrak{G}$ onto a group $\overline{\mathfrak{G}}$, all the elements of $\mathfrak{G}$ which are mapped onto the unit element of $\overline{\mathfrak{G}}$ form a normal subgroup $\mathfrak{E}$* of $\mathfrak{G}$, *called the kernel of σ. The factor group of $\mathfrak{G}$ with respect to $\mathfrak{E}$ is isomorphic to $\overline{\mathfrak{G}}$.*

Proof: All the elements of $\mathfrak{G}$ which are mapped on the unit element $\bar{e}$ of $\mathfrak{G}$ under σ form a subgroup $\mathfrak{E}$.

$\sigma(a\mathfrak{E}) = \sigma a$ gives a one-valued mapping.

$$\sigma(a\mathfrak{E} \cdot b\mathfrak{E}) = \sigma(ab\mathfrak{E}) = \sigma(ab) = \sigma a \cdot \sigma b = \sigma(a\mathfrak{E}) \cdot \sigma(b\mathfrak{E}).$$

From $\sigma(a\mathfrak{E}) = \sigma(\mathfrak{E}) = \sigma e$ it follows that $\sigma a = \sigma e$; and therefore

$$a \in \mathfrak{E},\ a\mathfrak{E} = \mathfrak{E}.$$

Therefore

$$a\mathfrak{E} \cdot b\mathfrak{E} = ab\mathfrak{E},$$

$\mathfrak{E}$ is a normal subgroup, and from the isomorphism it follows that the homomorphic image of $\mathfrak{G}$ has the same table as the factor group $\mathfrak{G}/\mathfrak{E}$. Therefore the question of the multiplication tables of homomorphic images will be resolved if we can give all the normal subgroups of the original group. (See Exercises 9, 10 at the end of Chap. I.)

SECOND ISOMORPHISM THEOREM: *If $\mathfrak{U}$ is a subgroup and $\mathfrak{N}$ is a normal subgroup of the group $\mathfrak{G}$, then the intersection $\mathfrak{U} \cap \mathfrak{N}$ is a normal subgroup of $\mathfrak{U}$ and*

$$\mathfrak{U}/\mathfrak{U} \cap \mathfrak{N} \simeq \mathfrak{U}\mathfrak{N}/\mathfrak{N}.$$

The isomorphism is obtained by means of the mapping:

$$U(\mathfrak{U} \cap \mathfrak{N}) \rightarrow U(\mathfrak{U} \cap \mathfrak{N}) \cdot \mathfrak{N} = U\mathfrak{N}.$$

Proof: The homomorphism $a \rightarrow a\mathfrak{N}$ of $\mathfrak{G}$ onto $\mathfrak{G}/\mathfrak{N}$ is again denoted by σ. Then $\mathfrak{U} \sim \sigma\mathfrak{U}$. Under the mapping of $\mathfrak{U}$ onto $\sigma\mathfrak{U}$, precisely the elements of $\mathfrak{U} \cap \mathfrak{N}$ map onto e; therefore $\mathfrak{U} \cap \mathfrak{N}$ is a normal subgroup of $\mathfrak{U}$, and $\mathfrak{U}/\mathfrak{U} \cap \mathfrak{N} \simeq \sigma\mathfrak{U}$. If we replace $\mathfrak{U}$ by $\mathfrak{U}\mathfrak{N}$ then the same argument shows that $\mathfrak{U}\mathfrak{N}/\mathfrak{N} \simeq \sigma\mathfrak{U}$. The theorem follows from both isomorphisms.

If $\mathfrak{H}$ is a normal subgroup of the group $\mathfrak{G}$, then for every homomorphism σ the group $\overline{\mathfrak{H}} = \sigma\mathfrak{H}$ is a normal subgroup of $\overline{\mathfrak{G}} = \sigma\mathfrak{G}$ since $\sigma x \cdot \overline{\mathfrak{H}}(\sigma x)^{-1} = \sigma(x\mathfrak{H}x^{-1}) = \sigma\mathfrak{H} = \overline{\mathfrak{H}}$. If, conversely, $\overline{\mathfrak{H}}$ is a normal subgroup of $\overline{\mathfrak{G}}$, then all the elements of $\mathfrak{G}$ whose image is in $\overline{\mathfrak{H}}$ form a normal subgroup $\mathfrak{H}$ of $\mathfrak{G}$, since

$$\sigma(x\mathfrak{H}x^{-1}) = \sigma(x)\overline{\mathfrak{H}}(\sigma x)^{-1} = \overline{\mathfrak{H}},$$

and therefore $x\mathfrak{H}x^{-1} \leq \mathfrak{H}$. Information on the relation between factor groups is given by the

THIRD ISOMORPHISM THEOREM: *Let σ be a homomorphic mapping of $\mathfrak{G}$ onto $\overline{\mathfrak{G}}$. Let $\mathfrak{E}$ be the normal subgroup composed of the elements of $\mathfrak{G}$ which map onto the unit element of $\overline{\mathfrak{G}}$; let $\overline{\mathfrak{H}}$ be a normal subgroup of $\overline{\mathfrak{G}}$, let $\mathfrak{H}$ be the group of elements in $\mathfrak{G}$ whose image falls in $\overline{\mathfrak{H}}$.*

Then $\mathfrak{H}$ is a normal subgroup of $\mathfrak{G}$, and

$$\mathfrak{G}/\mathfrak{H} \simeq \overline{\mathfrak{G}}/\overline{\mathfrak{H}} \simeq \mathfrak{G}/\mathfrak{E}/\mathfrak{H}/\mathfrak{E}.$$

Proof: We have $\mathfrak{G} \sim \overline{\mathfrak{G}}$ and $\overline{\mathfrak{G}} \sim \overline{\mathfrak{G}}/\overline{\mathfrak{H}}$. Under the second homomorphism exactly the elements of $\overline{\mathfrak{H}}$ map onto the identity coset $\overline{\mathfrak{H}}$. Under the first homomorphism precisely the elements of $\mathfrak{H}$ map onto $\overline{\mathfrak{H}}$. Therefore $\mathfrak{G}/\mathfrak{H} \simeq \overline{\mathfrak{G}}/\overline{\mathfrak{H}}$. If we set $\overline{\mathfrak{G}} = \mathfrak{G}/\mathfrak{E}$, then $\mathfrak{G}/\mathfrak{H} \simeq \mathfrak{G}/\mathfrak{E}/\mathfrak{H}/\mathfrak{E}$.

§ 2. Representation of Groups by Means of Permutations

We want to find the homomorphisms of given abstract groups onto permutation groups.

DEFINITION: A single-valued mapping $x \to \pi_x$ of the elements x of a group $\mathfrak{G}$ onto the permutations π_x of ω letters is called a *representation of $\mathfrak{G}$ (as a permutation group) of degree a* if

$$\pi_{xy} = \pi_x \cdot \pi_y.$$

All permutations π_x form a group $\mathfrak{P}$, the *representation group.*

A representation is said to be faithful if the homomorphy induced by the representation is an isomorphy.

Two representations Δ, Δ' by means of letters from $\mathfrak{M}_1$, $\mathfrak{M}_2$ respectively, are said to be *equivalent* if there is a 1-1 mapping $a \to a'$ of the letters of $\mathfrak{M}_1$ onto those of $\mathfrak{M}_2$ such that $(\pi_x a)' = \pi_x' a'$ for all x; in short if the representations are the same except for the naming of the letters.

If the permuted letters form a system of transitivity under $\mathfrak{P}$, then the permutation group $\mathfrak{P}$ *and* the representation of $\mathfrak{G}$ by $\mathfrak{P}$ are called *transitive,* otherwise they are called *intransitive.*

If $\mathfrak{T}$ is a system of transitivity of $\mathfrak{P}$, then $\pi_x' = \binom{t}{\pi_x t}$, where $t \in \mathfrak{T}$, is a permutation of letters in $\mathfrak{T}$, and the mapping $x \to \pi_x'$ gives a transitive representation $\Delta_{\mathfrak{T}}$ of $\mathfrak{G}$. The representation group belonging to $\Delta_{\mathfrak{T}}$ is called a *transitive component* of the original representation.

Since, clearly, every representation can be constructed from the transitive sub-representations, it is sufficient to investigate the transitive representations of a given group $\mathfrak{G}$.

Let a transitive representation Δ of degree ω of the group $\mathfrak{G}$ be given. We choose a letter a and consider two elements of $\mathfrak{G}$ to be in the same class if the corresponding permutations have the same effect on the letter a. With the help of this decomposition into classes, a normal congruence relation is definable. Moreover, $x \equiv y$ implies $\pi_x a = \pi_y a$, which implies $\pi_{zx} a = \pi_z \pi_x a = \pi_z \pi_y a = \pi_{zy} a$, and thus $zx \equiv zy$; therefore we have a right congruence with respect to the subgroup $\mathfrak{G}_a$ which consists of all elements of $\mathfrak{G}$ whose corresponding permutation leaves the letter a fixed. If we call the left coset consisting of all elements x for which $\pi_x a = b$, R_b, then

$$y R_b = R_{\pi_y b} \quad \text{or also} \quad \pi_y = \binom{R_b}{y R_b}.$$

The subgroups $\mathfrak{G}_a$, $\mathfrak{G}_b$, ... form a family of conjugate subgroups of $\mathfrak{G}$, since $x \mathfrak{U}_a x^{-1} = \mathfrak{G}_{\pi_x a}$. The same family of conjugate subgroups belongs to all equivalent transitive representations of $\mathfrak{G}$.

Conversely we assert: *If* $\mathfrak{U}$ *is a subgroup of* $\mathfrak{G}$ *and* $\mathfrak{G} = \sum_1^{\omega} R_i$ *is the decomposition of* $\mathfrak{G}$ *into left cosets with respect to* $\mathfrak{U}$, *then the mapping*

$$x \to \pi_x = \binom{R_i}{x R_i}$$

is a transitive representation of degree ω *of* $\mathfrak{G}$ *as a permutation group of left cosets.*

In fact $x R_i$ is also a coset; moreover

$$\pi_{xy} = \binom{R_i}{x y R_i} = \binom{R_i}{x R_i} \binom{R_i}{y R_i} = \pi_x \pi_y$$

and $\pi_e = \binom{R_i}{R_i} = \underline{1}$; therefore the π_x are permutations and the mapping $x \to \pi_x$ is a homomorphy. Transitivity follows from the remark that for every index pair i, k the equation $x R_i = R_k$ is solvable. The transitive representation just found is called *the representation belonging to* $\mathfrak{U}$. Equivalent representations belong to conjugate subgroups and to them only.

The degree of the representation belonging to $\mathfrak{U}$ is equal to the index of $\mathfrak{U}$ in $\mathfrak{G}$.

Under the representation belonging to $\mathfrak{U}$, exactly those elements in the intersection $\mathfrak{N}$ of all the subgroups conjugate to $\mathfrak{U}$ are mapped onto $\underline{1}$. Consequently the representation group is isomorphic to $\mathfrak{G}/\mathfrak{N}$ and the representation is faithful only when $\mathfrak{N} = e$.

We denote the corresponding representation group by $\mathfrak{G}_{\mathfrak{U}}$ and the image of a subset $\mathfrak{K}$ of $\mathfrak{G}$ by $\mathfrak{K}_{\mathfrak{U}}$.

If the subgroup $\mathfrak{U}$ is of finite index in $\mathfrak{G}$, then the representation group is finite, and conversely.

The left and right representative systems of $\mathfrak{G}$ with respect to $\mathfrak{U}$ go into left and right representative systems of $\mathfrak{G}_{\mathfrak{U}}$ with respect to $\mathfrak{U}_{\mathfrak{U}}$ and conversely; therefore Theorem 4 of Chap. I. holds for infinite groups $\mathfrak{G}$.

If we set $\mathfrak{U} = e$, we obtain the regular representation of $\mathfrak{G}$ known from Chap. I., Theorem 2. The degree of the regular representation is equal to the order of $\mathfrak{G}$.

The representation group $\mathfrak{G}_e$ is transitive and every permutation in $\mathfrak{G}_e$ either leaves every letter fixed, or leaves none fixed.

Permutation groups with the two preceding properties are called *regular permutation groups*. Regular permutation groups are their own regular representations. *Moreover, every transitive representation group of an abelian group is regular* (since every subgroup is a normal subgroup.)

The permutations π of a regular permutation group have the property that $\pi^i a = a$ implies $\pi^i = \underline{1}$.

Permutations with this last property are said to be *regular*. *A permutation of a finite number of letters is regular if and only if all its cycles are of the same length.*

A transitive permutation group consisting of regular permutations only is a regular permutation group.

How does the representation of a group $\mathfrak{H}$, properly between $\mathfrak{U}$ and $\mathfrak{G}$, look in the transitive representation group $\mathfrak{P} = \mathfrak{G}_{\mathfrak{U}}$?

We decompose $\mathfrak{G} = \sum_1^r G_i \mathfrak{H}$ and $\mathfrak{H} = \sum_1^s H_k \mathfrak{U}$ into left cosets and observe that the multiplication of left cosets of $\mathfrak{G}$ (with respect to $\mathfrak{U}$) by any x in $\mathfrak{G}$ permutes them in bundles: Either the cosets of $\mathfrak{G}(\mathfrak{U}r)$ in a complex $G_i\mathfrak{H}$ are left there or they are mapped onto a complex $G_k\mathfrak{H}$ disjoint from it.

Therefore there is a decomposition of the aggregate $\mathfrak{M}$ *of the permuted objects into mutually disjoint systems* $\mathfrak{J}_1, \ldots, \mathfrak{J}_r$, *each containing s elements and such that the permutations in* $\mathfrak{P}$ *permute the systems* $\mathfrak{J}_1, \ldots, \mathfrak{J}_r$ *and* $r>1$, $s>1$.

We call the system $\mathfrak{J}_1, \ldots, \mathfrak{J}_r$ a family of (conjugate) *systems of imprimitivity.*

If the letters permuted by a transitive permutation group $\mathfrak{P}$ can be decomposed into a family of systems of imprimitivity, then $\mathfrak{P}$ is said to be imprimitive. Otherwise $\mathfrak{P}$ is said to be *primitive.*

Correspondingly, the representation of $\mathfrak{G}$ by $\mathfrak{U}$ is said to be primitive or imprimitive according as the representation group $\mathfrak{G}_{\mathfrak{U}}$ is primitive or imprimitive.

To a decomposition of the totality $\mathfrak{M}$ *of permuted objects of a transitive representation group* $\mathfrak{P}=\mathfrak{G}_{\mathfrak{U}}$ *of the group* $\mathfrak{G}$ *into a family of systems* $\mathfrak{J}_1, \ldots, \mathfrak{J}_r$ *of imprimitivity, there belongs a group* $\mathfrak{H}$ *properly between* $\mathfrak{G}$ *and* $\mathfrak{U}$ *such that the left cosets of* $\mathfrak{G}(\mathfrak{U}r)$ *in* $\mathfrak{J}_i$ *form a left coset of* $\mathfrak{G}$ *with respect to* $\mathfrak{H}$.

Let us suppose that the left coset $\mathfrak{U}$ of $\mathfrak{G}(\mathfrak{U}r)$ is in $\mathfrak{J}_1$. Then we say two *elements* of $\mathfrak{G}$ are congruent if their corresponding permutations map $\mathfrak{J}_1$ onto the same $\mathfrak{J}_j$. This is a normal congruence relation. Furthermore it follows from $\pi \equiv \varrho$ that $\sigma\pi \equiv \sigma\varrho$, and therefore that we are considering a right congruence of $\mathfrak{G}$ with respect to a group $\mathfrak{H}$ which contains $\mathfrak{U}$ in any case. From the definition of imprimitivity and transitivity of $\mathfrak{P}$ it follows that $\mathfrak{H}$ is a group properly between $\mathfrak{U}$ and $\mathfrak{G}$, thus the theorem is proven.

The transitive representation $\mathfrak{G}_{\mathfrak{U}}$ *is primitive if and only if* $\mathfrak{U}$ *is a maximal subgroup of* $\mathfrak{G}$. For example transitive representations of prime order are primitive.

When is a letter system $\mathfrak{J}$ a system of imprimitivity? As a criterion for this we have: *If* $\mathfrak{J}$ *contains more than one, but not all, the permuted letters and if* $\pi\mathfrak{J} \leq \mathfrak{J}$ *whenever a permutation* π *in* $\mathfrak{P}$ *leaves a letter of* $\mathfrak{J}$ *in* $\mathfrak{J}$, *then* $\mathfrak{J}$ *is a system of imprimitivity.*

In any case the condition is necessary. If the condition is fulfilled, then two letters a and b are called congruent if there is a permutation π in $\mathfrak{P}$ which maps a and b into $\mathfrak{J}$. The symmetry of the congruence is clear. Since $\mathfrak{P}$ is transitive, the congruence is reflexive. If, moreover, $\pi a, \pi b, \varrho b, \varrho c$ are contained in $\mathfrak{J}$, then $\pi\varrho^{-1}(\varrho b) \in \mathfrak{J}$, and therefore by assumption $(\pi\varrho^{-1})\varrho c \in \mathfrak{J}$, $\pi c \in \mathfrak{J}$, so that the congruence relation is also transitive. Consequently the totality of permuted letters is decomposed into disjoint classes of congruent letters. Among these letter systems we find $\mathfrak{J}$, since two elements in $\mathfrak{J}$ are congruent and if, on the other hand, $a \in \mathfrak{J}$, πa, $\pi b \in \mathfrak{J}$, then it follows that $\pi^{-1}(\pi a) \in \mathfrak{J}$, and therefore by assumption $\pi^{-1}(\pi b) \in \mathfrak{J}$, $b \in \mathfrak{J}$. Obviously the letter systems found in this way are conjugate under $\mathfrak{P}$. It follows from the assumption that $\mathfrak{J}$ is a system of imprimitivity.

A letter system $\mathfrak{J}$ of a finite number of letters is a system of imprimitivity if it contains more than one, but not all of the permuted letters, and if for a fixed letter a_0 in $\mathfrak{J}$ and all permutations π in $\mathfrak{P}$, $\pi a_0 \in \mathfrak{J}$ implies $\pi \mathfrak{J} \leq \mathfrak{J}$.

Proof: For every letter a in $\mathfrak{J}$, there is a permutation ϱ in $\mathfrak{P}$ which maps a_0 onto a. It follows from the assumption that $\varrho\mathfrak{J} \leq \mathfrak{J}$, and, since $\mathfrak{J}$ is finite, $\varrho\mathfrak{J} = \mathfrak{J}$. If $\pi a \in \mathfrak{J}$ holds for a permutation π in $\mathfrak{P}$, then $\pi\varrho\, a_0 \in \mathfrak{J}$, $\pi\varrho\mathfrak{J} = \pi\mathfrak{J} \leq \mathfrak{J}$. Now we need merely apply the previous criterion.

THEOREM 1: *A normal subgroup $\mathfrak{N} \neq \mathfrak{1}$ of a primitive permutation group $\mathfrak{P}$ is transitive.*

Proof: Let $\mathfrak{T}$ be a system of transitivity of $\mathfrak{N}$ having at least two letters and let a_0 be a letter in $\mathfrak{T}$. If the permutation π in $\mathfrak{P}$ leaves the letter a_0 in $\mathfrak{T}$, then for any permutation ν in $\mathfrak{N}$, $\pi\nu\pi^{-1} = \begin{pmatrix} \pi a \\ \pi\nu a \end{pmatrix}$ lies in $\mathfrak{N}$ and therefore whenever πa_0 lies in $\mathfrak{T}$, $\pi\nu a_0$ lies in $\mathfrak{T}$ also. Since $\mathfrak{N}a_0 = \mathfrak{T}$, we have $\pi\mathfrak{T} \leq \mathfrak{T}$. Since this conclusion holds for every letter a_0 in $\mathfrak{T}$, $\mathfrak{T}$ is either a system of imprimitivity of $\mathfrak{P}$, or $\mathfrak{T}$ contains all the letters. It follows from the assumptions that $\mathfrak{N}$ is transitive.

As examples of primitive groups we have the *multiply transitive* permutation groups.

DEFINITION: A permutation group $\mathfrak{P}$ is said to be *k-tuply transitive* if the number of permuted letters is at least k and for any two (ordered) k-tuples of letters $(a_1, \ldots, a_k)$ and $(b_1, \ldots, b_k)$ there is a permutation π in $\mathfrak{P}$ which maps a_1 onto b_1, a_2 onto $b_2, \ldots, a_k$ onto b_k.

$\mathfrak{P}$ is called exactly k-tuply transitive if $\mathfrak{P}$ is k-tuply but not $k+1$-tuply transitive.

Every k-tuply transitive permutation group is transitive.

A permutation group $\mathfrak{P}$ is k-tuply transitive if for a fixed k-tuple $(1, \ldots, k)$ and every k-tuple $(a_1, \ldots, a_k)$ there is a permutation π in $\mathfrak{P}$ which maps 1 onto a_1, 2 onto $a_2, \ldots, k$ onto a_k. Then for any other k-tuple $(b_1, b_2, \ldots, b_k)$, there is a permutation ϱ in $\mathfrak{P}$ which maps 1 onto b_1,

2 onto $b_2, \ldots, k$ onto b_k, and therefore $\varrho\pi^{-1}$ maps a_1 onto b_1, a_2 onto $b_2, \ldots, a_k$ onto b_k.

For example, the symmetric permutation group is the only n-tuply transitive permutation group of n letters.

An $(n-1)$-tuply transitive permutation group of $n \geq 2$ letters is also n-tuply transitive and therefore symmetric.

The alternating permutation group of $n > 2$ letters is exactly $(n-2)$-tuply transitive, for one of the two permutations

$$\begin{pmatrix} 1\ 2 \ldots n-2\ n-1\ n \\ a_1\, a_2 \ldots a_{n-2}\ \ a_{n-1}\ a_n \end{pmatrix}, \quad \begin{pmatrix} 1\ 2 \ldots n-2\ n-1\ n \\ a_1\, a_2 \ldots a_{n-2}\ \ a_n\ a_{n-1} \end{pmatrix}$$

is always even.

A doubly transitive permutation group $\mathfrak{P}$ is also said to be *multiply transitive* and is primitive. Since a permutation can be found in $\mathfrak{P}$ which leaves a letter a fixed but maps a letter b different from a onto any letter $\neq a$, a lies in no system of imprimitivity.

A transitive permutation group $\mathfrak{P}$ is k-tuply transitive with $k > 1$ if the subgroup $\mathfrak{U}_a$ of all permutations which leave a fixed permutes the remaining letters in a $(k-1)$-tuply transitive manner. There is a permutation π in $\mathfrak{P}$ which maps a_1 onto a, a permutation ϱ which maps b_1 onto a and a permutation σ which leaves a fixed and maps πa_2 onto ϱb_2, πa_3 onto $\varrho b_3, \ldots,$ πa_k onto ϱb_k. But then $\varrho^{-1}\sigma\pi$ maps the letter a_1 onto b_1, a_2 onto $b_2, \ldots, a_k$ onto b_k.

There is a conjecture that any sextuply transitive permutation group of n-th degree (and, apart from a finite number of exceptions, even any quadruply transitive permutation group of n-th degree) contains the alternating permutation group of n-th degree.

The construction of all finite multiply transitive permutation groups is an interesting but still unsolved problem.

§ 3. Operators and Operator Homomorphies

DEFINITION: *A homomorphy of a multiplicative domain into itself is called an operator (or endomorphism).*

If the image of x is denoted by x^{Θ}, then an operator Θ is defined as a single-valued mapping of $\mathfrak{G}$ into itself with the properties

$$x^{\Theta} \in \mathfrak{G}, \quad (xy)^{\Theta} = x^{\Theta} \cdot y^{\Theta} \text{ for all } \quad x, y \in \mathfrak{G}$$

and the product of two operators Θ_1 and Θ_2 is defined by the equation

$$x^{\Theta_1 \Theta_2} = (x^{\Theta_2})^{\Theta_1}.$$

The identity mapping of $\mathfrak{G}$ onto itself is an operator with the property

$$\underline{1}\Theta = \Theta\underline{1} = \Theta$$

for all operators Θ.

All operators of a multiplicative domain form a semi-group with a unit element.

A semi-group with unit element denoted by $\{\Omega\}$ is generated by a complex Ω of operators by adjoining a unit element and forming all possible products of elements of Ω. Every semi-group of operators which contains $\underline{1}$ and Ω also contains the domain $\{\Omega\}$ of operators generated by Ω.

DEFINITION: *The multiplicative domains $\mathfrak{G}_1, \mathfrak{G}_2, \ldots$ have a common operator domain Ω if*

1) *a multiplication is defined in Ω,*

2) *to every element Θ in Ω, there corresponds an operator Θ of $\mathfrak{G}_i$,*

3) *to the product $\Theta_1\Theta_2$ of two elements in Ω, there corresponds the product of the operators Θ_1 and Θ_2 in $\mathfrak{G}_i$.*

In all the following considerations a fixed common operator domain is assumed unless something else is explicitly stated.

DEFINITION: A homomorphy σ of $\mathfrak{G}$ into $\mathfrak{G}^*$ is said to be an *operator homomorphy* if

$$\sigma(x^\Theta) = (\sigma x)^\Theta$$

for all x in $\mathfrak{G}$ and for all operators Θ in the common operator domain Ω.

This relation is transitive and reflexive.

If we talk of a homomorphy we shall, if nothing is said to the contrary, mean an operator homomorphy over Ω [1]. If other homomorphies are also considered, then we shall explicitly give an operator domain belonging to them. For example a $\underline{1}$-homomorphy means an ordinary homomorphy.

DEFINITION: For a given operator domain, a multiplicative subdomain $\mathfrak{U}$ of $\mathfrak{G}$ is said to be *admissible*[2] if for all Θ in Ω: $\mathfrak{U}^\Theta \subseteq \mathfrak{U}$.

Given two admissible multiplicative sub-domains, their intersection and the multiplicative sub-domain generated by them are also admissible.

An operator in Ω maps an admissible subdomain onto an admissible subdomain. Moreover, for an arbitrary complex $\mathfrak{K}$, the multiplicative subdomain $\{\mathfrak{K}^{\Theta_1} \cup \mathfrak{K}^{\Theta_2} \cup \ldots\}$ generated by the union of all $\mathfrak{K}^{\Theta_i}$ with $\Theta_i \in \Omega$ forms an admissible subdomain which contains $\mathfrak{K}$. It is called

---

[1] Also called an Ω-homomorphism. See JACOBSON, *Theory of Rings* (Amer. Math Soc.)
[2] Also called an Ω-subdomain. (*Ed.*)

the subdomain generated by $\mathfrak{K}$ over Ω and is denoted by $\{\mathfrak{K}\}_\Omega$. Every admissible subdomain which contains $\mathfrak{K}$ also contains $\{\mathfrak{K}\}_\Omega$.

We assume now that $\mathfrak{G}$ is a group.

Subgroups which are mapped into themselves by all operators of the entire group, are said to be *fully invariant subgroups.*

The unit element and the entire group are fully invariant subgroups.

An admissible subgroup is mapped homomorphically onto an admissible subgroup by an operator homomorphy.

This is because, for all x in $\mathfrak{U}$: $(\sigma x)^\Theta = \sigma(x^\Theta) \in \sigma\mathfrak{U}$, since $x^\Theta \in \mathfrak{U}$.

If $\overline{\mathfrak{U}}$ *is an admissible subgroup of the image domain* $\overline{\mathfrak{G}}$, *then all the elements in* $\mathfrak{G}$ *whose images lie in* $\overline{\mathfrak{U}}$ *form an admissible subgroup* $\mathfrak{U}$ *of* $\mathfrak{G}$ and

$$\overline{\mathfrak{U}} = \sigma\mathfrak{U}.$$

Every element in $\overline{\mathfrak{U}}$ is of the form σx with x in $\mathfrak{U}$; from $x \in \mathfrak{U}, \Theta \in \Omega$ it follows that

$$\sigma x \in \overline{\mathfrak{U}} \ni (\sigma x)^\Theta = \sigma(x^\Theta), \quad x^\Theta \in \mathfrak{U}.$$

In the following investigations it will be assumed that the subgroups used are admissible, if another operator domain is not explicitly assigned to the subgroup in question. For example a $\underline{1}$-group is an ordinary subgroup.

Let $\mathfrak{G}$ be a group with operator domain Ω. Given a homomorphism σ of $\mathfrak{G}$ we wish to consider the operator domain also as the operator domain of $\sigma\mathfrak{G}$ so that σ becomes an operator homomorphism.

Then $\sigma(x^\Theta) = (\sigma x)^\Theta$ for all Θ in Ω. In order that this mapping Θ be single-valued in $\sigma\mathfrak{G}$, the normal subgroup $\mathfrak{E}$ of all elements of $\mathfrak{G}$ which are mapped by σ onto σe must be admissible with respect to Ω.

If conversely the normal subgroup $\mathfrak{E}$ of all elements in the group $\mathfrak{G}$ which map onto σe under the homomorphism σ is admissible with respect to the operator domain Ω, then we define the *extension* of Ω to $\sigma\mathfrak{G}$ by the condition

$$(\sigma x)^\Theta = \sigma(x^\Theta)$$

for all Θ in Ω. Then Θ is an operator of $\sigma\mathfrak{G}$.

For it follows from $\sigma x = \sigma y$ that $x = a y$, where $a \in \mathfrak{E}$.

By assumption $a^\Theta \in \mathfrak{E}$, $(\sigma x)^\Theta = \sigma(x^\Theta) = \sigma(a^\Theta \cdot y^\Theta) = \sigma(y^\Theta)$. Moreover, $(\sigma x \cdot \sigma y)^\Theta = (\sigma(x y))^\Theta = \sigma((x y)^\Theta) = \sigma(x^\Theta \cdot y^\Theta) = \sigma(x^\Theta) \cdot \sigma(y^\Theta)$ $= (\sigma x)^\Theta \cdot (\sigma y)^\Theta$.

Ω is a common operator domain of $\mathfrak{G}$ and $\sigma\mathfrak{G}$ since $\Theta_1 \Theta_2 = \Theta_3$ in Ω over $\mathfrak{G}$ implies $(\sigma x^{\Theta_2})^{\Theta_1} = \sigma(x^{\Theta_2})^{\Theta_1} = \sigma((x^{\Theta_2})^{\Theta_1}) = \sigma x^{\Theta_3}$, and therefore $\Theta_3 = \Theta_1\Theta_2$ in $\sigma\mathfrak{G}$. From the definition of Ω over $\sigma\mathfrak{G}$ it follows that σ is an operator homomorphism.

The operator domain Ω of $\mathfrak{G}$ is extended to the factor group of $\mathfrak{G}$ with respect to the admissible normal subgroup $\mathfrak{N}$ by the condition

$$(a\mathfrak{N})^{\underline{\Theta}} = a^{\Theta}\mathfrak{N}$$

for all Θ in Ω.

If we regard $(a\mathfrak{N})^{\Theta}$ as a coset mod $\mathfrak{N}$ we can delete the dash without misunderstanding.

We apply the new concepts to a cyclic group $\mathfrak{G} = (A)$. *The homomorphic image of a cyclic group is cyclic.*

From $\sigma A^m = (\sigma A)^m$, it follows that $\sigma\mathfrak{G} = (\sigma A)$ and the above statement is proven. *Every subgroup of a cyclic group is admissible.*

This is because $A^{\Theta} = A^t$, and therefore

$$(A^m)^{\Theta} = (A^{\Theta})^m = A^{tm} \subseteq (A^m).$$

Every cyclic group has as many operators as it has elements.

This follows from the fact that an operator Θ is uniquely determined by its effect on A:

$$(A^m)^{\Theta} = (A^{\Theta})^m = A^{mt}.$$

Conversely, the mapping $(A^m)^{\Theta} = A^{mt}$ is an operator.

§ 4. On the Automorphisms of a Group

DEFINITIONS: An isomorphy of a group with itself is called a *meromorphy*.

An isomorphic mapping of a group onto itself is called an *automorphic mapping* (*automorphism*).

An isomorphic mapping of a group onto a subgroup is called a *meromorphic mapping* (*meromorphism*). The mapping is called a *proper* meromorphism if the subgroup is a proper subgroup.

If a group $\mathfrak{G}$ has a proper meromorphism σ, then $\mathfrak{G}$ contains the infinite and decreasing sequence of subgroups:

$$\mathfrak{G} \supset \sigma\mathfrak{G} \supset \sigma^2\mathfrak{G}, \ldots$$

A finite group, therefore, has no proper meromorphism. Every operator of the infinite cyclic group which is different from 0, ± 1, is a proper meromorphism.

The product of two proper meromorphisms is also a proper meromorphism.

All the automorphisms of a group $\mathfrak{G}$ form a group.

The group of automorphisms of a group $\mathfrak{G}$ without operators is denoted by $A_{\mathfrak{G}}$.

The group of automorphisms admissible over an operator domain Ω is denoted by $(A_{\mathfrak{G}})_\Omega$.

1. *Inner Automorphisms.*

The "transformation" of the elements of a group $\mathfrak{G}$ by a fixed element x is an automorphism. First we have the simple but important rule: $x a x^{-1} \cdot x b x^{-1} = x a b x^{-1}$; secondly, the transformations form a group with the identity automorphism as unit element.

We call the automorphism $\binom{a}{x a x^{-1}} = \binom{a}{a^{\underline{x}}}$ an *inner automorphism* of the group. All the inner automorphisms of $\mathfrak{G}$ form a group $J_{\mathfrak{G}}$.

We saw previously that the mapping $x \to \underline{x}$ defines a homomorphism between $\mathfrak{G}$ and $J_{\mathfrak{G}}$. Precisely those elements in the center of $\mathfrak{G}$ are mapped onto the identity automorphism, so that we have the isomorphy

(1) $$\mathfrak{G}/\mathfrak{z} \simeq J_{\mathfrak{G}}.$$

The group of inner automorphisms is a normal subgroup of the group of all automorphisms. This is because for every operator Θ

(2) $$(a^{\underline{x}})^\Theta = (x a x^{-1})^\Theta = x^\Theta a^\Theta (x^\Theta)^{-1},$$

and therefore $\Theta \underline{x} = \underline{x^\Theta} \Theta$; and if Θ is an automorphism, then

(3) $$\Theta \underline{x} \Theta^{-1} = \underline{x^\Theta}.$$

The factor group of $A_{\mathfrak{G}}$ over $J_{\mathfrak{G}}$ is called the group of *outer automorphisms.*

From formula (2) we see that:

An automorphism maps a series of complexes which are conjugate under $\mathfrak{G}$ onto a series of complexes which are again conjugate under $\mathfrak{G}$.

In particular, classes of conjugate elements go into classes of conjugate elements. But not every automorphism of a finite group which maps every class of conjugate elements onto itself is an inner automorphism (see Ex. 10, Appendix).

2. *Complete groups.*

A group is said to be complete if its center is e and every automorphism is an inner automorphism.

THEOREM 2: *The automorphism group of a simple non-abelian group is complete.*

Proof: Let $\mathfrak{G}$ be simple but non-abelian. Since the center of $\mathfrak{G}$ is a normal abelian subgroup, it must be e. Therefore $\mathfrak{G}$ is isomorphic to the group J of inner automorphisms. We may even identify $\mathfrak{G}$ with J so that in the full automorphism group A of $\mathfrak{G}$: $\alpha x \alpha^{-1} = x^\alpha$ for all x in J, and $x^\alpha = x$ for all x implies $\alpha = 1$. Therefore the center of A is 1, and the only element in A with which all the elements of J commute, is 1.

An automorphism σ of A onto itself maps J onto a normal subgroup J^σ of A. J^σ is isomorphic to J and therefore simple. The intersection of J and J^σ is a normal subgroup of J, and therefore $J = J^\sigma$ or $J \cap J^\sigma = 1$.

But in the latter case J and J^σ commute elementwise, since $\alpha \in J$, $\beta \in J^\sigma$ imply:

$$\alpha\beta\alpha^{-1} \in J^\sigma, \quad \beta\alpha^{-1}\beta^{-1} \in J,$$

$$\alpha\beta\alpha^{-1}\beta^{-1} \in J \cap J^\sigma, \quad \alpha\beta\alpha^{-1}\beta^{-1} = 1,$$

$$\alpha\beta = \beta\alpha.$$

Since $J \neq 1$, we must have $J = J^\sigma$. Consequently the mapping $x \rightarrow x^\sigma$ for all x in J is a certain automorphism σ' of J. We want to prove that the automorphism σ of A is an inner automorphism and may, for this purpose, replace σ by the automorphism $\tau = \sigma'^{-1}\sigma$ of A. Now $x^\tau = x$ for all x in J. Therefore, $\alpha x \alpha^{-1} = x^\alpha \in J$ implies $\alpha^\tau x (\alpha^\tau)^{-1} = (x^\alpha)^\tau = x^\alpha$, and therefore $\alpha^\tau = \alpha$. Thus τ is the identity automorphism of A, Q.E.D.

Are there any simple non-abelian groups? Since every subgroup of an abelian group is a normal subgroup, it follows that:

A simple abelian group $\neq e$ *is cyclic of prime order.*

Conversely a group of prime order is cyclic and simple.

Thus "*simple and non-abelian*" is equivalent to "*simple of composite order.*"

THEOREM 3: *The alternating permutation group on five letters is simple.*

Proof: A normal subgroup of $\mathfrak{A}_5$ can be divided into classes of elements conjugate under $\mathfrak{A}_5$ among which 1 must occur, and its order divides the order of $\mathfrak{A}_5$. The classes of $\mathfrak{A}_5$, as previously shown, are: (1) The identity permutation, (2) The 15 double transpositions, (3) The 20 3-cycles, (4) and (5) 12 5-cycles each. But no sum of two or more integers from the set 1, 15, 20, 12, 12 is a proper divisor of 60. Therefore $\mathfrak{A}_5$ contains no proper normal subgroup, as was to be proved.

3. *Characteristic Subgroups. Centralizers.*

A subgroup of a group $\mathfrak{G}$ which is mapped into itself by all automorphisms of $\mathfrak{G}$ is said to be a *characteristic subgroup.* $\mathfrak{G}$ and e are characteristic subgroups. The subgroups admissible under all inner automorphisms are precisely the normal subgroups. Consequently, characteristic subgroups are always normal subgroups.

The center of a group is a characteristic subgroup.

Proof: Since $\mathfrak{G}^{\alpha} = \mathfrak{G}$ for an automorphism α and $xz = zx$ for all x implies that $x^{\alpha}z^{\alpha} = z^{\alpha}x^{\alpha}$ for all x, we have z^{α} in the center for each z in the center.

The factor group over the center is likewise characteristic. We can even form a series of characteristic subgroups, the *ascending central series,* by defining recursively: $\mathfrak{z}_0 = e$, $\mathfrak{z}_1 = \mathfrak{z}(\mathfrak{G})$; if $\mathfrak{z}_i$ has already been defined as a characteristic subgroup, then $\mathfrak{z}_{i+1}/\mathfrak{z}_i$ will be the center of $\mathfrak{G}/\mathfrak{z}_i$.

A group is said to be *characteristically simple* if it does not contain any proper characteristic subgroup. The investigation of the structure of the finite characteristically simple groups will later be reduced to the investigation of the structure of simple groups.

If $\mathfrak{A}$ is a group of automorphisms of the group $\mathfrak{G}$, and $\mathfrak{N}$ is a normal subgroup admissible under $\mathfrak{A}$, then we can derive the structure of $\mathfrak{A}$ from the structure of certain groups of automorphisms of $\mathfrak{G}/\mathfrak{N}$ and $\mathfrak{N}$ in the following way: All the automorphisms of $\mathfrak{A}$ which leave the elements of $\mathfrak{N}$ fixed form a normal subgroup $\mathfrak{A}_1$ of $\mathfrak{A}$. All the automorphisms of $\mathfrak{A}$ which leave the elements of $\mathfrak{G}/\mathfrak{N}$ fixed form a normal sugroup $\mathfrak{A}_2$ of $\mathfrak{A}$. By the first isomorphy theorem, $\mathfrak{A}/\mathfrak{A}_1$ is isomorphic to a group of automorphisms of $\mathfrak{N}$ and $\mathfrak{A}/\mathfrak{A}_2$ is isomorphic to a group of automorphisms of $\mathfrak{G}/\mathfrak{N}$. $\mathfrak{A}_1 \cap \mathfrak{A}_2$ consists of all the automorphisms α in $\mathfrak{A}$ such that

1) $\nu^{\alpha} = \nu$ for all ν in $\mathfrak{N}$,

2) $x^{\alpha}x^{-1} \in \mathfrak{N}$ for all x in $\mathfrak{G}$.

See exercise 6 at the end of the chapter concerning this point.

Now we apply the first isomorphy theorem to the normalizer $N_{\mathfrak{U}}$ of a subgroup $\mathfrak{U}$.

The mapping $x \rightarrow \begin{pmatrix} U \\ xUx^{-1} \end{pmatrix}$ of the elements x in $N_{\mathfrak{U}}$ is a homomorphism onto a group of automorphisms of $\mathfrak{U}$.

Consequently all the elements of $\mathfrak{G}$ which commute with *every element* of $\mathfrak{U}$ form a subgroup $Z_{\mathfrak{U}}$. $Z_{\mathfrak{U}}$ is called the *centralizer* of $\mathfrak{U}$. *The centralizer of a subgroup is a normal subgroup of its normalizer and the factor group is isomorphic to a group of automorphisms of the subgroup.*

The centralizer of the whole group $\mathfrak{G}$ is equal to its center.

4. *The Φ - subgroup.*

An automorphism of the group $\mathfrak{G}$ is uniquely determined by its effect on the elements of a system of generators of $\mathfrak{G}$. In order to state this circumstance more sharply, we introduce the following concept.

DEFINITION: The set Φ of all the elements which may be deleted from every system of generators of a non-trivial group is a subgroup, *the Φ- subgroup of* $\mathfrak{G}$:

1) Since $\mathfrak{G} \neq e$, e is a member of Φ;

2) If $x \in \Phi$ and $y \in \Phi$, then it follows from $\mathfrak{G} = \{xy, \mathfrak{K}\}$ that $\mathfrak{G} = \{x, y, \mathfrak{K}\}$ and therefore $\mathfrak{G} = \{y, \mathfrak{K}\}$; therefore $\mathfrak{G} = \{\mathfrak{K}\}$; xy also belongs to Φ;

3) If $x \in \Phi$ then it follows from $\mathfrak{G} = \{x^{-1}, \mathfrak{K}\}$ that $\mathfrak{G} = \{x, \mathfrak{K}\}$ and therefore $\mathfrak{G} = \{\mathfrak{K}\}$; x^{-1} also belongs to Φ.

The Φ-subgroup is a characteristic subgroup (since every automorphism maps a system of generators onto a system of generators), hence also normal.

The Φ-subgroup is the intersection of the whole group with all of its maximal subgroups.

If $\mathfrak{D}$ is this intersection then we shall show 1) $\Phi \subseteq \mathfrak{D}$, in other words: If x does not lie in the maximal subgroup $\mathfrak{U}$ then it is not in Φ either. This is because $\{x, \mathfrak{U}\} = \mathfrak{G} \neq \mathfrak{U}$. 2) $\mathfrak{D} \subseteq \Phi$; in other words: If x does not lie in Φ, then neither does it lie in every maximal subgroup of $\mathfrak{G}$. If $\{x, \mathfrak{K}\} = \mathfrak{G} \neq \{\mathfrak{K}\}$, then x is not in $\{\mathfrak{K}\}$. By the theorem on maximal subgroups[1], there is a largest possible subgroup $\mathfrak{U}$ which contains $\mathfrak{K}$ but not x. $\mathfrak{U}$ is moreover a maximal subgroup of $\mathfrak{G}$, since any larger subgroup would contain x and $\mathfrak{K}$ and therefore also the group

$$\{x, \mathfrak{K}\} = \mathfrak{G}.$$

If every proper subgroup can be embedded in a maximal (proper) subgroup, then we have the

[1] Theorem 5, Chap. I.

Basis Theorem: *If a complex* $\mathfrak{K}$ *together with* Φ *generates* $\mathfrak{G}$, *then* $\mathfrak{K}$ *alone generates* $\mathfrak{G}$.

If $\{\mathfrak{K}\} \neq \mathfrak{G}$, then $\{\mathfrak{K}\}$ could be embedded in a maximal subgroup $\mathfrak{U}$. But then Φ would also lie in $\mathfrak{U}$. Therefore the group $\mathfrak{G}$ generated by $\mathfrak{K}$ and Φ would lie in $\mathfrak{U}$, which is a contradiction.

Now let $\mathfrak{G}$ be finite, $A_{\mathfrak{G}}$ the automorphism group of $\mathfrak{G}$, $\mathfrak{B}$ the normal subgroup consisting of all automorphisms which leave every coset of $\mathfrak{G}$ with respect to Φ fixed. Then $A_{\mathfrak{G}} : \mathfrak{B}$ is isomorphic to a group of automorphisms of $\mathfrak{G}/\Phi$. The factor group $\mathfrak{G}/\Phi$ has a finite number of generators $R_1, \ldots, R_d$. Since, by the basis theorem, a representative system $S_1, \ldots, S_d$ generates the whole group $\mathfrak{G}$, there are as many different systems conjugate to $S_1, \ldots, S_d$ under $\mathfrak{B}$ as there are elements in $\mathfrak{B}$. The $(\Phi : 1)^d$ representative systems decompose, therefore, into a certain number of classes of systems conjugate under $\mathfrak{B}$, such that the classes each contain $\mathfrak{B}:1$ elements. Thus we obtain the divisibility condition

(4) $$(A_{\mathfrak{G}}:1)\quad(\Phi:1)^d\cdot(A_{\mathfrak{G}/\Phi}:1).$$

5. *Normal and Central Operators.*

An operator is said to be a *normal operator* if

$$x y^{\Theta} x^{-1} = (x y x^{-1})^{\Theta}$$

for all x, y in $\mathfrak{G}$, i.e.:

An operator is normal if it commutes with all the inner automorphisms.

Therefore a normal operator maps a normal subgroup onto a normal subgroup.

If α is a normal automorphism, then $x^{\alpha} y^{\alpha} x^{-\alpha} = x y^{\alpha} x^{-1}$ or

$$x^{-1} x^{\alpha} y^{\alpha} = y^{\alpha} x^{-1} x^{\alpha}$$

for all x, y and, since $\mathfrak{G}^{\alpha} = \mathfrak{G}$, $x^{-1}x^{\alpha}$ is in the center of $\mathfrak{G}$, and conversely. An automorphism is normal if and only if it multiplies every element of $\mathfrak{G}$ by an element of the center.[1]

The mapping $x \to x^{-1}x^{\alpha}$ is an operator $-1 + \alpha$, since

$$(x y)^{-1+\alpha} = (x y)^{-1} (x y)^{\alpha} = y^{-1} x^{-1} x^{\alpha} y^{\alpha} = x^{-1} x^{\alpha} y^{-1} y^{\alpha} = x^{-1+\alpha} y^{-1+\alpha}.$$

An operator which maps every element of the group onto a center element is said to be a *central operator*. Every central operator is normal.

6. *The Holomorph of a Group.*

[1] Because of this property a normal automorphism is also called a *center automorphism*.

Is it possible to extend a given group $\mathfrak{G}$ to a group $\mathfrak{H}$ so that every automorphism of $\mathfrak{G}$ can be induced by a transformation by an element in $\mathfrak{H}$?

Let $\mathfrak{M}$ be any group of automorphisms of $\mathfrak{G}$, and consider the set $\mathfrak{H}$ of permutations $\binom{x}{y x^{\alpha}}$ with $x, y \in \mathfrak{G}$, $\alpha \in \mathfrak{M}$.

Then

$$\binom{x}{y x^{\alpha}}\binom{x}{z x^{\beta}} = \binom{x}{y z^{\alpha} x^{\alpha\beta}}.$$

The permutations $\pi_y = \binom{x}{y x}$ form a group $\overline{\mathfrak{G}}$ of permutations in $\mathfrak{H}$, and by I, Theorem 2, the mapping $y \to \pi_y$ gives the regular representation of $\mathfrak{G}$. $\overline{\mathfrak{G}}$ is therefore a regular permutation group which we may identify with $\mathfrak{G}$.

The permutations $\binom{x}{x^{\alpha}}$ form a permutation group $\overline{\mathfrak{M}}$ in $\mathfrak{H}$ isomorphic to $\mathfrak{M}$ and we identify $\overline{\mathfrak{M}}$ with $\mathfrak{M}$.

We can verify easily that

(5) $$\binom{x}{x^{\alpha}}\binom{x}{y x} = \binom{x}{y^{\alpha} x}\binom{x}{x^{\alpha}},$$

and consequently $\mathfrak{H} = \overline{\mathfrak{G}}\overline{\mathfrak{M}} = \overline{\mathfrak{M}}\overline{\mathfrak{G}}$ is a group of permutations.

According to (5), transforming by $\binom{x}{x^{\alpha}}$ in $\mathfrak{H}$ induces the automorphism α in $\mathfrak{G}$.

The permutation group just constructed is called the *holomorph of the automorphism group* $\mathfrak{M}$ *over* $\mathfrak{G}$. The holomorph of the group of all automorphisms over $\mathfrak{G}$ is called simply the *holomorph of* $\mathfrak{G}$.

Now we wish to start, in the reverse order, with a transitive permutation group $\mathfrak{G}$, and we form the group $\mathfrak{H}$ of all permutations that form $\mathfrak{G}$ onto itself.

Which automorphisms of the abstract group can be induced by transformation with elements of $\mathfrak{H}$?

Let $\mathfrak{G}_i$ be the subgroup of all permutations in $\mathfrak{G}$ which leave the letter i fixed. For a permutation π in $\mathfrak{H}$ it follows that

$$\pi\mathfrak{G}_i\pi^{-1} \subseteq \mathfrak{G}, \quad \pi\mathfrak{G}_i\pi^{-1}(\pi i) = \pi i,$$

and therefore $$\pi\mathfrak{G}_i\pi^{-1} \subseteq \mathfrak{G}_{\pi i},$$

likewise $$\pi^{-1}\mathfrak{G}_{\pi i}\pi \subseteq \mathfrak{G}_i$$

$$\mathfrak{G}_{\pi i} \subseteq \pi\mathfrak{G}_i\pi^{-1},$$

therefore

$$\pi \mathfrak{G}_i \pi^{-1} = \mathfrak{G}_{\pi i}.$$

Conversely, let α be an automorphism of $\mathfrak{G}$ which maps $\mathfrak{G}_1$ onto $\mathfrak{G}_i$. We wish to show that there is a permutation π in $\mathfrak{H}$ such that

$$\pi x \pi^{-1} = x^{\alpha}$$

for all x in $\mathfrak{G}$. In order to prove this we look for a permutation y in $\mathfrak{G}$ which maps i onto 1. (There are such, since $\mathfrak{G}$ is transitive.) Then $\mathfrak{G}_1^{y\alpha} = \mathfrak{G}_1$, and so without loss of generality we may replace $y\alpha$ by a new α with

$$\mathfrak{G}_1^{\alpha} = \mathfrak{G}_1$$

Let R_i be the left coset of $\mathfrak{G}$ over $\mathfrak{G}_1$ consisting of all permutations which map 1 onto i. The mapping $\begin{pmatrix} R_i \\ R_i^{\alpha} \end{pmatrix}$ is a permutation $\begin{pmatrix} R_i \\ R_{\pi i} \end{pmatrix}$ of the left cosets since $\mathfrak{G}_1^{\alpha} = \mathfrak{G}_1$.

Then
$$\begin{pmatrix} R_i \\ (x R_i)^{\alpha} \end{pmatrix} = \begin{pmatrix} R_i \\ x^{\alpha} R_i^{\alpha} \end{pmatrix} = \begin{pmatrix} R_i \\ x^{\alpha} R_{\pi i} \end{pmatrix},$$

and therefore
$$\pi x i = x^{\alpha} \pi i,$$
$$\pi x \pi^{-1} = x^{\alpha}.$$

We have as a result:

THEOREM 4: *Let $\mathfrak{H}$ be the group of all automorphisms, of a transitive group $\mathfrak{G}$ which permute the subgroups $\mathfrak{G}_i$ (previously described and belonging to the given transitive representation of $\mathfrak{G}$). Then this group $\mathfrak{H}$ is precisely that induced by all transformations of $\mathfrak{G}$ by the elements of the normalizer of $\mathfrak{G}$ in the group of all permutations on the letters of $\mathfrak{G}$.*

We determine which permutations π in $\mathfrak{H}$ are elementwise commutative with $\mathfrak{G}$. Let $\pi^{-1}1 = i$. By assumption $\pi R_i = R_i \pi$ and also $\pi R_i 1 = 1$; therefore $1 = R_i \pi 1$. If we set $R_i = x \mathfrak{G}_1$, then it follows that

$$\begin{aligned} x \mathfrak{G}_1 x^{-1} 1 &= x \mathfrak{G}_1 x^{-1} \cdot x \pi 1 \\ &= R_i \pi 1 = 1, \end{aligned}$$

therefore
$$\mathfrak{G}_i = x \mathfrak{G}_1 x^{-1} \subseteq \mathfrak{G}_1.$$

Since $\pi i = 1$, we find through similar considerations that: $\mathfrak{G}_1 \subseteq \mathfrak{G}_i$, and therefore $\mathfrak{G}_i = \mathfrak{G}_1$.

Conversely, let $\mathfrak{G}_i = x\mathfrak{G}_1 x^{-1} = \mathfrak{G}_1$ with $x \in R_i$. Since $\mathfrak{G}_1 x = x\mathfrak{G}_1$, the mapping $\binom{R_i}{R_i x}$ is a permutation $\binom{R_i}{R_{\bar{x}i}}$ of the left cosets of $\mathfrak{G}$ over $\mathfrak{G}_1$.

Since

$$\binom{R_i}{yR_i} \cdot \binom{R_i}{R_i x} = \binom{R_i}{R_i x}\binom{R_i}{yR_i} = \binom{R_i}{yR_i x},$$

$\bar{x}$ commutes with all the permutations in $\mathfrak{G}$. The mapping $x \rightarrow \bar{x}^{-1}$ gives a homomorphy, between the normalizer of $\mathfrak{G}_1$ in $\mathfrak{G}$ and the group of all permutations commuting elementwise with $\mathfrak{G}$, under which $\mathfrak{G}_1$ is mapped onto $\underline{1}$.

If however π commutes with $\mathfrak{G}$ elementwise, and $\pi 1 = 1$, then

$$\pi i = \pi R_i 1 = R_i \pi 1 = R_i 1 = i,$$

and therefore $\pi = \underline{1}$.

We obtain as the result:

THEOREM 5: *The centralizer of a transitive permutation group $\mathfrak{G}$ in the group of all permutations is isomorphic to the factor group $N_{\mathfrak{G}_1}/\mathfrak{G}_1$ of the normalizer $N_{\mathfrak{G}_1}$ in $\mathfrak{G}$ of a subgroup $\mathfrak{G}_1$ which belongs to the transitive representation of $\mathfrak{G}$. It consists wholly of regular permutations.*

As a special case we obtain the THEOREM OF JORDAN: *The centralizer of a group $\mathfrak{G}$ in its holomorph consists of the permutations*

$$\varrho_y = \binom{x}{xy^{-1}} = \binom{x}{y^{-1}x^y},$$

which form a regular permutation group isomorphic to $\mathfrak{G}$.

Moreover it follows that the center of a primitive permutation group is $\underline{1}$, or the group consists of the powers of a cycle whose length is a prime.

A transitive permutation group $\mathfrak{H}$ which contains a regular normal subgroup $\mathfrak{G}$ is, by what has just been proven, the holomorph of the group $\mathfrak{M}$ of all permutations in $\mathfrak{H}$ which leave a letter fixed over the group $\mathfrak{G}$.

Thus the holomorph of a group $\mathfrak{G}$ is primitive if and only if the group is characteristically simple[1], since a system of imprimitivity which contains e consists of the elements of a non-trivial characteristic subgroup of $\mathfrak{G}$.

THEOREM 6: *If the holomorph of a finite group is doubly transitive, then $\mathfrak{G}$ is abelian and there is a prime integer p such that the p-th power of every element in $\mathfrak{G}$ is equal to e.*

---

[1] i.e., has no proper characteristic subgroups.

Proof: It follows from the hypothesis that the automorphisms of $\mathfrak{G}$ permute transitively all the elements $\neq e$ of $\mathfrak{G}$. Therefore all the elements $\neq e$ of $\mathfrak{G}$ have the same order p. Since $\mathfrak{G} \neq e$, there are elements of prime order in $\mathfrak{G}$ and therefore p is a prime. Moreover all the normalizers of elements $\neq e$ in $\mathfrak{G}$ have the same order $\frac{\mathfrak{G}:1}{h}$. Thus there are h elements in each class of conjugate elements $\neq e$ in $\mathfrak{G}$. If there are $r+1$ classes it follows that

$$\mathfrak{G} : 1 = rh + 1.$$

On the other hand h is a divisor of $\mathfrak{G}:1$; therefore $h = 1$, i.e., $\mathfrak{G}$ is abelian, Q.E.D.

THEOREM 7: *If the holomorph of a group $\mathfrak{G}$ consisting of more than three elements is triply transitive, then $\mathfrak{G}$ is abelian and the square of every element is e.*

Proof: It follows from the hypothesis that the automorphisms of $\mathfrak{G}$ permute the elements $\neq e$ in $\mathfrak{G}$ in a doubly transitive manner. If, for an x in $\mathfrak{G}$: $x^2 \neq e$, $x^3 \neq e$, then there is an automorphism which maps x onto x^2 but leaves x^3 fixed. But then $(x^2)^3 = x^3$ and therefore $x^3 = e$. If, however, $x^2 \neq e$, $x^3 = e$, then by hypothesis there is an element y in $\mathfrak{G}$ which does not lie in (x) and we can find an automorphism of $\mathfrak{G}$ which maps x^2 onto y but leaves x fixed. But then $x^2 = y$ which is a contradiction. Consequently for all x in $\mathfrak{G}$, $x^2 = e$, i.e., $x = x^{-1}$. From this it follows that $xy = x^{-1}y^{-1} = (yx)^{-1} = yx$. Therefore $\mathfrak{G}$ is abelian, Q.E.D.

If the holomorph $\mathfrak{H}$ of a group $\mathfrak{G}$ is quadruply transitive, then $\mathfrak{G}$ must consist of exactly four elements:

By the previous theorem the square of every element in $\mathfrak{G}$ is equal to e. Moreover $\mathfrak{G}$ contains at least four elements e, x, y, xy. If there were a fifth element z in $\mathfrak{G}$ then an automorphism could be found which leaves x and y fixed but maps xy onto z and this is a contradiction, which establishes the above. There does, in fact, exist a group of four elements whose holomorph is quadruply transitive. In the symmetric permutation group of four letters, the three double transpositions (12) (34), (13) (24), (14) (23) together with 1 form a regular normal subgroup $\mathfrak{V}_4$ of four elements, as is easily seen. ($\mathfrak{V}_4$ is called the *Klein Four Group.*)

§ 5. Normal Chains and Normal Series

Let $\mathfrak{G}$ be a group with operators.

A *normal chain of length* r is a chain of $(r+1)$ subgroups:

$$\mathfrak{G} = \mathfrak{G}_0 \geq \mathfrak{G}_1 \geq \mathfrak{G}_2 \geq \cdots \geq \mathfrak{G}_r = 1,$$

which begins with $\mathfrak{G}$ and terminates with e, and is such that every member of the chain is a normal subgroup of the preceding member. The factor groups $\mathfrak{G}_i/\mathfrak{G}_{i+1}$ $(i = 0, 1, 2, \ldots, r-1)$ are called the *factors* of the chain.

A normal chain in which successive members are different is said to be a *normal chain without repetitions.*

A normal chain is said to be a *refinement* of a given normal chain if the members of the given chain are among the members of the new chain.

THEOREM 8 (Jordan - Hölder - Schreier) : *Two given normal chains can be refined so that the series of factors of the two new chains are identical up to order and isomorphism.*

In order to carry out the proof, we ask not only if a refinement process can be found, but still more, namely :

Are there convenient methods for constructing the refinement?

Let the two given chains be :

$$\mathfrak{G} = \mathfrak{G}_0 \geq \mathfrak{G}_1 \geq \mathfrak{G}_2 \geq \cdots \geq \mathfrak{G}_r = e$$

and
$$\mathfrak{G} = \mathfrak{H}_0 \geq \mathfrak{H}_1 \geq \mathfrak{H}_2 \geq \cdots \geq \mathfrak{H}_s = e.$$

The following example shows that in general at least s–1 groups must be inserted between adjacent members of the first chain and similarly that at least r–1 groups must be inserted between adjacent members of the second chain.

Let p_{ik} $(i=1, 2, \ldots, r;\ k=1, 2, \ldots, s)$ be $r.s$ distinct prime numbers. Let $\mathfrak{G}$ be the cyclic group of order $n = \prod_{i,k} p_{ik}$. Then set

$$d_i = \prod_{k=1}^{s} p_{ik}, \qquad e_k = \prod_{i=1}^{r} p_{ik},$$

and $\mathfrak{G}_0 = \mathfrak{G}$; let $\mathfrak{G}_i$ be the subgroup of order $n/\prod_1^i d_\nu$ $(i=1, \ldots, r)$; similarly let $\mathfrak{H}_0 = \mathfrak{G}$ and let $\mathfrak{H}_k$ be the subgroup of order $n/\prod_1^k e_\mu$ $(k = 1, 2, \ldots, s)$. By inserting s–1 groups and r–1 between the adjacent members of the first and second chain respectively, both given chains can be refined so that the orders of the new factors run through all the primes p_{ik}. Since there is only one group of a given prime order, the resulting refinements are isomorphic. On the other hand isomorphic refinements between $\mathfrak{G}_{i-1}$ and $\mathfrak{G}_i$ (or $\mathfrak{H}_{k-1}$ and $\mathfrak{H}_k$) can not contain a factor whose order is divisible by two primes, since only common factor groups of order p_{ik} or 1 can lie between $\mathfrak{G}_{i-1}, \ldots, \mathfrak{G}_i$ (and $\mathfrak{H}_{k-1}, \ldots, \mathfrak{H}_k$). The intersection of two admissible subgroups and the product of an admissible normal subgroup with an admissible subgroup are again admissible subgroups. Consequently, multiplication of an intersection lying in $\mathfrak{G}_{i-1}$ with the normal subgroup $\mathfrak{G}_i$ yields a group between $\mathfrak{G}_{i-1}$ and $\mathfrak{G}_i$. In what follows, it will be shown that the intercalation of the $s-1$ groups

$$\mathfrak{G}_{i,k} = \mathfrak{G}_i \cdot (\mathfrak{G}_{i-1} \cap \mathfrak{H}_k) \qquad (k = 1, 2, \ldots, s-1)$$

between $\mathfrak{G}_{i-1}$ and $\mathfrak{G}_i$, and of the $r-1$ groups

$$\mathfrak{H}_{i,k} = \mathfrak{H}_k \cdot (\mathfrak{H}_{k-1} \cap \mathfrak{G}_i) \qquad (i = 1, 2, \ldots, r-1)$$

between $\mathfrak{H}_{k-1}$ and $\mathfrak{H}_k$, refines the given chains isomorphically.

$\mathfrak{G}_{i,k}$ and $\mathfrak{H}_{i,k}$ are defined for $i = 1, 2, \ldots, r-1$; $k=1, 2, \ldots, s-1$ by the above formulae. Moreover set

$$\mathfrak{G}_{i,0} = \mathfrak{G}_{i-1}, \ \mathfrak{H}_{0,k} = \mathfrak{H}_{k-1}; \quad \mathfrak{G}_{i,s} = \mathfrak{G}_i, \ \mathfrak{H}_{r,k} = \mathfrak{H}_k .$$

If it is shown that $\mathfrak{G}_{i,k}$ is a normal subgroup of $\mathfrak{G}_{i,k-1}$ $(k=1, 2, \ldots, s)$, then the $\mathfrak{G}_{i,k}$ form a refinement of $\mathfrak{G}_i$. Correspondingly for the $\mathfrak{H}_{i,k}$. If it is shown that

$$\frac{\mathfrak{G}_{i,k-1}}{\mathfrak{G}_{i,k}} \simeq \frac{\mathfrak{H}_{i-1,k}}{\mathfrak{H}_{i,k}} \qquad \begin{pmatrix} i = 1, 2, \ldots, r \\ k = 1, 2, \ldots, s \end{pmatrix},$$

then the refinements are isomorphic. The desired results are given by the following theorem concerning four groups:

If a subgroup $\mathfrak{u}$ *is a normal subgroup of the subgroup* $\mathfrak{U}$ *of* $\mathfrak{G}$, *and the subgroup* $\mathfrak{v}$ *is a normal subgroup of the subgroup* $\mathfrak{V}$ *of* $\mathfrak{G}$, *then* $\mathfrak{u}(\mathfrak{U} \cap \mathfrak{v})$ *is a normal subgroup of* $\mathfrak{u}(\mathfrak{U} \cap \mathfrak{V})$, *and* $\mathfrak{v}(\mathfrak{V} \cap \mathfrak{u})$ *is a normal subgroup of* $\mathfrak{v}(\mathfrak{V} \cap \mathfrak{U})$; *and*

$$\frac{\mathfrak{u}(\mathfrak{U} \cap \mathfrak{V})}{\mathfrak{u}(\mathfrak{U} \cap \mathfrak{v})} \simeq \frac{\mathfrak{v}(\mathfrak{V} \cap \mathfrak{U})}{\mathfrak{v}(\mathfrak{V} \cap \mathfrak{u})}.$$

Proof: By the second isomorphism theorem $\mathfrak{u} \cap \mathfrak{V}$ is a normal subgroup of $\mathfrak{U} \cap \mathfrak{V}$ and

$$\frac{\mathfrak{U} \cap \mathfrak{V}}{\mathfrak{u} \cap \mathfrak{V}} \simeq \frac{\mathfrak{u}(\mathfrak{U} \cap \mathfrak{V})}{\mathfrak{u}}.$$

Since $\mathfrak{u} \cap \mathfrak{V}$ is a normal subgroup of $\mathfrak{U} \cap \mathfrak{V}$, so is $\mathfrak{v} \cap \mathfrak{U}$, and therefore

$(\mathfrak{u} \cap \mathfrak{V})(\mathfrak{v} \cap \mathfrak{U})$ is also a normal subgroup of $\mathfrak{U} \cap \mathfrak{V}$. Under the above isomorphy, $(\mathfrak{u} \cap \mathfrak{V})(\mathfrak{v} \cap \mathfrak{U})$ is mapped onto $\mathfrak{u}(\mathfrak{u} \cap \mathfrak{V})(\mathfrak{v} \cap \mathfrak{U}) = \mathfrak{u}(\mathfrak{v} \cap \mathfrak{U})$. Therefore by the third isomorphy theorem $\mathfrak{u}(\mathfrak{U} \cap \mathfrak{v})$ is a normal subgroup of $\mathfrak{u}(\mathfrak{U} \cap \mathfrak{V})$, and

$$\frac{\mathfrak{U} \cap \mathfrak{V}}{(\mathfrak{u} \cap \mathfrak{V})(\mathfrak{v} \cap \mathfrak{U})} \simeq \frac{\mathfrak{u}(\mathfrak{U} \cap \mathfrak{V})}{\mathfrak{u}(\mathfrak{U} \cap \mathfrak{v})}.$$

Since the hypotheses are symmetric, it follows likewise that $\mathfrak{v}(\mathfrak{V} \cap \mathfrak{u})$ is a normal subgroup of $\mathfrak{v}(\mathfrak{V} \cap \mathfrak{U})$ and that

$$\frac{\mathfrak{U} \cap \mathfrak{V}}{(\mathfrak{u} \cap \mathfrak{V})(\mathfrak{v} \cap \mathfrak{U})} \simeq \frac{\mathfrak{v}(\mathfrak{V} \cap \mathfrak{U})}{\mathfrak{v}(\mathfrak{V} \cap \mathfrak{u})},$$

from which we obtain the desired isomorphy.

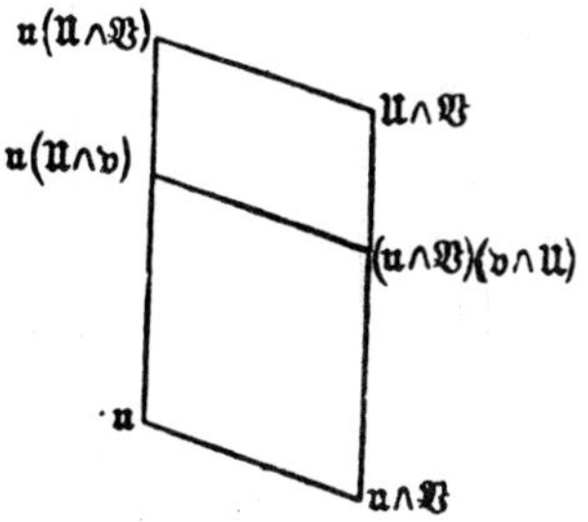

The method of proof can be made more meaningful by means of a diagram which shows the position of the groups occurring in the proof. In order to do this let a line between two groups, one of which is above the other, mean that the group at the upper end contains the group at the lower end of the line.[1]

The given method of refinement, applied for a second time, gives no new refinement of the first refinement. Nevertheless it may refine isomorphic chains still further.

Example: Let $\mathfrak{G}$ be cyclic of order 12. Let $\mathfrak{G}_1$ be the subgroup of order 6, $\mathfrak{H}_1$ the one of order 2. $\mathfrak{G}_2 = \mathfrak{H}_2 = e$. Then

$$\mathfrak{G}_{2,1} = \mathfrak{H}_1, \quad \mathfrak{H}_{1,1} = \mathfrak{G}_1.$$

A refinement is said to be a *proper refinement* if a new subgroup of $\mathfrak{G}$ actually occurs in the new chain.

A normal chain is said to be a *normal series* if it has no proper refinements.

If $\mathfrak{G}$ has a normal series then by the theorem of Jordan - Hölder - Schreier it follows that every normal chain can be refined so as to give a normal series. The series of factors in different normal series is identical up to sequence and isomorphy.

The existence of a normal series is assured if the *Double chain condition of group theory* holds:

[1] Often called a *Hasse diagram.*

1. *Minimal condition*: In every decreasing sequence of subgroups $\mathfrak{U}_1 \supseteq \mathfrak{U}_2 \supseteq \cdots$ there is an index after which all the members are equal. Equivalent to this is:

1. a) In every set of subgroups there is a subgroup which contains no other subgroup of the set.

2. *Maximal condition*: In every increasing sequence of groups $\mathfrak{U}_1 \subseteq \mathfrak{U}_2 \subseteq \mathfrak{U}_3 \subseteq \cdots$ there is an index after which all the members are equal.

2. a) In every set of subgroups there is a subgroup which is contained in no other subgroup of the set.

From the maximal condition it follows that $\mathfrak{G}$ contains a largest normal subgroup $\mathfrak{G}_1$ or is equal to e, that $\mathfrak{G}_1$ contains a largest normal subgroup $\mathfrak{G}_2$ or is equal to e, etc. It follows from the minimal condition that this sequence terminates at e after a finite number of steps. The normal chain thus obtained is a normal series.

If, conversely, every admissible subgroup is a normal subgroup of $\mathfrak{G}$, then the double chain theorem follows from the existence of a normal series.

The normal series in group theory have received different names, depending on the underlying domain of operators:

1. *Composition Series*: Every member of the chain

$$\mathfrak{G} = \mathfrak{G}_0 \supseteq \mathfrak{G}_1 \supseteq \cdots \supseteq \mathfrak{G}_{r-1} \supseteq \mathfrak{G}_r = e$$

which is different from $\mathfrak{G}$ is a maximal normal subgroup of the previous member.

2. *Principal Series*: Every member of the chain different from $\mathfrak{G}$ is a normal subgroup of $\mathfrak{G}$, maximal in the set of all proper subgroups of the preceding member.

3. *Characteristic Series*: Every member of the chain different from $\mathfrak{G}$ is a characteristic subgroup of $\mathfrak{G}$, maximal in the set of proper subgroups of the preceding member.

As an example we shall determine the *structure of the symmetric and the alternating permutation groups.*

The following theorem is useful when investigating the simplicity of a group.

THEOREM 9: *A transitive and primitive permutation group $\mathfrak{G}$ which contains no proper regular normal subgroups, and in which the subgroup of all permutations which leave a letter fixed is simple, must itself be simple.*

Proof: By Theorem 1 of this chapter, any normal subgroup $\mathfrak{N}$ other than $\mathfrak{1}$, of the primitive permutation group $\mathfrak{G}$ is transitive. Let $\mathfrak{G}_1$ be the group of all permutations of $\mathfrak{G}$ which leave the letter 1 fixed. Since $\mathfrak{N}$ is assumed not to be regular, the intersection $\mathfrak{N}_1$ of $\mathfrak{N}$ with the group $\mathfrak{G}_1$ is distinct from $\mathfrak{1}$. According to the 2nd isomorphy theorem $\mathfrak{N}_1$ is a normal subgroup of $\mathfrak{G}_1$, and inasmuch as $\mathfrak{G}_1$ is simple by hypothesis, we must have $\mathfrak{N}_1 = \mathfrak{G}_1$. Since $\mathfrak{N}$ is transitive,

$$\mathfrak{G} = \mathfrak{N}\mathfrak{G}_1 = \mathfrak{N}\mathfrak{N}_1 = \mathfrak{N},$$

as was to be proved.

THEOREM 10:[1] *The alternating permutation group of* $n \neq 4$ *letters is simple.*

Proof: $\mathfrak{A}_1$, $\mathfrak{A}_2$, $\mathfrak{A}_3$ are of orders 1, 1, 3, and are therefore simple. By Theorem 3, $\mathfrak{A}_5$ is simple. Now assume that we know that $\mathfrak{A}_{n-1}$ is simple and $n>5$. Then $\mathfrak{A}_n$ is quadruply transitive, therefore primitive, and according to the remark following Theorem 7, $\mathfrak{A}_n$ contains no regular normal subgroups.

The subgroup of all permutations in $\mathfrak{A}_n$ which leave the letter n fixed permutes the remaining letters $1, \ldots, n-1$ as $\mathfrak{A}_{n-1}$ does, and thus is simple by the induction assumption. By the preceding theorem $\mathfrak{A}_n$ itself is simple, as was to be shown.

THEOREM 11: *If* $n \neq 4$, $n > 2$, *then the symmetric permutation group* $\mathfrak{S}_n$ *has exactly the one composition series*

$$\mathfrak{S}_n \supset \mathfrak{A}_n \supset e.$$[2]

Proof: If $n > 2$, then $\mathfrak{S}_n$ is doubly transitive and therefore primitive; consequently a proper normal subgroup $\mathfrak{N}$ of $\mathfrak{S}_n$ is transitive. By the second isomorphy theorem $\mathfrak{N} \cap \mathfrak{A}_n$ is a normal subgroup of $\mathfrak{A}_n$; moreover $\mathfrak{N} : \mathfrak{N} \cap \mathfrak{A}_n / 2$, while the transitivity of $\mathfrak{N}$ implies $\mathfrak{N} \cap \mathfrak{A}_n \neq \mathfrak{1}$. If moreover $n \neq 4$, then because of the previously proven simplicity of $\mathfrak{A}_n$

$$\mathfrak{N} \cap \mathfrak{A}_n = \mathfrak{A}_n,$$

and therefore $\mathfrak{N} = \mathfrak{A}_n$, as was to be shown.

By the same method of proof, it follows that for a proper normal subgroup $\mathfrak{N}$ of $\mathfrak{A}_4$, the intersection $\mathfrak{N}_1 = \mathfrak{N} \cap \mathfrak{A}_4$ is a normal subgroup $\neq \mathfrak{1}$ of $\mathfrak{A}_4$. Since $\mathfrak{A}_4$ is doubly transitive, $\mathfrak{N}_1$ is transitive. Therefore the order of $\mathfrak{N}_1$ is divisible by 4. Either $\mathfrak{N}_1$ contains 3-cycles so that $\mathfrak{N}_1 = \mathfrak{A}_4$, or $\mathfrak{N}_1$ consists of double transpositions and $\mathfrak{1}$. $\mathfrak{A}_4$ actually con-

---

[1] With the help of Exercise 9 at the end of the chapter, the reader can develop the usual proof of Theorem 10.

[2] Naturally this is also the principal series, indeed the characteristic series.

tains the transitive normal subgroup $\mathfrak{V}_4$ which consists of the three double transpositions and $\mathfrak{1}$. The subgroup $\mathfrak{S}_3$ of all permutations which leave the letter 4 fixed can be taken as a representative system of $\mathfrak{S}_4$ over $\mathfrak{V}_4$ and so we finally obtain: *Every composition series of* $\mathfrak{S}_4$ *begins with* $\mathfrak{S}_4 \supset \mathfrak{A}_4 \supset \mathfrak{V}_4$.

Since the abelian group $\mathfrak{V}_4$ contains three proper subgroups, $\mathfrak{S}_4$ has three different compositions series.

Close inspection of the proof of the Jordan-Hölder-Schreier theorem shows that its validity in a given group depends more on the relation between the subgroups of the group than on the behavior of the individual elements. This observation lead Oystein Ore[1] to consider problems of this type from an abstract point of view, and for this purpose he defined a new algebraic system which he called a *structure.* In more recent times the theory of structures has been developed into a new branch of mathematics.

To begin with, we define a *partially ordered set* (*poset*) as a set S in which a binary relation is defined between certain of its elements; this may be denoted by

$$a \subseteq b \qquad \text{or by} \qquad b \supseteq a$$

or, in case it should be necessary to indicate the poset to which a and b belong, by

$$a \underset{S}{\subseteq} b \qquad \text{and} \qquad b \underset{S}{\supseteq} a,$$

respectively. This relation is required to be subject to the conditions of reflexivity and transitivity:

(1) $$a \subseteq a$$

(2) $$\text{if} \quad a \subseteq b \quad \text{and} \quad b \subseteq c \qquad \text{then} \quad a \subseteq c.$$

Example: The set $\Sigma(S)$ consisting of all subsets of a given set S is a poset when we define the binary relation as set-theoretical inclusion.

Every subset T of a poset S is a poset if we take as its binary relation the one induced by the binary relation of S. We call two elements a, b of a poset S *equivalent*, if $a \subseteq b$ and $b \subseteq a$. This equivalence is normal. Furthermore it satisfies the law of substitution:

If a is equivalent to a' and b equivalent to b' then from $a \subseteq b$ it follows that $a' \subseteq b'$.

[1] Many results and problems have been anticipated by Dedekind.

Therefore we are able to define uniquely the relation between the classes of equivalent elements of S by the following rule:

$$\bar{a} \leq \bar{b} \textit{ if and only if } a \leq b,$$

where $\bar{x}$ denotes the class of all elements of S equivalent to the element x of S.

It follows that the classes of equivalent elements of a partially ordered set S themselves form a partially ordered set $\bar{S}$ satisfying the additional rule:

If $\bar{a} \leq \bar{b}$ *and* $\bar{b} \leq \bar{a}$ *then* $\bar{a} = \bar{b}$.

Usually in dealing with partially ordered sets we assume that equivalence amounts to equality. This certainly holds for the posets $\Sigma(S)$ formed by all the subsets of a set S. At any rate, the fundamental concepts remain invariant if equivalent elements are substituted.

Let us observe that reflexivity and transitivity are *self-dual concepts* inasmuch as a poset S is carried over into another poset S^* if we introduce the new relation

$$a \underset{S^*}{\leq} b \textit{ if and only if } a \underset{S}{\geq} b.$$

S^* is called the *dual poset* of S. The dual poset of S^* is S. For any concept employing the symbol $\leq$ we obtain the *dual concept* by employing the symbol $\geq$ instead.

The *principle of duality* states that every theorem concerning posets remains true if the symbols $\leq$, $\geq$ are interchanged and every derived concept is replaced by the corresponding dual concept.

1. *Homomorphisms and anti-homomorphisms.*

A one-to-one correspondence between two posets that preserves the binary relation in both directions is called an *isomorphism* between the two posets. Naturally, the fundamental concepts of the theory of structures are so chosen that they remain invariant under isomorphisms.

We define more generally: A single-valued mapping φ of the elements of a poset S onto a certain subset of a poset S_1 is called a *homomorphy* if from $a \leq b$ in S it always follows that $\varphi a \leq \varphi b$ in S_1 and if from $\varphi c \leq \varphi d$ in S_1 it always follows that there exist two elements a, b of S such that $\varphi a = \varphi c$ and $\varphi b = \varphi d$ and $a \leq b$. This defines a *homomorphic mapping* or a *homomorphism* of the poset S onto the poset φS consisting of all the images $\varphi a (a \varepsilon S)$.

The most important poset for the theory of groups is the poset $S(\mathfrak{G})$ formed by all the subgroups of a group $\mathfrak{G}$, where $\leq$ is taken as set-theoretical inclusion. We observe that a homomorphism of a group $\mathfrak{G}$ onto another group $\mathfrak{H}$ induces a homomorphism of $S(\mathfrak{G})$ onto $S(\mathfrak{H})$.

Every homomorphy φ of a poset S gives rise to a normal congruence relation $R\varphi$ in S defined by:

$$a \equiv b \qquad (R\varphi)$$

if and only if $\varphi a = \varphi b$. This congruence relation is *isotonic*: If $a \leq b$, $b' \leq c$ and $b' \equiv b(R\varphi)$ then there are elements a', c' in S satisfying $a' \equiv a(R\varphi)$, $c' \equiv c(R\varphi)$ and $a' \leq c'$. For any isotonic normal congruence relation R in S, denote by $a(R)$ the residue class represented by a modulo R. These classes themselves form a poset S/R in which the binary relation is defined by:

$$a(R) \leq b(R) \tag{3}$$

if and only if there are elements a', b' in S satisfying $a' \equiv a(R)$, $b' \equiv b(R)$ and $a' \leq b'$.

The mapping $a \to a(R)$ defines a homomorphism between S and S/R which we may call the natural homomorphism between S and S/R. For a homomorphism φ of S we find that the mapping $a(R\varphi) \to \varphi a$ defines an isomorphism between the posets $S/R\varphi$ and φS which we may call the *natural isomorphism* between $S/R\varphi$ and φS. Also, φ induces the natural homomorphism between S and $S/R\varphi$. Two homomorphisms, say φ mapping poset S into poset S_1 and ψ mapping poset S_2 into poset S_3 may be multiplied if and only if ψS_2 is part of S. The rule of multiplication is given by

$$(\varphi\psi)a = \varphi(\psi a) \qquad (a \varepsilon S_2) \tag{4}$$

and $\varphi\psi$ turns out to be a homomorphism of S_2 into S_1.

This multiplication is associative. A left identity for the homomorphisms into a poset S is given by the identity 1_S of S. This also acts as right identity for all homomorphisms of the poset S into another poset.

A homomorphism φ of the poset S onto the poset φS is an isomorphism if and only if it is one-to-one. In this case the inverse mapping φ^{-1} of φS onto S characterized by

$$\varphi^{-1}\varphi = 1_S$$

is an isomorphism between φS and S. Its inverse is φ itself:

$$\varphi\varphi^{-1} = 1_{\varphi S}.$$

If there exists an isomorphism between two posets then the posets are called *isomorphic*. If there is given a number of isomorphic posets, then all the isomorphisms between any two of them form a groupoid under multiplication, with the identity mappings of the members of the system acting as units.

The isomorphisms of a poset with itself are called the *automorphisms* of the poset. They form a group under multiplication. Isomorphic posets have isomorphic groups of automorphisms. A single-valued mapping φ of the poset S onto the poset φS will be said to be an *anti-homomorphism* if φ induces a homomorphism between the poset S and the dual poset of φS. Two single-valued mappings φ, ψ which are either anti-homomorphisms (denote by a) or homomorphisms (denote by h) are multiplied according to the same rule as was given for the multiplication of homomorphisms, and the outcome is either a homomorphism or an anti-homomorphism as given by the following "multiplication table" of a and h:

	a	h
a	h	a
h	a	h

A one-to-one anti-homomorphism is called *anti-isomorphism*. E.g., the one-to-one correspondence by which an element a of a poset S corresponds to itself in the dual poset S^* defined in the introduction is an anti-isomorphism.

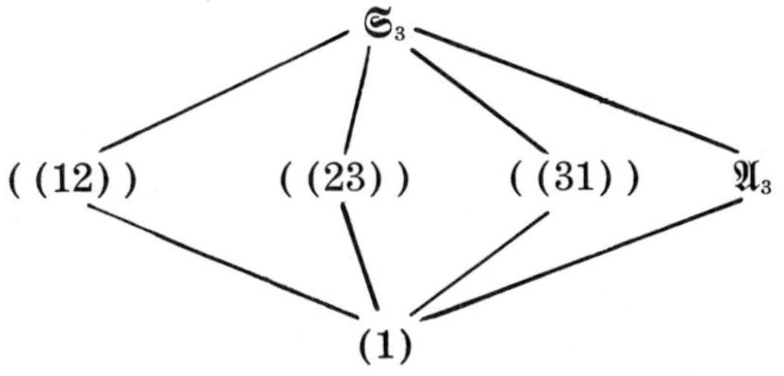

An anti-isomorphism of a poset S with itself is an *anti-automorphism*. A poset S which admits an anti-automorphism is called *self-dual*. In this case all the automorphisms and anti-automorphisms of S together form a group containing the group of automorphisms of S as normal subgroup of index 2. An example of a self-dual poset is provided by $S(\mathfrak{S}_3)$, the Hasse diagram of which is exhibited just above. The automorphism group of $S(\mathfrak{S}_3)$ is isomorphic to $\mathfrak{S}_4$. There are 24 anti-automorphisms one of which may be obtained by simply interchanging $\mathfrak{S}_3$ and (1), leaving the other subgroups invariant. Only six of the automorphisms of $S(\mathfrak{S}_3)$ are induced by the six automorphisms of the group $\mathfrak{S}_3$, which is the largest number possible because $\mathfrak{A}_3$, as the only non-trivial normal subgroup, is necessarily invariant under all automorphisms of the group.

The correspondence between any automorphism α of a group $\mathfrak{G}$ and the automorphism α_S induced by α on $S(\mathfrak{G}) = S$ provides a homomorphism between the group $A_{\mathfrak{G}}$ of all automorphisms of $\mathfrak{G}$ and a subgroup $A_{\mathfrak{G},S}$ of the group $A_{S(\mathfrak{G})}$ of all automorphisms of $S(\mathfrak{G})$. The kernel of this homomorphism consists of all automorphisms of $\mathfrak{G}$ that map every subgroup of $\mathfrak{G}$ onto itself (not necessarily elementwise).

2. *Meet and join.*

Let S be a poset. We describe sets formed by some elements of S by taking a non-empty index set A and assigning to each α in A an element a_α of S. We define: The element x of S is the *meet* of the elements a_α ($\alpha \varepsilon A$) if and only if $x \subseteq a_\alpha$ for all α in A and if from $y \subseteq a_\alpha$ for all α in A it always follows that $y \subseteq x$. We write $x = \bigcap_{\alpha \varepsilon A} a_\alpha$. We also write $x = a_1 \cap a_2 \cap \ldots \cap a_n$ if the index set is finite.

It may happen that the meet does not exist. *E.g.*, if no inclusions other than the trivial ones $a \subseteq a$ hold in S, then it is impossible to form the meet of two or more different elements. But if the meet does exist then it is uniquely determined up to equivalence: If $x = \bigcap_{\alpha \varepsilon A} a_\alpha$, $y = \bigcap_{\alpha \varepsilon A} a_\alpha$ then $x \subseteq a_\alpha$ and $y \subseteq a_\alpha$ for all α in A and thus $x \subseteq y$, $y \subseteq x$. Hence we are allowed to deal with $\bigcap_{\alpha \varepsilon A} a_\alpha$ as with an element of S.

We always have:

(5) $$\bigcap_{\alpha \varepsilon A} a_\alpha = a \qquad \text{if } a_\alpha = a \text{ for all } \alpha \varepsilon A.$$

(6) $$a \cap b = b \qquad \text{if } b \subseteq a, \text{ and conversely.}$$

Furthermore,

$$\bigcap_{\alpha \in A} a_\alpha \leq \bigcap_{\alpha \in A} b_\alpha \qquad \text{if } a_\alpha \leq b_\alpha \text{ for } \alpha \in A, \tag{7a}$$

$$\bigcap_{\alpha \in B} a_\alpha \geq \bigcap_{\alpha \in A} a_\alpha \tag{7b}$$

for a subset B of A provided both sides of the inequality can be formed. (7b) turns into an equality if each element a_α that is removed also occurs among the remaining elements, i.e. if $\alpha \in A - B$, $a_\alpha = a_\beta$ for some $\beta \in b$. Finally, we have

$$\bigcap_{\beta \in B} b_\beta = \bigcap_{\alpha \in A} a_\alpha, \tag{8}$$

where B denotes a second set of indices and to each β of B there corresponds a subset A_β of A such that $b_\beta = \bigcap_{\alpha \in A_\beta} a_\alpha$ can be formed and A is the union of all subsets A_β, provided at least one of the two sides of (8) can be formed.

The previous rules imply the rules

$$a \wedge a = a \tag{9}$$

$$a \wedge b = b \wedge a \tag{10}$$

$$a \wedge (b \wedge c) = (a \wedge b) \wedge c \tag{11}$$

provided all but one of the meets involved can be formed.

In the poset $S_1 = \Sigma(S)$ of all subsets of a set S the meet coincides with the set-theoretical intersection. This remains true for the poset $S(\mathfrak{G})$ of all subgroups of a group $\mathfrak{G}$. But, for a poset S_2 formed from certain subsets of a set S, we may by no means infer that meet and intersection coincide. The reason is simply that the intersection of certain subsets a_α of S $(\alpha \in A)$ belonging to S_2 may not belong to S_2. *E.g.*, Let S be the set of all real numbers, and let S_2 be the set of all open intervals $-\varepsilon < x < \varepsilon$ $(\varepsilon > 0)$ together with the empty subset. The meet of these intervals taken in S_2 will be the empty subset, but the intersection will be the origin.

More generally, let S_2 be a subposet of the poset S_1 and assume that $\bigcap^{S_2}_{\alpha \in A} a_\alpha$ exists ($a_\alpha \in S_2$ for $\alpha \in A$). Then $\bigcap^{S_1}_{\alpha \in A} a_\alpha$ may not exist at all, or it may exist and be different from $\bigcap^{S_2}_{\alpha \in A} a_\alpha$. We only can infer the rule

$$\bigcap^{S_2}_{\alpha \in A} a_\alpha = \bigcup^{S_1}_{\substack{a_\alpha \geq u \in S_2 \\ \text{for } \alpha \in A}} u \tag{12}$$

The dual operation of meet is called *join*. The element x of the poset S is called the *join* of the elements a_α of $S\,(\alpha \in A)$ if for all α in A, $a_\alpha \subseteq x$ and if, for all $y \in S$ satisfying $a_\alpha \subseteq y$ for all α in A, $x \subseteq y$. We write $x = \bigcup_{\alpha \in A} a_\alpha$. The same considerations as for the meet hold for the join if we replace the symbol $\wedge$ by the symbol $\vee$.

Meet and join are interdependent in the following way. If for the subset U of all elements u of S satisfying $a_\alpha \subseteq u$ for all α of A the meet $\bigcap_{u \in U} u$ can be formed, then we have

$$\bigcap_{a_\alpha \subseteq u \text{ for all } \alpha \in A} u = \bigcup_{\alpha \in A} a_\alpha \tag{13}$$

Similarly,

$$\bigcup_{a_\alpha \supseteq u \text{ for all } \alpha \in A} u = \bigcap_{\alpha \in A} a_\alpha \tag{14}$$

if one of the two sides can be formed. In fact, since $a_\alpha \subseteq u$ for all $u \in U$ it follows that $a_\alpha \subseteq \bigcap_{u \in U} u$ and this holds for all α in A. But if $a_\alpha \subseteq b$ for all α in A, then we have $b \in U$, $\bigcap_{u \in U} u \subseteq b$, and hence (13).

A poset in which the meet and the join can always be formed, is called *complete*. E.g., the poset $\Sigma(S)$ of all subsets of a given set S is complete. If the sum of all elements of a poset can be formed then the result will be an *all element* characterized as an element including every other element of the poset. If there is an all element and if the meet always can be formed, then from (13) it follows that the join always can be formed, and hence the poset is complete. This happens, for example, for the poset $S(\mathfrak{G})$ of all subgroups of a group $\mathfrak{G}$ in which $\mathfrak{G}$ is all element and in which the intersection always can be formed. The join of any number of subgroups coincides with the intersection of all subgroups of $\mathfrak{G}$ containing each subgroup of the given system of subgroups. Hence the join coincides with the subgroup generated by all the subgroups of the given system.

If the meet of all elements of a poset exists, then it is called *null element* of S. It is characterized up to equivalence as an element of S included in every element of S.

If there should not be a null element of S then we enlarge S artificially by the addition of an element m subject to the inequalities $m \subseteq a$ for all $a \in S$ and $m \subseteq m$. The enlarged set is a poset with null element m. Similarly, we may add an all element if there is none.

3. *Lattices.*

A *lattice* or a *structure* is a set L with two binary operations $\wedge$ and $\vee$ such that for any two elements a, b there are always uniquely defined elements $a \cap b$, $a \cup b$ in L. The operations are called *meet* and *join* respectively and are subject to the rules

(15) (idempotency)

a) $a \cap a = a$

b) $a \cup a = a$

(16) (commutativity)

a) $a \cap b = b \cap a$

b) $a \cup b = b \cup a$

(17) (associativity)

a) $a \cap (b \cap c) = (a \cap b) \cap c$

b) $a \cup (b \cup c) = (a \cup b) \cup c$

(18) a) $a \cap (a \cup b) = a$

b) $a \cup (a \cap b) = a.$

For example, a poset S in which equivalence amounts to equality and in which the meet and the join of any two elements can always be formed, is a lattice. According to (6) and the dual rule, we have

(19) $a \subseteq b$ if and only if $a \cap b = a,$

(20) $a \subseteq b$ if and only if $a \cup b = b.$

Conversely, in an arbitrary lattice we may define a binary relation by (19). From (16a) follows $a \subseteq a$. If $a \subseteq b, b \subseteq c$, then $a \cap b = a$, $b \cap c = b$, $a \cap c = (a \cap b) \cap c = a \cap (b \cap c) = a \cap b = a$ so that $a \subseteq c$. If $a \subseteq b, b \subseteq a,$ then $a \cap b = a, b \cap a = b, a = a \cap b = b \cap a = b.$

Hence a lattice determines a poset in which equivalence amounts to equality. Furthermore we have $(a \cap b) \cap a = (b \cap a) \cap a = b \cap (a \cap a) = b \cap a = a \cap b$, hence $a \cap b \subseteq a$; similarly $a \cap b \subseteq b$. If $x \subseteq a, x \subseteq b$, then $x \cap a = x, x \cap b = x, x \cap (a \cap b) = (x \cap a) \cap b = x \cap b = x$, so that $x \subseteq a \cap b$. Hence $a \cap b$ coincides with the meet of a and b formed in the poset L. Using (18) we prove the equivalence of (19) with (20). By dual arguments it follows now that $a \cup b$ coincides with the join of a, b formed in the poset L.

We have found that lattices may be defined as posets in which equivalence amounts to equality and in which the meet and join of any two elements always can be formed.

From the definition it follows that the concept of a lattice is a self-dual concept.

A subset of a lattice is called a *sublattice* if it is closed under both lattice operations. A sublattice is itself a lattice.

Trivially, L is a sublattice of L. For any pair of elements a, b of L satisfying $b \subseteq a$ there is defined the *factor lattice* a/b consisting of all elements x of L which satisfy $b \subseteq x \subseteq a$ ("x is between b and a"). This is a sublattice of L. Another sublattice of L is the lattice L/a of all elements x of L which satisfy $a \subseteq x$; the dual notion $a/0$ is defined to be the sublattice consisting of all elements x of L satisfying $x \subseteq a$. In case there is an element M, it follows that $M/a = L/a$. If there is a null element m then we have $a/m = a/0$.

A complete poset in which equivalence amounts to equality is a lattice. Not every lattice is complete—e.g, the lattice consisting of all open non-empty intervals containing the origin on the real axis is not complete. However every finite lattice is complete. More generally, a lattice satisfying the *maximal condition* that in every non-empty subset of the lattice there is an element which is not included in any other element of the subset, is a complete lattice provided that there is an all element and a null element. In fact, for any subset U of the lattice the set V of all elements of the lattice included in every element of U will not be empty; hence it contains an element a not included by any other element of V. But the join of a and any other element of V also belongs to V and so includes a; hence it must coincide with a. Thus a includes every element of V, so that a is the meet of the elements of U.

4. *Projections and antiprojections.*

For any lattice L and any element a of L there is defined a homomorphism φ_a of the poset L onto the poset $a/0$ by the rule

$$x \to \varphi_a x = a \cap x;$$

This homomorphism may be called the *projection* of L with respect to a or onto $a/0$. In fact, we even have

$$\varphi_a(x \cap y) = \varphi_a x \cap \varphi_a y, \tag{21a}$$

as follows from (17a) and (18a).

The dual concept is the *antiprojection* of L with respect to the element a of L which is defined as the homomorphic mapping φ^a given by

$$x \to \varphi^a x = a \cup x$$

of the poset L onto the poset L/a. This homomorphism preserves the join operation

(21b) $$\varphi^a(x \cup y) = \varphi^a x \cup \varphi^a y,$$

as follows from (17b) and (18b).

The projection φ_a induces in any factor lattice b/c a homomorphism of b/c into $(a \cap b)/(a \cap c)$ such that $\varphi_a b = a \cap b$, $\varphi_a c = a \cap c$. But we can by no means infer that φ_a maps b/c onto $(a \cap b)/(a \cap c)$.

E.g., in $S(\mathfrak{S}_4)$ let $a = \{(123), (12)\}$, $b = \mathfrak{S}_4$, $c = ((1234))$. There is exactly one subgroup x between c and b other than c and b itself viz. $x = \{(1234), (24)\}$ but we have $a \cap b = a$, $a \cap c = a \cap x = m = (1)$; hence the subgroup $y = ((123))$ of a/m is not in $\varphi_a(b/c)$. The six subgroups involved form a sublattice whose Hasse diagram is given below. (Hexagon lattice.)

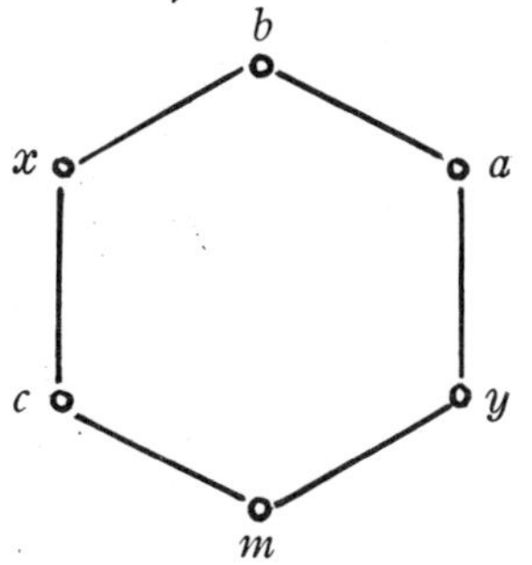

Generally we have

(22) $$\varphi_a x = \varphi_a(\varphi_a x \cup c) \qquad \text{if } c \leq x,$$

because

$$\varphi_a x \leq x, \ \varphi_a x \cup c \leq x \cup c = x,$$
$$\varphi_a(\varphi_a x \cup c) \leq \varphi_a x \leq a \cap \varphi_a x \leq a \cap (\varphi_a x \cup c) = \varphi_a(\varphi_a x \cup c).$$

Hence a necessary and sufficient condition that φ_a induce an isomorphism of b/c into $a/0$ is that $x = \varphi_a x \cup c$ whenever $c \leq x \leq b$, or

(23) $$\varphi^c \varphi_a x = x \qquad \text{for all elements } x \text{ of } b/c.$$

This condition guarantees that φ_a induces an isomorphism between b/c and some subset of $(a \cap b)/(a \cap c)$, but it still may happen that there are elements in $(a \cap b)/(a \cap c)$ not belonging to $\varphi_a(b/c)$ as is the case for the *pentagon* which one obtains from the last figure by omitting x, and which looks as follows:

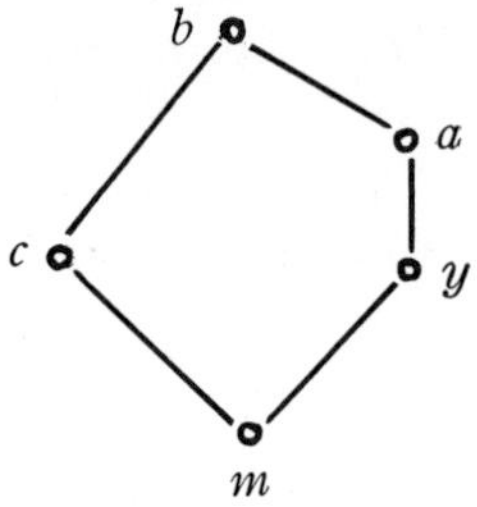

If φ_a induces an isomorphism of b/c into $a/0$ then its inverse is induced by φ^c according to (23). Hence in order to be sure that φ_a induces an isomorphism between b/c and $(a \cap b)/(a \cap c)$ we have to amplify (23) by

$$\varphi_a \varphi^c y = y \qquad \text{for all } y \text{ in } (a \cap b)/(a \cap c). \tag{24}$$

DEFINITION: The factor lattice b/c is *projective* with the factor lattice d/e if φ_a induces an isomorphism of b/c onto d/e or what is equivalent, if φ^c induces an isomorphism of d/e onto b/c.

This relation between b/c and d/e is described properly by the formulas:

$$c \cup d = b, \qquad c \cap d = e \tag{25}$$

and the identities

$$((x \cup c) \cap d) \cup c = (x \cup c) \cap (c \cup d) \tag{26}$$

$$((y \cap d) \cup c) \cap d = (y \cap d) \cup (c \cap d) \tag{27}$$

for all x and y of L.

The relation is reflexive: φ_b induces the identity isomorphism between b/c and b/c.

The relation is transitive: If b/c is projective with d/e and d/e projective with f/g, then we have $f \subseteq d$, and hence from the associative law

$$\varphi_f \varphi_d = \varphi_f.$$

It follows that φ_f induces an isomorphism between b/c and f/g, viz., the product of the isomorphism induced by φ_f between d/e and f/g and the isomorphism induced by φ_d between b/c and d/e.

Moreover if b/c is projective with d/e and if also b/c is projective with f/g such that $f \subseteq d$, then d/e is projective with f/g. (*Cancellation Law*).

Proof: φ_d induces an isomorphism $\bar{\varphi}_d$ between b/c and d/e, φ_f induces an isomorphism $\bar{\varphi}_f$ between b/c and f/g. Since $f \subseteq d$, we find that $\varphi_f = \bar{\bar{\varphi}}_f \varphi_d$ where $\bar{\bar{\varphi}}_f$ denotes the homomorphism induced by $\bar{\varphi}_f$ on d/e. It follows that $\bar{\bar{\varphi}}_f$ is an isomorphism between d/e and f/g.

By duality we find: If b/c is projective with f/g and if d/e is projective with f/g and if in addition the inequality $c \supseteq e$ holds, then b/c is projective with d/e.

Finally, let us notice the rule of inclusion: If b/c is projective with d/e and if $c \subseteq c' \subseteq b' \subseteq b$, then b'/c' is projective with $(b' \cap d)/(c' \cap d)$. And the dual rule: If b/c is projective with d/e and if $e \subseteq e' \subseteq d' \subseteq d$, then $(d' \cap c)/(e' \cap c)$ is projective with d'/e'.

The first of the two rules follows from the fact that φ_d induces an isomorphism between b'/c' and $(b' \cap d)/(c' \cap d)$ which according to the associative law also is induced by $\varphi_{x \cap d}$. The dual rule follows by duality.

An important application of these concepts can be made to the lattice $S(\mathfrak{G})$ of all subgroups of a given group $\mathfrak{G}$. First, $\mathfrak{G}$ is the all element of $S(\mathfrak{G})$, (1) the null element of $S(\mathfrak{G})$. Second, if $\mathfrak{B}, \mathfrak{C} \in S(\mathfrak{G})$, $\mathfrak{C} \subseteq \mathfrak{B}$, then $\mathfrak{C}/\mathfrak{B}$ will be the lattice formed by all subgroups of $\mathfrak{C}$ between $\mathfrak{C}$ and $\mathfrak{B}$. If $\mathfrak{C}$ is a normal subgroup of $\mathfrak{B}$, then for brevity we introduce the notation

$$\mathfrak{C} \lhd \mathfrak{B}$$

for this relation between $\mathfrak{C}$ and $\mathfrak{B}$.

If $\mathfrak{C} \lhd \mathfrak{B}$ and if for a certain subgroup $\mathfrak{A}$ of $\mathfrak{G}$ it holds that $\{\mathfrak{A}, \mathfrak{C}\} = \mathfrak{B}$ then it follows from the second isomorphy theorem that there is the isomorphism $\varphi_{\mathfrak{A},\mathfrak{B},\mathfrak{C}}$ between $\mathfrak{B}/\mathfrak{C}$ and $\mathfrak{A} \cap \mathfrak{B}/\mathfrak{A} \cap \mathfrak{C}$ defined by the formula

$$\varphi_{\mathfrak{A},\mathfrak{B},\mathfrak{C}}(a\mathfrak{C}) = a(\mathfrak{A} \cap \mathfrak{C}) \tag{28}$$

for all a of $\mathfrak{A}$, which may be called the *projection* of $\mathfrak{B}/\mathfrak{C}$ onto $\mathfrak{A} \cap \mathfrak{B}/\mathfrak{A} \cap \mathfrak{C}$. The group-theoretical isomorphism $\varphi_{\mathfrak{A},\mathfrak{B},\mathfrak{C}}$ induces an isomorphism between the corresponding lattices $\mathfrak{B}/\mathfrak{C}$ and $(\mathfrak{A} \cap \mathfrak{B})/(\mathfrak{A} \cap \mathfrak{C})$ and from (28) it becomes clear that the group-theoretical projection induces a lattice-theoretical projection.

5. *The Dedekind elements of a lattice.*

A. Kurosh has generalized the concept of normality to all lattices by making the following definition:

The element c of the lattice L is called a *Dedekind element* of L if and only if for each element a the mapping φ_c induces an isomorphism of $(a \cup c)/c$ onto $a/(a \cap c)$. This amounts to the identities (26) and (27) for all a, x, y of L. Or, using the terminology introduced by O. Ore: $(a \cup c)/c$ is projective with $a/(a \cap c)$ for all elements a of L. We write $c \mathrm{K} L$ if c is a Dedekind element of L. Furthermore, we write $c \mathrm{K} b$ if c is a Dedekind element of the lattice $b/0$. It is convenient to call the relation $c \mathrm{K} b$ *Kurosh invariance.*

We have seen in the last section that the relation $\mathfrak{C} \lhd \mathfrak{B}$ between two subgroups of a group $\mathfrak{G}$ implies the Kurosh invariance $\mathfrak{C} \mathrm{K} \mathfrak{B}$. But the converse is not true in general, as may be seen from $S(\mathfrak{S}_3)$ where every element is Kurosh invariant but not every element is normal.

Kurosh invariance has the following five properties proved by Kurosh:

I. $$c \mathrm{K} c.$$

II. If $c \mathrm{K} L$, then $c \mathrm{K} a \cup c$ for all a of L.

Proof: For $x \epsilon L$ we find that $(x \cup c)/c$ is projective with $x/(x \cap c)$, and this holds *a fortiori* if $x \epsilon (a \cup c)/0$

Applying II to $b/0$ we find

IIa. If $c \mathrm{K} b$ then $c \mathrm{K} ((a \cap b) \cup c)$ for all a of L.

III. If $c \mathrm{K} L$ it follows that $a \cap c \mathrm{K} a$.

Proof: If $a' \epsilon a/0$, then

$$a' \leq a,\ a \cap c \leq (a' \cup (a \cap c)) \cap c \leq (a \cup (a \cap c)) \cap c = a \cap c,$$
$$(a' \cup (a \cap c)) \cap c = a \cap c,$$

hence, since $c \mathrm{K} L$, $((a' \cup (a \cap c)) \cup c)/c$ is projective with

$$(a' \cup (a \cap c))/(a \cap c).$$

But
$$(a' \cup (a \cap c)) \cup c = a' \cup ((a \cap c) \cup c) = a' \cup c;$$

hence $(a' \cup c)/c$ is projective with $(a' \cup (a \cap c))/(a \cap c)$. We also have that $(a' \cup c)/c$ is projective with $a'/(a' \cap c)$. Observing that $a' \cap c = a' \cap (a \cap c)$ and using the cancellation rule proved at the end of the last section, we find that $(a' \cup (a \cap c))/(a \cap c)$ is projective with

$$a'/(a' \cap (a \cap c)).$$

Applying III to $b/0$ we find

IIIa. If $c \,\mathrm{K}\, b$, then $c \cap a \,\mathrm{K}\, b \cap a$ for all a of L.

IV. If $c \,\mathrm{K}\, L$ and if $x \,\mathrm{K}\, L/c$, then $x \,\mathrm{K}\, L$. Conversely, if $c \subseteq x$ and $x \,\mathrm{K}\, L$, then $x \,\mathrm{K}\, L/c$.

Applying IV to $b/0$ we find

IVa. If $c \,\mathrm{K}\, b$ and if $c \subseteq x \subseteq b$, then $x \,\mathrm{K}\, b$ if and only if $x \,\mathrm{K}\, b/c$.

V. If $c_1 \,\mathrm{K}\, L$, $c_2 \,\mathrm{K}\, L$, then $c_1 \cup c_2 \,\mathrm{K}\, L$.

Applying V to $b/0$ we find:

Va. If $c_1 \,\mathrm{K}\, b$, $c_2 \,\mathrm{K}\, b$, then $c_1 \cup c_2 \,\mathrm{K}\, b$.

Proof: Let $x \,\mathrm{K}\, L/c$, $a \varepsilon L$. It follows that $(x \cup (a \cup c))/x$ is projective with $(a \cup c)/(x \cap (a \cup c))$. Furthermore, from $c \,\mathrm{K}\, L$ it follows that $(a \cup c)/c$ is projective with $a/(a \cap c)$, hence $(a \cup c)/(x \cap (a \cup c))$ is projective with $a/(a \cap (x \cap (a \cup c)))$. Observing that

$$x \cup (a \cup c) = (x \cup c) \cup a = x \cup a,$$

$$a \cap (x \cap (a \cup c)) = (a \cap (a \cup c)) \cap x = a \cap x$$

we find that $(x \cup a)/x$ is projective with $(a \cup c)/(x \cap (a \cup c))$ and $(a \cup c)/(x \cap (a \cup c))$ is projective with $a/(a \cap x)$. From the transitivity of the 'projective' relation it follows that $(x \cup a)/x$ is projective with $a/(a \cap x)$. Since this holds for all a of L it follows that $x \,\mathrm{K}\, L$. Conversely if $x \,\mathrm{K}\, L$, then for $c \subseteq a$ it follows that $(x \cup a)/x$ is projective with $a/(x \cap a)$; hence we have $x \,\mathrm{K}\, L/c$.

Let $c_1 \subseteq x$. Since $c_2 \,\mathrm{K}\, L$, it follows that $(c_2 \cup x)/c_2$ is projective with $x/(c_2 \cap x)$. Hence $(c_2 \cup x)/(c_2 \cup c_1)$ is projective with $x/((c_2 \cup c_1) \cap x)$. Since $c_1 \subseteq x$, it follows that $c_2 \cup x = c_2 \cup (c_1 \cup x) = (c_1 \cup c_2) \cup x$. Hence $((c_1 \cup c_2) \cup x)/(c_1 \cup c_2)$ is projective with $x/(x \cap (c_1 \cup c_2))$. Hence $c_1 \cup c_2 \,\mathrm{K}\, L/c_1$. From IV it follows that $c_1 \cup c_2 \,\mathrm{K}\, L$.

Finally we observe that Kurosh-invariance is not a self-dual concept, which may be seen from the pentagon (p. 72). We have $a \mathsf{K} b$, $y \not\mathsf{K} b$, but there is an anti-automorphism interchanging a and y.

6. *The Jordan-Hölder-Schreier Theorem for lattices.*

As minimum requirements of any concept of normality in a lattice we lay down the following rules:

1) Between certain pairs of elements c, b of the lattice L there is defined a relation: c is normal with respect to (*or*: under) b. We write $c \mathsf{N} b$.

2) If $c \mathsf{N} b$, then $c \mathsf{K} b$ and hence in particular $c \subseteq b$.

3) $c \mathsf{N} c$.

4) If $c \mathsf{N} b$, $a \varepsilon L$, then $a \cap c \mathsf{N} a \cap b$.

5) If $c \mathsf{N} a \cup c$, $y \mathsf{N} a$, then $c \cup y \mathsf{N} c \cup a$.

6) If $c_1 \mathsf{N} b$, $c_2 \mathsf{N} b$, then $c_1 \cup c_2 \mathsf{N} b$.

From 4) follows:

4a) If $c \mathsf{N} b$, $c \subseteq x \subseteq b$, then $c \mathsf{N} x$.

These rules are satisfied by Kurosh-invariance

Proof of 5): From IIIa follows $a \cap c \mathsf{K} a$. From Va follows $y \cup (a \cap c) \mathsf{K} a$. From IVa it follows that $y \cup (a \cap c) \mathsf{K} a/(a \cap c)$. Applying the isomorphism induced by γ^c we find that $y \cup c \mathsf{K} (a \cup c)/c$. From IVa it follows that $y \cup c \mathsf{K} a \cup c$. They are also satisfied by the concept of 'normal subgroup' in the lattice of all subgroups of a given group. We proceed to prove the Jordan-Hölder-Schreier Theorem for lattices. First we prove the

LEMMA ON FOUR ELEMENTS: *If* $u \mathsf{N} U$, $v \mathsf{N} V$, *then we have*

$$u \cup (v \cap U) \mathsf{N} u \cup (V \cap U)$$

$$v \cup (u \cap V) \mathsf{N} v \cup (U \cap V)$$

and

(29) $$(u \cup (V \cap U))/(u \cup (V \cap U)) \simeq (v \cup (U \cap V))/(v \cup (u \cap V))$$

In fact, we have:

$$u \cap V \; \mathsf{N} \; U \cap V, \; v \cap U \; \mathsf{N} \; U \cap V \qquad \text{from rule 4)},$$

$$(u \cap V) \cup (v \cap U) \; \mathsf{N} \; U \cap V \qquad \text{from rule 6)},$$

$$u \; \mathsf{N} \; u \cup (U \cap V) \qquad \text{from rule 4a)},$$

$$u \cup (v \cap U) \; \mathsf{N} \; u \cup (U \cap V) \qquad \text{from rule 5)},$$

$(u \cup (U \cap V)/u$ is projective with $(U \cap V)/(u \cap (U \cap V))$ from rule 2).

Observing that

$$u \cup ((u \cap V) \cup (v \cap U)) = u \cup (u \cap V) \cup (v \cap U) = u \cup (v \cap U)$$

and

$$u \cap (U \cap V) = (u \cap U) \cap V = u \cap V \leq (u \cap V) \cup (v \cap U)$$

we conclude that also

$$(u \cup (U \cap V))/(u \cup (U \cap v)) \text{ is projective with}$$
$$(U \cap V)/((u \cap V) \cup (v \cap U))$$

Hence it follows that

$$(u \cup (U \cap V))/(u \cup (U \cap v)) \simeq (U \cap V)/((u \cap V) \cup (v \cap U))$$

and from arguments of symmetry

$$(v \cup (V \cap U))/(v \cup (u \cap V)) \simeq (U \cap V)/((u \cap V) \cup (v \cap U)).$$

Since the right-hand sides coincide, comparison of the left-hand sides gives us (29).

In a lattice L with all element M and null element m and with a normality relation a normal chain of length r is a chain of $r+1$ elements

$$M = a_0 \; \mathsf{N} \; a_1 \; \mathsf{N} \; a_2 \ldots \mathsf{N} \; a_r = m.$$

The factor lattices a_i/a_{i+1} are called the *factors* of the chain. If successsive numbers are different, then the normal chain is said to be a normal chain *without repetitions*. A normal chain is said to be a *refinement* of a given normal chain if the members of the given chain are among the members of the new chain. Now Theorem 8 can be reformulated as follows: Two given normal chains of a lattice with normality relations and extreme elements can be refined so that the series of factors of the two new chains are identical up to order and isomorphisms.

The proof is carried out as before by basing it on the lemma. The applications to the situations in groups gives us back Theorem 8.

§ 6. Commutator Groups and Commutator Forms

We saw on page 9 that there are groups in which the commutative law $ab=ba$ does not hold. If we wish nevertheless to calculate in an arbitrary group $\mathfrak{G}$ in commutative fashion we must create a multiplicative normal congruence relation between its elements for which the condition

$$ab \equiv ba \tag{1}$$

holds. By page 22 the congruence relation sought is the congruence in $\mathfrak{G}$ with respect to a normal subgroup $\mathfrak{G}'$, consisting of all elements which are to be congruent to e. From the proposed congruence, we conclude, upon multiplication by the congruence $(ba)^{-1} \equiv (ba)^{-1}$, that all elements $ab(ba)^{-1}$ must be in $\mathfrak{G}'$.

Definition: The element $ab\,a^{-1}b^{-1}$ is called the *commutator* of the elements a, b and is denoted by (a, b).

According to the defining equation,

$$ab = (a, b)\,ba \tag{2}$$

the commutator indicates the deviation from the commutative law.

The subgroup generated by all the commutators is called the *commutator subgroup* of $\mathfrak{G}$, and is denoted by $D\,\mathfrak{G}$ or by $\mathfrak{G}'$.

If we actually wish to calculate commutatively in $\mathfrak{G}$, then we must look upon two elements as congruent if their quotient lies in the commutator group. However if we do this, we calculate in an abelian manner, since from (2) it follows that

$$ab \equiv ba\ (D\mathfrak{G}l)$$

and from $a \equiv b,\ c \equiv d$ it follows that

$$ac \equiv bc \equiv cb \equiv db \equiv bd\ (D\mathfrak{G}l).$$

The commutator group is the smallest normal subgroup with an abelian factor group.

The commutator group is invariant under every operator of the given group, since

(3) $$(a, b)^{\Theta} = (a b a^{-1} b^{-1})^{\Theta} = a^{\Theta} b^{\Theta} (a^{\Theta})^{-1} (b^{\Theta})^{-1} = (a^{\Theta}, b^{\Theta}),$$

and therefore $\mathfrak{G}'^{\Theta}$ lies in $\mathfrak{G}'$.

We now define higher commutator groups ("higher derivatives") recursively, setting

$$D^0 \mathfrak{G} = \mathfrak{G}$$

$$D^1 \mathfrak{G} = D \mathfrak{G} = \mathfrak{G}',$$

$$D^2 \mathfrak{G} = D \mathfrak{G}' = \mathfrak{G}'',$$

$$\cdots\cdots\cdots\cdots$$

$$D^r \mathfrak{G} = D(D^{r-1} \mathfrak{G}).$$

It is clear that the r-th commutator group $D^r \mathfrak{G}$ is a fully invariant subgroup of $\mathfrak{G}$ and that the successive factor groups of the normal subgroup chain

$$\mathfrak{G} = D^0 \mathfrak{G} \supseteq \mathfrak{G}' = D^1 \mathfrak{G} \supseteq D^2 \mathfrak{G} \ldots \supseteq D^r \mathfrak{G}$$

are abelian.

For a subgroup $\mathfrak{U}$ of $\mathfrak{G}$ it follows from the definition of the commutator group that

$$D\mathfrak{U} \subseteq D\mathfrak{G}$$

and by induction

$$D^r \mathfrak{U} \subseteq D^r \mathfrak{G}.$$

For the factor group over a normal subgroup $\mathfrak{N}$ we have

$$D^r (\mathfrak{G}/\mathfrak{N}) = (D^r \mathfrak{G})\, \mathfrak{N}/\mathfrak{N}.$$

The usefulness of these concepts is obvious from the following

DEFINITION: A group $\mathfrak{G}$ is said to be *solvable*, if the series of higher commutator groups terminates with e.

To a solvable group $\mathfrak{G} \neq e$ there corresponds a uniquely defined number k such that $D^k \mathfrak{G} = e$, $D^{k-1} \mathfrak{G} \neq e$. Since in the normal chain, $\mathfrak{G} \supset D^1 \mathfrak{G} \supset D^2 \mathfrak{G} \supset \cdots \supset D^k \mathfrak{G} = e$ just k abelian factor groups different from e appear, we say that the group $\mathfrak{G}$ is ***k*-step metabelian**.

The group consisting of only the unity element is said to be 0-step metabelian. The 1-step metabelian groups are exactly the abelian groups $\neq e$.

It follows immediately from our remarks above that *every subgroup and every factor group of a k-step metabelian group is itself at most k-step metabelian.*

If the group $\mathfrak{G}$ has a normal chain

$$\mathfrak{G} = \mathfrak{G}_0 \supseteq \mathfrak{G}_1 \supseteq \mathfrak{G}_2 \supseteq \cdots \supseteq \mathfrak{G}_k = e$$

which has only abelian factor groups $\mathfrak{G}_i/\mathfrak{G}_{i+1}$, then $\mathfrak{G}$ is at most k-step metabelian since $\mathfrak{G}' \subseteq \mathfrak{G}_1$, because $\mathfrak{G}/\mathfrak{G}_1$ is abelian, and it follows by induction that $D^r\mathfrak{G} \subseteq \mathfrak{G}_r$, $D^k\mathfrak{G} = e$.

Since the higher derivatives are fully invariant in $\mathfrak{G}$, and the subgroups and factor groups of an abelian group are themselves abelian, it follows from the Jordan - Hölder - Schreier theorem that the factor groups of a normal series of a solvable group are abelian.

Since an abelian group is simple if and only if it is of prime order, it follows that:

A solvable group has a composition series if and only if it is finite. A finite group is solvable if and only if its composition factors are cyclic of prime order.

The following rules hold for calculation with commutators:

(4) $$(a, b) = e$$

is equivalent to $$ab = ba,$$

and in particular, we have

(4a) $$(e, a) = (a, e) = e$$

(4b) $$(a, a) = e.$$

(5) $$(a, b)\,(b, a) = e,$$

(6) $$aba^{-1} = b^a = (a, b)\,b,$$

$$(a, b) = a^{1-b} = b^{a-1},$$

(7) $$(ab, c) = (b, c)^a\,(a, c)$$

(8) $$(a, bc) = (a, b)\,(a, c)^b.$$

If the commutator group lies in the center, then rules (7) and (8) can be simplified to

(7a) $$(ab, c) = (a, c)\,(b, c)$$

(8a) $$(a, bc) = (a, b)\,(a, c),$$

in particular

(9) $$(a^n, c) = (a, c^n) = (a, c)^n.$$

From this we can derive the useful power rule:

For (a, b) in the center of the group,

$$(ab)^n = (b, a)^{\frac{1}{2}n(n-1)} a^n b^n \tag{10}$$

Proof: For $n = 0$ the rule is trivially true. Now let $n > 0$ and assume we have already proven

$$(ab)^{n-1} = (b, a)^{\frac{1}{2}(n-1)(n-2)} a^{n-1} b^{n-1}.$$

Now

$$(ab)^n = (ab)^{n-1} \cdot ab$$

$$a^{n-1} b^{n-1} ab = a^{n-1} (b^{n-1}, a) a b^n$$

$$= a^{n-1} (b, a)^{n-1} a b^n \quad \text{by (9)}$$

$$= (b, a)^{n-1} a^n b^n, \quad \text{since } (b, a) \text{ is in the center}$$

and therefore

$$(ab)^n = (b, a)^{\frac{1}{2}(n-1)(n-2)} (b, a)^{n-1} a^n b^n,$$

from which the rule follows for positive n. For negative exponents the rule follows from the equations

$$(ab)^{-n} = (b^{-1} a^{-1})^n, \quad a^n b^n = (a^n, b^n) b^n a^n$$
$$= (a, b)^{n^2} b^n a^n = (b, a)^{-n^2} b^n a^n.$$

The *mutual commutator group* $(\mathfrak{U}, \mathfrak{V})$ of two complexes $\mathfrak{U}$ and $\mathfrak{V}$ of given group $\mathfrak{G}$ is the subgroup generated by all the commutators (U, V) where $U \in \mathfrak{U}$, $V \in \mathfrak{V}$.[1] $\mathfrak{U}$ commutes with $\mathfrak{V}$ elementwise if and only if $(\mathfrak{U}, \mathfrak{V}) = e$..

From (5) we have

$$(\mathfrak{U}, \mathfrak{V}) = (\mathfrak{V}, \mathfrak{U}). \tag{11}$$

If $\mathfrak{U}$ and $\mathfrak{V}$ are normal subgroups of $\mathfrak{G}$ then it follows from (3) and (6) that $(\mathfrak{U}, \mathfrak{V})$ is a normal subgroup of $\mathfrak{G}$ and is contained in $\mathfrak{U} \cap \mathfrak{V}$. Then by (7) and (8), for an arbitrary complex $\mathfrak{K}$:

$$(\mathfrak{K}\mathfrak{U}, \mathfrak{V}) = (\mathfrak{U}, \mathfrak{V})(\mathfrak{K}, \mathfrak{V}), \tag{12}$$

$$(\mathfrak{U}, \mathfrak{K}\mathfrak{V}) = (\mathfrak{U}, \mathfrak{K})(\mathfrak{U}, \mathfrak{V}). \tag{13}$$

Let $\mathfrak{K}_1$ and $\mathfrak{K}_2$ be two complexes, $\mathfrak{U}_1$, $\mathfrak{U}_2$ the subgroups generated by them.

THEOREM 12: *The normal subgroup $\mathfrak{N}$ of $\mathfrak{G}$ generated by $(\mathfrak{K}_1, \mathfrak{K}_2)$ is equal to the normal subgroup $\mathfrak{N}_1$ of $\mathfrak{G}$ generated by $(\mathfrak{U}_1, \mathfrak{U}_2)$.*

Proof: In any case $\mathfrak{N}$ is contained in $\mathfrak{N}_1$. We must show that

[1] If confusion with the commutator is to be feared then we write $\{(\mathfrak{U}, \mathfrak{V})\}$.

$\mathfrak{N}=e$ implies $\mathfrak{N}_1=e$[1]. In fact, then $\mathfrak{K}_1$, $\mathfrak{K}_2$ commute elementwise and from (4), (7), (8) it follows that $\mathfrak{U}_1$, $\mathfrak{U}_2$ commute elementwise, as was to be shown.

It is useful to introduce higher commutators,e.g.,

$$(a, b, c) = (a, (b, c)) \tag{14}$$

$$(a_1, a_2, \ldots, a_n) = (a_1, (a_2, \ldots, a_n)) \tag{15}$$

$$(a, b; c, d) = ((a, b), (c, d)). \tag{16}$$

Rules (7) and (8) can now be written

$$(ab, c) = (a, b, c)\,(b, c)\,(a, c) \tag{7b}$$

$$(a, bc) = (a, b)\,(b, a, c)\,(a, c). \tag{8b}$$

In order to understand these multiple commutators completely we define recursively a "linear expression of weight w and type s", in symbols $x_1, x_2, \ldots, x_w$. The linear expression of weight 1 in x is the symbol x itself. Let this correspond to type 0. As a separating symbol of the first type we use a comma; for the second type, a semi-colon; for the third type, a triple point $\vdots$; and in general for the sth type the symbol Ⓢ is used. Now let $w > 1$ and assume that all the linear expressions of weight $< w$ are defined, and let a type correspond to each of them. Then we define expressions $(f_1 \text{ Ⓢ } f_2)$ as linear expressions of weight w and of type $s > 0$ where f_1 is of weight w_1 in $x_1, x_2, \ldots, x_{w_1}$ and of type s_1, and f_2 is of weight w_2 in $x_{w_1+1}, x_{w_1+2}, \ldots, x_{w_1+w_2}$ and of type s_2, such that $w = w_1 + w_2$, $s = \text{Max}\,(s_1 + 1, s_2)$. The weight is therefore simply the number of symbols in the "linear expression" and the type is equal to the highest type of separation symbol.

If $f\,(x_1, \ldots, x_w)$ is a linear expression of weight w and type s, then for arbitrary elements $G_1, \ldots, G_w$ in the group $\mathfrak{G}$ we define:

$f\,(G_1, \ldots, G_w)$ is a *commutator of weight w and type s in the G_i*. Moreover for arbitrary subgroups $\mathfrak{U}_1, \mathfrak{U}_2, \ldots, \mathfrak{U}_w$ we define: $f(\mathfrak{U}_1, \mathfrak{U}_2, \ldots, \mathfrak{U}_w)$ is a *commutator form of weight w and of type s in the $\mathfrak{U}_i$*.

In the successive construction of corresponding linear expressions the separation symbols are to indicate commutator formation. E.g.,

$D\mathfrak{G} = (\mathfrak{G}, \mathfrak{G})$ is of weight 2 and type 1,

$D^2\mathfrak{G} = (\mathfrak{G}, \mathfrak{G}; \mathfrak{G}, \mathfrak{G})$ is of weight 4 and type 2,

$D^r\mathfrak{G} = (D^{r-1}\mathfrak{G} \text{ ⓡ } D^{r-1}\mathfrak{G})$ is of weight 2^r and type r

in the components $\mathfrak{G}, \mathfrak{G}, \ldots, \mathfrak{G}$.

[1] We calculate in the factor group $\mathfrak{G}/\mathfrak{N}$!

If the complexes $\mathfrak{K}_1, \ldots, \mathfrak{K}_w$ are transformed into themselves by every element in $\mathfrak{G}$, then every commutator form formed from them is a normal subgroup of $\mathfrak{G}$, since from (3) it follows, by induction on the weight w, that for each operator Θ of $\mathfrak{G}$:

(3a) $$(f(G_1, G_2, \ldots, G_w))^\Theta = f(G_1^\Theta, G_2^\Theta, \ldots, G_w^\Theta).$$

If the subgroup generated by the complex $\mathfrak{K}_i$ is equal to the normal subgroup $\mathfrak{N}_i$ of $\mathfrak{G}$ then we actually have:

THEOREM 13: *The commutator form $f(\mathfrak{N}_1, \mathfrak{N}_2, \ldots, \mathfrak{N}_w)$ is equal to the subgroup $\mathfrak{N}$ of $\mathfrak{G}$, $\mathfrak{N}$ being generated by all the elements $f(N_1, \ldots, N_w)$ with $\mathfrak{N}_i$ in $\mathfrak{K}_i$.*

Proof: For $w = 1$ the theorem is clear. Let $w > 1$ and assume that the theorem is proven for commutator forms of lower weight.

$f = (f_1 \circledS f_2)$, where the weights w_1, w_2 of f_1, f_2 are lower than w. By the induction hypothesis,

$$f_1(\mathfrak{N}_1, \mathfrak{N}_2, \ldots, \mathfrak{N}_{w_1}) \text{ is generated by all } f_1(N_1, \ldots, N_{w_1}),$$
$$f_2(\mathfrak{N}_{w_1+1}, \mathfrak{N}_{w_1+2}, \ldots, \mathfrak{N}_w) \text{ is generated by all } f_2(N_{w_1+1}, \ldots, N_w).$$

Now the statement of the theorem follows from Theorem 12. From the previous definitions and the last theorem the following "substitution principle" follows immediately:

If $f(y_1, y_2, \ldots, y_\omega)$ is a linear expression of weight ω and if $\varphi_i(x_1^{(i)}, x_2^{(i)}, \ldots, x_{w_i}^{(i)})$ $(i=1, 2, \ldots, \omega)$ are linear expressions of weight w_i, then $f(\varphi_1, \varphi_2, \ldots, \varphi_\omega)$ is a linear expression g of weight $w = w_1 + w_2 + \ldots + w_\omega$ in $x_1^{(1)}, \ldots, x_{w_1}^{(1)}, \ldots, x_{w_\omega}^{(\omega)}$[1]*. For normal subgroups*

$$\mathfrak{N}_\nu^{(i)} \; (\nu = 1, 2 \ldots w_i;\; i = 1, 2 \ldots \omega)$$

of a group $\mathfrak{G}$, we have

$$g(\mathfrak{N}_\nu^{(i)}) = f(\mathfrak{M}_1, \mathfrak{M}_2 \ldots \mathfrak{M}_\omega),$$

where $\mathfrak{M}_i = \varphi_i(\mathfrak{N}_1^{(i)}, \ldots, \mathfrak{N}_{w_i}^{(i)})$.

In a group $\mathfrak{G}$ with abelian commutator group $\mathfrak{G}'$ we have the following important rule:

(17) $$(a, b, c)\,(b, c, a)\,(c, a, b) = e,$$

which we derive in the following way:

---

[1] The type of the separating symbols in f may have to be raised by the substitution.

$$c(a,b)c^{-1} = (a,b)^c = (c,a,b)\,(a,b)$$
$$= cac^{-1}\cdot cbc^{-1}\cdot (cbc^{-1}\cdot cac^{-1})^{-1}$$
$$= (c,a)a\cdot (c,b)b\cdot ((c,b)b\,(c,a)a)^{-1}$$
$$= (c,a)\,(a,c,b)\,(c,b)ab\,((c,b)\,(b,c,a)\,(c,a)ba)^{-1}.$$

$$(17\text{a})\quad (c,a,b)\,(a,b) = (c,a)\,(a,c,b)\,(c,b)\,(a,b)\,(c,a)^{-1}(b,c,a)^{-1}(c,b)^{-1}.$$

Since $\mathfrak{G}'$ is abelian we have

$$(c,a,b) = (a,c,b)\,(b,c,a)^{-1},$$

and moreover by (8) $\quad (a,e) = e = (a,b,c)\,(a,c,b),$

and therefore finally we have (17). Now we can prove the following important theorem.

THEOREM 14: *In a group $\mathfrak{G}$ with the three normal subgroups $\mathfrak{A}, \mathfrak{B}, \mathfrak{C}$, each of the three normal subgroups $(\mathfrak{A}, \mathfrak{B}, \mathfrak{C})$, $(\mathfrak{B}, \mathfrak{C}, \mathfrak{A})$, $(\mathfrak{C}, \mathfrak{A}, \mathfrak{B})$ is contained in the product of the two others.*

Proof: We may assume that $(\mathfrak{A}, \mathfrak{B}, \mathfrak{C})\cdot(\mathfrak{B}, \mathfrak{C}, \mathfrak{A}) = e$, and must then prove that $(\mathfrak{C}, \mathfrak{A}, \mathfrak{B}) = e$. By Theorem 13, $(\mathfrak{C}, \mathfrak{A}, \mathfrak{B})$ is generated by all (c,a,b) where $a \in \mathfrak{A}$, $b \in \mathfrak{B}$, $c \in \mathfrak{C}$, so that we must prove $(c,a,b) = e$. In any case $(\mathfrak{A}, \mathfrak{B}, \mathfrak{C}) = (\mathfrak{B}, \mathfrak{C}, \mathfrak{A}) = e$, and therefore also $(\mathfrak{A}, \mathfrak{C}, \mathfrak{B}) = e$. In formula (17a) we may insert $(a,c,b) = (b,c,a) = e$, so that

$$(18)\qquad (c,a,b)\,(a,b) = (c,a)\,(c,b)\,(a,b)\,(c,a)^{-1}\,(c,b)^{-1}.$$

Since $\mathfrak{A}$ is a normal subgroup, $(\mathfrak{A}, \mathfrak{B}) \leq \mathfrak{A}$, and therefore, $(\mathfrak{A}, \mathfrak{B}; \mathfrak{B}, \mathfrak{C}) = e$ and $(\mathfrak{A}, \mathfrak{C}; \mathfrak{B}, \mathfrak{C}) = e$. Since $\mathfrak{B}$ is a normal subgroup, $(\mathfrak{A}, \mathfrak{B}) \leq \mathfrak{B}$, $(\mathfrak{A}, \mathfrak{B}; \mathfrak{C}, \mathfrak{A}) = e$. The factors on the right of (18) may be permuted so that we finally obtain $(c,a,b) = e$.

§ 7. On the Groups of an Algebra

In this paragraph we give a short survey of the groups occurring in an algebra and of groups with operators.

1. *Modules.*

A commutative group in which the symbol of combination is written as the $(+)$ symbol, is called a **module**.

Consequently, the *sum* of two *summands* a and b is denoted by $a+b$. The following laws must then be valid for this *addition*:

I. $a+(b+c) = (a+b)+c$.

II. There is a null element 0 with the property $0+a=a$ for all a.

III. The equation $x+a=b$ is solvable for all pairs a, b.

IV. $a+b=b+a$.

As we saw earlier, it follows from this that in a sum of a finite number of summands the order and parenthesizing can be changed arbitrarily, without altering the value of the sum.

A sum consisting of the n summands $a_1, \ldots, a_n$ is written

$$a_1+a_2+\ldots+a_n \text{ or } \sum_1^n a_i .$$

Addition has a unique inverse, i.e., the equation $x+a=b$ has exactly one solution for each pair a, b. By the commutative law, the equation $a+x=b$ is equivalent to $x+a=b$. Zero has the property: $0+a=a$, $a+0=a$ and it is uniquely defined by any of these equations. The solution of $x+a=0$ is denoted by $-a$ and is uniquely determined. We have

$$a+-a=-a+a=0$$

and therefore $$-(-a)=a$$
The *difference* $a+(-b)$ of a and b will be denoted by $a-b$.

The sum of n equal summands a is denoted by $n\,a$. $0a$ is defined as 0 and $(-n)\,a$ is defined as $-(n\,a)$.

Then we have rules analogous to the power rules:

(1) $$n\,(a+b)=n\,a+n\,b\ ,$$

(2) $$(n+m)\,a=n\,a+m\,a\ ,$$

(3) $$(n\,m)\,a=n\,(m\,a),$$

for all rational integers n, m. Consequently the mapping $a \to n\,a$ is an operator $\underline{n}$ of the module such that the rules for calculation

(4) $$\underline{n+m}=\underline{n}+\underline{m}$$

(5) $$\underline{n\,m}=\underline{n}\,\underline{m}$$

are valid. $\underline{1}$ leaves each element fixed; $\underline{0}$ maps each element onto 0.

For example, the rational integers $0, \pm 1, \pm 2, \ldots$ form a module $\mathfrak{o}$. $\mathfrak{o}$ is additively generated by 1 and is therefore cyclic; moreover $\mathfrak{o}$ is infinite.

By I § 5, the *submodules* of $\mathfrak{o}$ are exactly the modules (n), consisting of all multiples $m\,n$ of the non-negative rational integer n.

Two numbers are said to be *congruent* mod (n) if their difference is in (n) and therefore divisible by n. The number of residue classes of $\mathfrak{o}$ with respect to (n) is n if $n>0$. They form a module $\mathfrak{o}(n)$, the *factor-module* of $\mathfrak{o}$ with respect to (n).

For an arbitrary module $\mathfrak{M}$, all the rational integers m for which $\underline{m}=\underline{0}$ form a submodule of $\mathfrak{o}$, the so-called *exponential module* of $\mathfrak{M}$. The non-negative rational integer generating the exponential module is called the *characteristic* of the module. For example, the factor-module $\mathfrak{o}/(n)$ is of characteristic n.

The *sum* $\mathfrak{m}_1 + \mathfrak{m}_2$ of two submodules $\mathfrak{m}_1, \mathfrak{m}_2$ of $\mathfrak{M}$ consists of all sums $a_1 + a_2$ with $a_i \in \mathfrak{m}_i$. It is a submodule.

The sum of the two submodules (n) and (m) of $\mathfrak{o}$ is generated by the *greatest common divisor* (g. c. d.) (n, m) of n and m. The intersection of (n) and (m) is generated by the *least common multiple* of n and m (l. c. m.) From the second isomorphy theorem it follows that

$$n. m = (n, m). (\text{l. c. m. } (n, m))$$

The *maximal submodules* of $\mathfrak{o}$ are exactly the submodules generated by the prime natural numbers.

2. *Rings.*

DEFINITION: A *ring* is a module in which besides addition, a multiplication of elements is defined such that

1. $a\ (b\ c) = (a\ b)\ c$ (associative law)
2. $a\ (b + c) = a\ b + a\ c$ (left distributive law)
   $(b + c)\ a = b\ a + c\ a$ (right distributive law).

Thus a ring is an abelian group in which a right and a left operator is associated with each element.

In particular,

$$a. 0 = 0. a = 0,\ a. -b = -a. b = -(a. b).$$

DEFINITION: The admissible subgroups of a ring are said to be *ideals.*

A *right ideal* is a submodule in which ϱa is contained in the submodule if ϱ is in the submodule; similarly a left ideal is a submodule which contains $a\lambda$ if it contains λ, where a in each case runs through all the elements of the ring $\mathfrak{S}$.

A submodule which is at the same time a right and left ideal is said to be a *two-sided ideal.*

As the *product* $\mathfrak{m}_1\mathfrak{m}_2$ of two submodules $\mathfrak{m}_1, \mathfrak{m}_2$ of a ring $\mathfrak{S}$ we define the set of all finite sums

$$a_1b_1 + a_2b_2 + \cdots + a_rb_r\ ,$$

where $a_i \in \mathfrak{m}_1$, $b_i \in \mathfrak{m}_2$, r arbitrary.

With this definition, $\mathfrak{m}_1\mathfrak{m}_2$ first becomes a submodule of $\mathfrak{S}$.

The sum, intersection and product of two ideals of the same sort are also ideals of this same sort.

The residue classes (cosets) over a right– (left–) ideal have $\mathfrak{S}$ as a right– (left–) domain of operators. The residue classes with respect to a two-sided ideal form a ring, the factor-ring, where the residue class R_{ab} is defined as the product of the residue classes R_a and R_b.

The ring $\mathfrak{f}$ is said to be *homomorphic* to the ring $\mathfrak{S}$ if there is a single-valued mapping σ of $\mathfrak{S}$ on $\mathfrak{f}$ such that $\sigma(a+b) = \sigma a + \sigma b$, $\sigma(ab) = \sigma a \cdot \sigma b$. If the mapping is one-one, then $\mathfrak{f}$ is said to be isomorphic to $\mathfrak{S}$.

Here the first isomorphy theorem reads:

A ring $\mathfrak{f}$ homomorphic to the ring $\mathfrak{S}$ is isomorphic to the residue class ring of $\mathfrak{S}$ with respect to the two-sided ideal consisting of all the elements of $\mathfrak{S}$ which are mapped onto 0 *by the homomorphic mapping of $\mathfrak{S}$ onto $\mathfrak{f}$.*

An example of a ring is the operator domain of a module $\mathfrak{M}$. An operator Θ of $\mathfrak{M}$ is a single-valued mapping of $\mathfrak{M}$ into itself such that $\Theta(a+b) = \Theta a + \Theta b$. The product of two operators is defined by $(\Theta_1\Theta_2)a = \Theta_1(\Theta_2 a)$ which we encountered previously; on the other hand the sum is defined by $(\Theta_1 + \Theta_2)a = \Theta_1 a + \Theta_2 a$. One can easily show that the operators of $\mathfrak{M}$ form a ring with the unit element $\underline{1}$.

3. *Division Rings,*[1] *Commutative Rings, and Fields.*

DEFINITION: A ring in which the elements different from zero form a multiplicative group is said to be a *division ring*.

This would follow from the additional conditions:

3. There are at least two different elements;

4. The equations $a.x=b$ and $y.a=b$ are solvable if $a \neq 0$. (If $a \neq 0$, $b \neq 0$, then the equations $ae=a$ and $bx=e$ are solvable and give $abx = ae = a \neq 0$, therefore $ab \neq 0$. Thus the non-zero elements form a semi-group which is actually a group because of 4.)

A ring is said to be a *commutative ring* if the commutative law for multiplication holds in it.

A commutative ring which is at the same time a division ring is called a *field*.

For example all the rational numbers, as well as the domain of all real numbers, form a field.

DEFINITION: The *center* of a ring is the (commutative) ring of all elements which commute with every element of the ring.

The center of a division ring is a field.

[1] Also called skew fields, s-fields, non-commutative fields, etc.

In a commutative ring with unit element, the residue class ring with respect to an ideal is a field if and only if the ideal is maximal.

For example, the set of all the rational integers form a commutative ring $\mathfrak{o}$. Every submodule of $\mathfrak{o}$ is also an ideal of $\mathfrak{o}$. The residue class ring of $\mathfrak{o}$ with respect to the ideal (n) is a field if and only if n is a prime. Therefore for every prime p we obtain a field k_p of p elements.

In an arbitrary division ring K, all the elements obtainable from 1 by combinations of the four operations form a sub-field k, the *prime field* of K. Either none of the sums $1, 1+1, 1+1+1, \ldots$ is equal to zero in K, in which case k is isomorphic to the field of rational numbers, or a sum $1+1+\ldots+1$ is equal to zero, in which case k is isomorphic to a field k_p of p elements.

The characteristic of a division ring is equal to the characteristic of its prime field and is therefore zero or a natural prime, since from $n\,a=0$, $a\neq 0$ *it follows that* $n=0$.

4. $\mathfrak{S}$-*Modules.*

DEFINITION: A module with a ring $\mathfrak{S}$, as operator domain is said to be an $\mathfrak{S}$-*module.*

We define in greater detail:

The module $\mathfrak{M}$, given a ring $\mathfrak{S}$, is said to be a *left* $\mathfrak{S}$-*module* if a multiplication of elements α in $\mathfrak{S}$ with elements u in $\mathfrak{M}$ is uniquely defined so that

1. $$\alpha u \in \mathfrak{M},$$
2. $$\alpha(u+v) = \alpha u + \alpha v,$$
3. $$(\alpha+\beta)u = \alpha u + \beta u$$
4. $$(\alpha\beta)u = \alpha(\beta u).$$

We also speak of an $\mathfrak{S}$-module in the case where $\mathfrak{S}$ is only a semi-group in which case requirement 3. becomes meaningless.

$\mathfrak{M}$ is said to be a *proper* $\mathfrak{S}$-module if 5. $\mathfrak{S}\mathfrak{M}=\mathfrak{M}$; 6. $\alpha\mathfrak{M}=0$, $\mathfrak{M}\neq 0$, imply $\alpha=0$.

If $\mathfrak{S}$ contains a unit element 1, then condition 5. is equivalent to condition 5a): $1.\mathfrak{M}=\mathfrak{M}$. 5a) is equivalent to 5b) : $1.u=u$ for all u.

If $\mathfrak{S}$ is a division ring, then condition 5. suffices to make $\mathfrak{M}$ a proper $\mathfrak{S}$-module: for from $\alpha\mathfrak{M}=0$ and $\alpha\neq 0$ we would have $\mathfrak{S}\alpha\mathfrak{M}=0$, and since $\mathfrak{S}\alpha=\mathfrak{S}$, then $\mathfrak{S}\mathfrak{M}=0$.

The concept of a right $\mathfrak{S}$-module is defined similarly, the module being multiplied on the right.

An $\mathfrak{S}$-module $\mathfrak{M}$ is said to be a *finite $\mathfrak{S}$-module* if $\mathfrak{M}$ can be generated over $\mathfrak{S}$ by a finite number of its elements, and therefore if there are a finite number of elements $u_1, \ldots, u_n$ in $\mathfrak{M}$ such that every element in $\mathfrak{M}$ is of the form

$$u = \alpha_1 u_1 + \alpha_2 u_2 + \cdots + \alpha_n u_n$$

with $\alpha_i \in \mathfrak{S}$.

Examples of $\mathfrak{S}$-modules are the *n-dimensional $\mathfrak{S}$-vector module* consisting of all ordered n-tuples $(\alpha_1, \alpha_2, \ldots, \alpha_n)$ (*vectors*) with *components* α in $\mathfrak{S}$, among which only a finite number are different from zero[1], and with the calculation rules

$$(\alpha_1, \alpha_2, \ldots, \alpha_n) + (\beta_1, \beta_2, \ldots, \beta_n) = (\alpha_1 + \beta_1, \alpha_2 + \beta_2, \ldots, \alpha_n + \beta_n)$$

$$\alpha(\alpha_1, \alpha_2, \ldots, \alpha_n) = (\alpha\alpha_1, \alpha\alpha_2, \ldots, \alpha\alpha_n).$$

If $\mathfrak{S}$ contains a unit element 1, then the n-dimensional $\mathfrak{S}$-vector module over $\mathfrak{S}$ is generated by the *n unit vectors*

$$u_i = (0, \ldots, 0, 1, 0, \ldots, 0) \qquad (i = 1, 2, \ldots, n).$$

The u_i are then called a *basis* of the $\mathfrak{S}$-module.

In an arbitrary $\mathfrak{S}$-module $\mathfrak{M}$, the expression $\alpha_1 u_1 + \alpha_2 u_2 + \cdots + \alpha_r u_r$ is called a *linear combination* of the u_i. The elements $u_1, u_2, \ldots, u_r$ are said to be *linearly independent* if

1. $u_i \neq 0$ $(i = 1, \ldots, r)$
2. $\alpha_1 u_1 + \alpha_2 u_2 + \cdots + \alpha_r u_r = 0$ implies $\alpha_i u_i = 0, i = 1, 2, \ldots, r.$

Now we generalize the definition of basis. A system $\mathfrak{B}$ of elements $u_1, u_2, \ldots, u_\omega$ is said to be an *$\mathfrak{S}$-basis system* if each element of $\mathfrak{M}$ is of the form $u = \alpha_{\nu_1} u_{\nu_1} + \cdots + \alpha_{\nu_r} u_{\nu_r}$ and every finite set of elements $u_{\nu_1}, u_{\nu_2}, \ldots, u_{\nu_r}$ is linearly independent $(\nu_1 < \nu_2 < \cdots < \nu_r)$.

If a module $\mathfrak{M}$ over a division ring K has a finite basis, then it is a vector module.[2]

Since $K\mathfrak{M} = \mathfrak{M}$, $\mathfrak{M}$ is a proper K-module. If, moreover, $u_1, u_2, \ldots, u_n$ is the basis, then it follows from $\alpha_1 u_1 + \alpha_2 u_2 + \cdots + \alpha_n u_n = 0$ that $\alpha_i u_i = 0$. If we were to have $\alpha_i \neq 0$ then $\alpha_i^{-1}(\alpha_i u_i) = 1 u_i = u_i = 0$. Therefore we must have $\alpha_i = 0$ $(i = 1, 2, \ldots, n)$. Every element of $\mathfrak{M}$ can be represented in only one way as $\alpha_1 u_1 + \alpha_2 u_2 + \cdots + \alpha_n u_n$.

---

[1] The number n can be any ordinal number whatsoever.

[2] More precisely: is operator-isomorphic to an $\mathfrak{S}$-vector module.

THEOREM 15: *For any division ring K, every proper K-module $\mathfrak{M} \neq 0$ has a basis.*

Proof: With the help of transfinite induction we shall give a method of construction only. The reader can carry out the proof himself without difficulty.

We first pick a system of generators $v_1, v_2, \ldots, v_\omega$ of $\mathfrak{M}$ over K, for example, $\mathfrak{M}$ itself, for which we can assume that $v_1 \neq 0$ and that the indices $1, 2, \ldots, \omega$ are well ordered. Let $\mathfrak{M}_\mu$ be the K-module of all linear combinations of the elements $v_1, v_2, \ldots, v_\mu$. We wish to define a basis system $\mathfrak{B}_\mu$ of $\mathfrak{M}_\mu$. In order to do this, let $\mathfrak{B}_1$ be the set consisting of v_1 alone. Moreover let $\nu > 1$ and let $\mathfrak{B}_\mu$ be defined for all $\mu < \nu$. Let Σ_ν be the union of all $\mathfrak{B}_\mu$ with $\mu < \nu$, and let $\mathfrak{m}_\nu$ be the K-module union of all $\mathfrak{M}_\mu$ with $\mu < \nu$. One can show easily that Σ_ν is a K-basis for $\mathfrak{m}_\nu$. Now we define: $\mathfrak{B}_\nu = \Sigma_\nu$ if $v_\nu \in \mathfrak{m}_\nu$ but $\mathfrak{B}_\nu$ = the union of Σ_ν and v_ν if $v_\nu \notin \mathfrak{m}_\nu$.

Then $\mathfrak{B}_\omega$ is the desired basis system.

It is shown in the theory of linear algebra that for a division ring K the dimension of a finite K-vector module $\mathfrak{M}$ is uniquely determined (also see Chap. III. § 2).

The dimension of $\mathfrak{M}$ over K is denoted by $[\mathfrak{M}/K]$ or simply by $[\mathfrak{M}]$.

The dimension of a K-module in $\mathfrak{M}$ is at most equal to the dimension of $\mathfrak{M}$. If the dimension of $\mathfrak{M}$ is finite then a K-module in $\mathfrak{M}$ is identical with $\mathfrak{M}$ if and only if their dimensions are equal. Consequently, then, the double chain theorem holds in $\mathfrak{M}$ over the operator domain K.

If k is a division subring in K, then K can be conceived of as a proper left k-module. The dimension of K over k is called the *degree* of K over k. If K is a finite k-module, then K is said to be a *finite extension* of k.

In this case, the elements of K can be represented uniquely in the form

$$\Theta = \lambda_1 u_1 + \lambda_2 u_2 + \cdots \lambda_n u_n$$

where $\lambda_i \in k$ and $u_1, u_2, \ldots, u_n$ is a basis of K over k. If k contains q elements, then K contains q^n elements.

5. *Semi-modules, semi-rings, quasi-rings, $\mathfrak{S}$-rings, algebras.*

Many concepts defined for modules can be extended to additive semi-groups. For example, a sum of $a_1, a_2, \ldots, a_n$, in this order, is written as $a_1 + a_2 + \ldots + a_n$ or $\sum_{i=1}^{n} a_i$. A zero element of an additive semi-group $\mathfrak{a}$ is defined to be an element 0 satisfying $0 + x = x + 0$ for all x of $\mathfrak{a}$. There is at most one zero element, since $0 + 0' = 0 = 0'$ for any two zero elements $0, 0'$. If there is an element x such that $a + x = x + a = 0$

then x is called the *negative* of a, and it is denoted by $-a$. If a has a negative, then for any element b the two equations $a+x=b$, $y+a=b$ are uniquely solvable, the solutions being $x=(-a)+b$, $y=b+(-a)$, which is abbreviated as $x=-a+b$, $y=b-a$. For any natural number n, the sum of n equal summands a is denoted by na. For natural numbers the rules (2), (3) are satisfied. If there is a zero element then we define $0a=0$. If $\mathfrak{a}$ is an additive group, then for negative integers $-n$ we define $(-n)a=-(na)$, which is abbreviated as $-na$. The rules (2), (3) remain valid for all rational integers.

A *normal divisor* of an additive semi-group $\mathfrak{a}$ is defined as an additive sub-semigroup $\mathfrak{b}$ of $\mathfrak{a}$ which has the property that for $a_1 \epsilon \mathfrak{a}$, $a_2 \epsilon \mathfrak{a}$, $x \epsilon \mathfrak{b}$,

$$a_1 + x + a_2 \epsilon \mathfrak{b} \text{ if and only if } a_1 + a_2 \epsilon \mathfrak{b},$$
$$a_1 + x \epsilon \mathfrak{b} \text{ if and only if } a_1 \epsilon \mathfrak{b},$$
$$x + a_2 \epsilon \mathfrak{b} \text{ if and only if } a_2 \epsilon \mathfrak{b}.$$

This definition is in agreement with the one given in Exercise 25 at the end of Chap. I. The congruence modulo $\mathfrak{b}$ is the normal additive congruence relation generated from

$$a+b \equiv a\,,\quad b+a \equiv a \quad \text{for } a \epsilon \mathfrak{a},\ b \epsilon \mathfrak{b}.$$

The congruence modulo a normal divisor is additive in the sense that for $a_1 \equiv a_2$ (modulo $\mathfrak{b}$), $b_1 \equiv b_2$ (modulo $\mathfrak{b}$) we have $a_1+b_1 \equiv a_2+b_2$ (modulo $\mathfrak{b}$). The residue classes modulo a normal divisor are added according to the rule: $\bar{a}+\bar{b}=\overline{a+b}$, where $\bar{x}$ denotes the residue class modulo $\mathfrak{b}$ that is represented by the element x of $\mathfrak{a}$.

The residue classes of $\mathfrak{a}$ modulo the normal divisor $\mathfrak{b}$ form an additive semi-group which is called the *factor semi-group* of $\mathfrak{a}$ over $\mathfrak{b}$ and is denoted by $\mathfrak{a}/\mathfrak{b}$. The mapping $a \to \bar{a}$ establishes a homomorphism of $\mathfrak{a}$ onto $\mathfrak{a}/\mathfrak{b}$ called the natural homomorphism of $\mathfrak{a}$ onto $\mathfrak{a}/\mathfrak{b}$. The elements of $\mathfrak{b}$ form a residue class $\bar{\mathfrak{b}}$ which is the zero element of $\mathfrak{a}/\mathfrak{b}$.

For a homomorphism Θ of the additive semi-group $\mathfrak{a}$ onto the additive semi-group $\Theta\mathfrak{a}$ having a zero element, all those elements of $\mathfrak{a}$ mapped onto the zero element of $\Theta\mathfrak{a}$ form a normal divisor $\mathfrak{a}_\Theta$ of $\mathfrak{a}$ which is called the kernel of Θ. The homomorphism Θ induces the homomorphism $\bar{\Theta}$ of $\mathfrak{a}/\mathfrak{b}$ onto $\Theta\mathfrak{a}$ defined by $\bar{\Theta}\bar{a}=\Theta a$ having $\mathfrak{a}_\Theta$ as its kernel. If $\mathfrak{a}/\mathfrak{a}_\Theta$ is an additive group, then $\bar{\Theta}$ is an isomorphism and the First Isomorphism Theorem

$$\mathfrak{a}/\mathfrak{a}_{\Theta} \simeq \Theta\,\mathfrak{a}$$

applies.

DEFINITION: An additive abelian semi-group is called a *semi-module.* A semi-module is called a halfmodule if it can be embedded into a module.

The natural numbers, for example, form a halfmodule. The rule (1) for natural numbers as multipliers holds true in semi-modules.

DEFINITION: An additive *sub-semigroup* of an additive semigroup $\mathfrak{a}$ is a non-empty subset $\mathfrak{b}$ of $\mathfrak{a}$ which is closed under addition. $\mathfrak{b}$ is itself an additive semi-group. It is called a sub-semimodule if it is commutative.

DEFINITION: An element s of the additive semi-group $\mathfrak{a}$ is called a *subtrahend,* if

(1) $$s + a = a + s \text{ for all } a \varepsilon \mathfrak{a}$$

(2) $$\text{from } s + a = s + b \text{ it follows that } a = b.$$

Exercise 1: All subtrahends of $\mathfrak{a}$ form a sub-semimodule $S(\mathfrak{a})$ satisfying the cancellation laws of addition, provided there is at least one subtrahend in $\mathfrak{a}$.

Exercise 2: All elements of $\mathfrak{a}$ together with the formal differences $a - s$ $(a \varepsilon \mathfrak{a}, s \varepsilon S(\mathfrak{a}))$ and the symbol 0 form an additive semi-group $\delta(\mathfrak{a})$ (*difference semi-group*) containing $\mathfrak{a}$ as additive sub-semigroup if equality is defined as follows:

$a = b$ as in $\mathfrak{a}$; $a = b - s$ if $a + s = b$; $a_1 - s_1 = a_2 - s_2$ if $a_1 + s_2 = a_2 + s_1$; $0 = a$, $a = 0$ if a is zero element of $\mathfrak{a}$ and if addition is defined as follows:

$a + b$ as in $\mathfrak{a}$; $(b - s) + a = (b + a) = s$; $a + (b - s) = (a + b) - s$; $(a - s) + (b - t) = (a + b) - (s + t)$; $s - s = 0$, $0 + a = a + 0 = a$; $0 + (b - s) = (b - s) + 0 = b - s$; $0 + 0 = 0$.

Exercise 3: The subtrahends of $\delta(\mathfrak{a})$ form a module $S\delta(\mathfrak{a})$ which coincides with $\delta S(\mathfrak{a})$. Prove also that the formal differences $a - s$ occurring in the construction of $\delta(\mathfrak{a})$ are actual differences between a and s as defined above.

Exercise 4: $d(\mathfrak{a})$ is a module if and only if $\mathfrak{a}$ is a semi-module satisfying the cancellation laws of addition. This is the case if and only if $\mathfrak{a}$ is a halfmodule.

DEFINITION: A semi-module for which a multiplication is uniquely defined is called a *semi-ring* if the distributive laws

$$a(b+c)=ab+ac$$
$$(b+c)a=ba+ca$$

are satisfied.

A sub-semimodule of a semi-ring closed under multiplication is called a *sub-semiring*. A sub-semiring is itself a semi-ring.

Exercise 5: The operators of a semi-module $\mathfrak{a}$ form a semi-ring $O(\mathfrak{a})$. Here an operator of $\mathfrak{a}$ is a unique mapping Θ of $\mathfrak{a}$ into $\mathfrak{a}$ satisfying $\Theta(a+b)=\Theta a+\Theta b$. The addition and multiplication of two operators Θ_1, Θ_2 of $\mathfrak{a}$ is defined as usual by the rules

$$(\Theta_1+\Theta_2)a=\Theta_1 a+\Theta_2 a, \quad \Theta_1\Theta_2 a=\Theta_1(\Theta_2 a).$$

The following three exercises show that we may interpret semi-rings as a special case of semi-modules with operators.

Exercise 6: For a semi-ring $\mathfrak{S}$ the mapping

$$a \to a_l = \binom{x}{a\,x}(x \varepsilon \mathfrak{S})$$

of the elements a of $\mathfrak{S}$ onto the *left multiplications* as well as the mapping

$$a \to a_r = \binom{x}{x\,a}(x \varepsilon \mathfrak{S})$$

of the elements of $\mathfrak{S}$ onto the *right multiplications* establishes homomorphic mappings of the semi-module $\mathfrak{S}$ onto the sub-semimodule $\mathfrak{S}_l$, and $\mathfrak{S}_r$ of $O(\mathfrak{S})$ respectively.

Exercise 7: To each homomorphic mapping $a \to a_l$ of a semi-module $\mathfrak{S}$ into the operator-semi-ring $O(\mathfrak{S})$ of $\mathfrak{S}$ there belongs one semi-ring defined over the semi-module $\mathfrak{S}$ as follows:

$$ab = a_l(b) \quad \text{for } a \varepsilon \mathfrak{S},\ b \varepsilon \mathfrak{S}.$$

The associative case is treated in

Exercise 8: The associative law of multiplication in a semi-ring $\mathfrak{S}$ is equivalent to each of the following three statements:

1. The correspondence $a \to a_l$ maps the semi-ring $\mathfrak{S}$ homomorphically onto a sub-semiring $\mathfrak{S}_l$ of $O(\mathfrak{S})$ such that $a_l + b_l = (a+b)_l$, $a_l b_l = (ab)_l$.

2. The correspondence $a \to a_r$ maps the semi-ring $\mathfrak{S}$ homomorphically onto a sub-semiring $\mathfrak{S}_r$ of $O(\mathfrak{S})$ such that $a_r + b_r = (a+b)_r$, $a_r b_r = (ab)_r$.

3. The two sub-semimodules $\mathfrak{S}_l$ and $\mathfrak{S}_r$ of $O(\mathfrak{S})$ are elementwise permutable, i.e., $a_l b_r = b_r a_l$ for $a \in \mathfrak{S}$, $b \in \mathfrak{S}$.

DEFINITION: The normal divisors of the semi-module formed by the elements of a semi-ring under addition which are invariant under $\mathfrak{S}_l$, $\mathfrak{S}_r$, $\mathfrak{S}_l \cup \mathfrak{S}_r$ are called, respectively, left ideals, right ideals, and two-sided ideals of $\mathfrak{S}$.

It is clear that the sum and the intersection of two ideals of one kind is itself an ideal of the same kind. The factor semi-module $\mathfrak{S}/\mathfrak{b}$ of $\mathfrak{S}$ over a two-sided ideal $\mathfrak{b}$ becomes a semi-ring with the introduction of the multiplication $\bar{a}\bar{b} = \overline{ab}$, where $\bar{x}$ denotes the residue class modulo $\mathfrak{b}$ represented by the element x of $\mathfrak{S}$. The residue class $\bar{\mathfrak{b}}$ consisting of the elements of $\mathfrak{b}$ is the zero element, and multiplication by zero always yields zero. The natural homomorphism of $\mathfrak{S}$ onto $\mathfrak{S}/\mathfrak{b}$ also preserves multiplication. Conversely, for a homomorphic mapping Θ of $\mathfrak{S}$ onto a semi-ring $\Theta\mathfrak{S}$ having a zero element such that multiplication by zero yields zero, the kernel $\mathfrak{S}_\Theta$ of Θ is a two-sided ideal. Θ induces a homomorphism Θ of the semi-ring $\mathfrak{S}/\mathfrak{S}_\Theta$ onto the semi-ring $\Theta\mathfrak{S}$ mapping only the residue class $\mathfrak{S}_\Theta$ onto the zero element of $\Theta\mathfrak{S}$. If $\mathfrak{S}/\mathfrak{S}_\Theta$ is a module, then Θ is an isomorphism and $\mathfrak{S}/\mathfrak{S}_\Theta \simeq \Theta\mathfrak{S}$.

For any two sub-semimodules $\mathfrak{m}_1, \mathfrak{m}_2$ of the semi-ring $\mathfrak{S}$ the set of products $x_1 x_2$ $(x_i \in \mathfrak{m}_i)$ is not necessarily closed under addition. It is customary to denote by $\mathfrak{m}_1\mathfrak{m}_2$ the set of all expressions

$$\sum_{i=1}^{r} x_{1i}x_{2i} \quad (x_{1i} \in \mathfrak{m}_1,\ x_{2i} \in \mathfrak{m}_2;\ r \text{ any natural number}),$$

that is, the smallest sub-semimodule of $\mathfrak{S}$ containing all products $x_1 x_2$ with $x_1 \in \mathfrak{m}_1$, $x_2 \in \mathfrak{m}_2$.

Exercise 9: If the subtrahends of a semi-ring $\mathfrak{S}$ form a two-sided ideal $S(\mathfrak{S})$ of $\mathfrak{S}$, then the difference semi-module becomes a semi-ring (*difference semi-ring*) if multiplication is defined as follows:

$$ab \text{ as in } \mathfrak{S},\ a(b-s)=ab-as,\ (b-s)a$$
$$=ba-bs,\ (a-s)(b-t)=(ab+st)-(at+sb).$$

This is the only possibility for extending the multiplication so that $\delta(\mathfrak{S})$ becomes a semi-ring.

DEFINITION: A *halfring* is a semi-ring which is sub-semiring of a ring. The natural numbers, for example, form a halfring.

Exercise 10: Show that a semi-ring is a halfring if and only if it is associative and satisfies the cancellation laws of addition. In other words, the axioms defining a halfring are obtained from the ring axioms by weakening it through the replacement of existence of subtraction by the cancellation law of addition.

DEFINITION: A *quasi-ring* is a semi-ring which under addition is a module. In other words, the axioms of a quasi-ring are obtained from the ring axioms by omitting the associative law of multiplication.

A sub-semiring of a quasi-ring closed under subtraction is called a *subring*. Each subring of a quasi-ring is a quasi-ring. E.g., we obtain a commutative quasi-ring $J(\mathfrak{R})$ from a ring $\mathfrak{R}$ according to the new multiplication rule $a * b = ab + ba$. This ring is called the *Jordan ring* belonging to $\mathfrak{R}$.

The *Lie-ring* $L(\mathfrak{R})$ belonging to $\mathfrak{R}$ is obtained by replacing the rule of multiplication given in $\mathfrak{R}$ by the Lie-multiplication

$$a \circ b = ab - ba \quad (a, b \in \mathfrak{R})$$

The Lie product of a and b can be interpreted as a measure of the non-commutativity of a and b in terms of the ring $\mathfrak{R}$ since

$$a \circ b = 0$$

is equivalent to the statement

$$ab = ba.$$

Besides the axioms of a quasi-ring the following axioms are satisfied by the Lie multiplication:

$$a \circ a = 0 \tag{6}$$

$$\text{(Jacobi identity)}\quad a\circ(b\circ c) + b\circ(c\circ a) + c\circ(a\circ b) = 0, \tag{7}$$

which follow from obvious computations.

Generalizing this remark we obtain the following definition.

DEFINITION: *A Lie-ring* is a quasi-ring in which multiplication satisfies the rules (6), (7), where it is customary to denote the product of a, b by $a\circ b$.[1]

Each subring of a Lie-ring is a Lie-ring. From the distributive laws and from (6) follows the *anti-commutative law* of multiplication

$$b\circ a = -a\circ b \tag{6a}$$

as follows:

$$b\circ a = b\circ a + a\circ a + b\circ b - (a+b)\circ(a+b) =$$
$$b\circ a - (a\circ b + b\circ a) = -a\circ b.$$

Conversely, if in a quasi-ring the anticommutative law holds and if from $x + x = 0$ it always follows that $x = 0$, then (6) holds, i.e.,

$$a\circ a = -a\circ a,$$
$$a\circ a + a\circ a = 0,$$
$$a\circ a = 0.$$

From the anticommutative law it follows that each ideal of a Lie-ring is two-sided. Any factor ring is itself a Lie-ring. Also

$$m_1\circ m_2 = m_2\circ m_1 \tag{8}$$

for any two submodules m_1 m_2 of a Lie-ring L. From the Jacobi identity we derive

[1] Formerly written also as $[a, b]$, (a, b), or (ab).

(9) $$m_1 \circ (m_2 \circ m_3) \subseteq m_2 \circ (m_3 \circ m_1) + m_3 \circ (m_1 \circ m_2)$$

for any three submodules m_1, m_2, m_3 of L. From (8), (9) for $m_1 = L$, and m_2, m_3 ideals of L, we obtain

$$L \circ (m_2 \circ m_3) \subseteq m_2 \circ (m_3 \circ L) + m_3 \circ (L \circ m_2) \subseteq m_2 \circ m_3 + m_3 \circ m_2 \subseteq m_2 \circ m_3,$$

which means that the product of any two ideals of a Lie-ring is itself an ideal.

In the last part of this section we discuss briefly rings and quasi-rings over coefficient rings.

DEFINITION: Let $\mathfrak{M}$ be a semi-ring. An operator Θ of $\mathfrak{M}$ considered as a semi-module is called a *left scalar* if $(\Theta a) b = \Theta(ab)$. In other words,

$$(\Theta a)_l = \Theta \cdot a_l.$$

From this definition it follows that all the left scalars form an associative semi-ring. Similarly, the *right scalars*, i.e., the members ψ of $O(\mathfrak{M})$ satisfying $a(\psi b) = \psi(ab)$, i.e., $(\psi b)_r = b_r \psi$ for all $b \in \mathfrak{M}$, form an associative semi-ring. The intersection of the two semi-rings just defined is an associative semi-ring whose elements we will call *scalars*. These are the mappings $x \to \varphi x$ of $\mathfrak{M}$ into itself satisfying

$$\varphi(x+y) = \varphi x + \varphi y$$

$$\varphi(xy) = (\varphi x) y = x(\varphi y).$$

Each left scalar Θ_1 commutes with each right scalar Θ_2 for $\mathfrak{M}\mathfrak{M}$ since

$$\Theta_1 \Theta_2 (x_1 x_2) = \Theta_1 (\Theta_2 (x_1 x_2) = \Theta_1 (x_1 (\Theta_2 x_2)) = \Theta_1 x_1 \, . \, \Theta_2 x_2$$
$$= \Theta_2 ((\Theta_1 x_1) x_2) = \Theta_2 (\Theta_1 (x_1 x_2)) = \Theta_2 \Theta_1 (x_1 x_2).$$

This fact explains why one usually only defines the notion of an $\mathfrak{S}$-ring for *commutative* rings of coefficients. However a generalization of the narrower concept is possible by using the concept of left and right scalars as follows.

DEFINITION: Let $\mathfrak{S}$ be an associative semi-ring. The semi-ring $\mathfrak{M}$ is called an $\mathfrak{S}$-semi-ring if for each element λ of $\mathfrak{S}$ and each element a of $\mathfrak{M}$ there are defined unique products λa and $a\lambda$ also in $\mathfrak{M}$ such that

1) $$\lambda(a+b)=\lambda a+\lambda b,\quad (a+b)\lambda=a\lambda+b\lambda$$

2) $$(\lambda+\mu)a=\lambda a+\mu a,\quad a(\lambda+\mu)=a\lambda+a\mu$$

3) $$(\lambda\mu)a=\lambda(\mu a),\quad (a\lambda)\mu=a(\lambda\mu)$$

4) $$\lambda(ab)=(\lambda a)b,\quad (ab)\lambda=a(b\lambda),\quad (a\lambda)b=a(\lambda b)$$

5) $$(\lambda a)\mu=\lambda(a\mu)$$

The $\mathfrak{S}$- semi-ring $\mathfrak{M}$ is called *proper* if

6) $$\mathfrak{S}\mathfrak{M}=\mathfrak{M}=\mathfrak{M}\mathfrak{S}.$$

For example, if $\mathfrak{S}$ is a sub-semiring of the associative semi-ring $\mathfrak{M}$, then the multiplication of the elements of $\mathfrak{M}$ by elements of $\mathfrak{S}$ as defined in $\mathfrak{M}$ itself defines $\mathfrak{M}$ as an $\mathfrak{S}$-semi-ring. To give an illustration, the *linear transformations* of an $\mathfrak{S}$-module $\mathfrak{M}$, $\mathfrak{S}$ being an arbitrary ring, are defined as operators of $\mathfrak{M}$ which are permutable with the elements of $\mathfrak{S}$ as applied to $\mathfrak{M}$. In other words, a linear transformation is a mapping Θ of $\mathfrak{M}$ into $\mathfrak{M}$ satisfying

$$\Theta(u+v)=\Theta u+\Theta v,\ \Theta(\lambda u)=\lambda(\Theta u)\ \text{ for } u, v\,\epsilon\,\mathfrak{M}, \lambda\,\epsilon\,\mathfrak{S}.$$

The linear transformations of the $\mathfrak{S}$-module $\mathfrak{M}$ form a ring $L(\mathfrak{M}, \mathfrak{S})$. If $\mathfrak{S}$ is a commutative ring and if $\mathfrak{M}$ is a proper $\mathfrak{S}$-module then $L(\mathfrak{M}, \mathfrak{S})$ becomes an $\mathfrak{S}$-ring if the product of a linear transformation Θ and an element λ of $\mathfrak{S}$ is defined as the linear transformation $\lambda\Theta=\Theta\lambda=\binom{u}{\lambda(\Theta u)}$ of $\mathfrak{M}$. In fact, since $\mathfrak{M}$ is a proper $\mathfrak{S}$-module and since $\mathfrak{S}$ is commutative, it follows that the mapping $\lambda\rightarrow\binom{u}{\lambda u}$ provides an isomorphism between $\mathfrak{S}$ and a subring of $L(\mathfrak{M}, \mathfrak{S})$ which may take the place of $\mathfrak{S}$ in defining $L(\mathfrak{M}, \mathfrak{S})$ as an $\mathfrak{S}$-ring.

We remark that 6) is equivalent to

6a) $$1a=a=a1\ \text{ for } a\,\epsilon\,\mathfrak{M}$$

if $\mathfrak{S}$ contains a unit element 1. For a commutative ring $\mathfrak{S}$ of coefficients the concept of a *quasi-ring* over the coefficient ring $\mathfrak{S}$ is ordinarily defined as a quasi-ring $\mathfrak{M}$ which is an $\mathfrak{S}$-module subject to the further conditions

4a) $$\lambda(ab) = (\lambda a)b = a(\lambda b) \text{ for } \lambda \epsilon \gamma;\ a, b \epsilon \mathfrak{M},$$

6a) $$\mathfrak{S}\mathfrak{M} = \mathfrak{M}.$$

However, this becomes a symmetric $\mathfrak{S}$-quasi-ring in the more general sense defined above by the definition

7) $$\lambda a = a\lambda$$

where 7) simply expresses the symmetry.

DEFINITION: Let $\mathfrak{S}$ be a ring with a unit element and let $\mathfrak{M}$ be an $\mathfrak{S}$-quasi-ring. A subset $\mathfrak{B}$ of $\mathfrak{M}$ is called a *basis* of $\mathfrak{M}$ over $\mathfrak{S}$ if

a) $$\lambda b = b\lambda \text{ for } \lambda \epsilon \mathfrak{S},\ b \epsilon \mathfrak{B}$$

b) We have for $a \epsilon \mathfrak{M}$,

$$a = \sum_{b \epsilon B} \lambda_b b$$

where only a finite number of $\lambda_b b \neq 0$ and a is the sum of this finite number of elements. Such a sum is called a linear combination of the basis elements. Since such sums are finite, we can multiply one of them by a scalar and add two of them term by term.

c) A basis is linearly independent in the strict sense: A linear combination vanishes if and only if each coefficient vanishes.

It follows that for any two basis elements x, y there are equations

(8) $$xy = \sum_{b \epsilon B} \gamma_{x,y,b} b$$

with uniquely determined "*combination constants*" $\gamma_{x,y,b}$ which are contained in $\mathfrak{S}$ and are such that for every fixed pair of basis elements x, y all but a finite number of the combination constants vanishes. Furthermore, because of the distributive laws, the laws 4), and the condition a), the multiplication rule in $\mathfrak{M}$ is given as

(9) $$\sum_{x \in B} \alpha_x x \,.\, \sum_{y \in B} \beta_y y = \sum_{b \in B} \alpha_x \beta_y \gamma_{x,y,b} b$$

and the rule 4) finds its expression in

(10) $$\lambda \gamma_{x,y,b} = \gamma_{x,y,b} \lambda$$

i.e., the multiplication constants belong to the center of the coefficient ring.

Conversely, if a set of elements $\gamma_{x,y,b}$ of $\mathfrak{S}$ satisfying the previous conditions is given, then an $\mathfrak{S}$-quasi-ring $\mathfrak{M}$ with the given set $\mathfrak{B}$ as basis can be constructed as the set of all formal linear combinations

$$\sum_{b \in B} \alpha_b b$$

of the elements of $\mathfrak{B}$ over the coefficient ring $\mathfrak{S}$, where all but a finite number of the coefficients α_x vanish. The set $\mathfrak{M}$ of all those formal linear combinations is subject to the rules

$$\sum_{b \in B} \alpha_b b = \sum_{b \in B} \beta_b b \text{ if and only if } \alpha_b = \beta_b \text{ for } b \in \mathfrak{B},$$

(11) $$\sum_{b \in B} \alpha_b b + \sum_{b \in B} \beta_b b = \sum_{b \in B} (\alpha_b + \beta_b) b$$

(12) $$\lambda \sum_{b \in B} \alpha_b b = \sum_{b \in B} (\lambda \alpha_b) b, \ \ (\sum_{b \in B} \alpha_b b) \lambda = \sum_{b \in B} (\alpha_b \lambda) b$$

and (9). We may consider each element x of $\mathfrak{B}$ as an element of $\mathfrak{M}$ after identification of x with that linear combination of the elements of $\mathfrak{B}$ which bears the coefficient 1 in front of x and zero elsewhere. After this identification each formal linear combination is the actual linear combination of the elements of $\mathfrak{B}$ with the same set of coefficients, and

$\mathfrak{B}$ turns out to be a basis of $\mathfrak{M}$. The $\mathfrak{S}$-quasi-ring $\mathfrak{M}$ is associative, if and only if the *associativity relations*

$$(13) \qquad \sum_{b\in B} \gamma_{x,y,b}\gamma_{b,z,u} = \sum_{b\in B} \gamma_{x,b,u}\gamma_{y,z,b}$$

are true for all quadruples x, y, z, u. The $\mathfrak{S}$-quasi-ring $\mathfrak{M}$ is a Lie-ring if and only if

$$(14) \qquad \gamma_{x,x,b} = 0, \quad \gamma_{x,y,b} + \gamma_{y,x,b} = 0$$

$$(15) \qquad \sum_{b\in B} (\gamma_{x,b,u}\gamma_{y,z,b} + \gamma_{y,b,u}\gamma_{z,x,b} + \gamma_{z,b,u}\gamma_{x,y,b}) = 0.$$

for $x, y, z, u \in \mathfrak{B}$.

For example, let $\mathfrak{M}$ be the $\mathfrak{S}$-vector module with the basis elements $x^0 = 1$, $x^1 = x$, x^2, x^3, . . . and with the rule of combination $x^n \cdot x^m = x^{n+m}$. The ring defined by this is called the *polynomial domain of one variable* x over $\mathfrak{S}$ and is denoted by $\mathfrak{S}[x]$. Every element of $\mathfrak{S}[x]$ is uniquely of the form

$$(16) \qquad f(x) = \alpha_n x^n + \alpha_{n-1} x^{n-1} + \cdots + \alpha_0,$$

with $\alpha_i \in \mathfrak{S}$ and $\alpha_n \neq 0$, if $n > 0$. The number n is called the *degree of the polynomial* $f(x)$ if $f(x) \neq 0$.

The $\mathfrak{S}$-matrix rings are other examples of $\mathfrak{S}$ rings.

Let $\mathfrak{M}$ be the n-dimensional left vector module with basis $u_1, u_2, \ldots, u_n$. We wish to find all the operators σ of $\mathfrak{M}$ which map $\mathfrak{M}$ into itself operator-homomorphically with respect to $\mathfrak{S}$. Accordingly we define: A *linear transformation* σ of $\mathfrak{M}$ is a single-valued mapping $u \to u\sigma$ of $\mathfrak{M}$ into itself such that

$$(17) \qquad (u + v)\sigma = u\sigma + v\sigma,$$

$$(18) \qquad (\alpha u)\sigma = \alpha(u\sigma).$$

Therefore we have

(19) $$u_i\sigma = \alpha_{i1}u_1 + \alpha_{i2}u_2 + \cdots + \alpha_{in}u_n$$

and for $$u = \sum_1^n \lambda_i u_i,$$

(20) $$u\sigma = \sum_{k=1}^n (\sum_{i=1}^n \lambda_i \alpha_{ik}) u_k.$$

Conversely, every system of elements α_{ik} ($i, k=1, 2, \ldots, n$) in $\mathfrak{S}$ defines a linear transformation of $\mathfrak{M}$ uniquely by means of the above formulae.

We order the n^2 elements α_{ik} in a square configuration

$$A = \begin{pmatrix} \alpha_{11}\,\alpha_{12} \ldots \alpha_{1n} \\ \alpha_{21}\,\alpha_{22} \ldots \alpha_{2n} \\ \cdots\cdots\cdots \\ \alpha_{n1}\alpha_{n2} \ldots \alpha_{nn} \end{pmatrix} = (\alpha_{ik})$$

and call this configuration *the matrix of n-th degree associated with* σ.

According to the earlier definitions and statements, the linear transformations of $\mathfrak{M}$ form a ring.

If $$u_i\sigma = \sum_{k=1}^n \alpha_{ik}u_k \qquad (i=1,2,\ldots,n)$$

$$u_i\tau = \sum_{k=1}^n \beta_{ik}u_k, \qquad (i=1,2,\ldots,n)$$

then we defined

(21) $$u_i(\sigma+\tau) = u_i\sigma + u_i\tau = \sum_{k=1}^n (\alpha_{ik}+\beta_{ik})u_k,$$

(22) $$u_i(\sigma\tau) = (u_i\sigma)\tau = (\sum_{\nu=1}^n \alpha_{i\nu}u_\nu)\tau = \sum_{\nu=1}^n \alpha_{i\nu}(u_\nu\tau) = \sum_{k=1}^n(\sum_{\nu=1}^n \alpha_{i\nu}\beta_{\nu k})u_k.$$

Accordingly we define the sum and product of the matrices

$$A = (\alpha_{ik}),\quad B = (\beta_{ik}) \ :$$
$$A + B = (\alpha_{ik} + \beta_{ik}) \quad ,$$
$$AB = (\sum_{\nu=1}^n \alpha_{i\nu}\beta_{\nu k}).$$

All the matrices of the n-th degree with coefficients in the ring $\mathfrak{S}$ *with unit element form a ring* M_n *isomorphic to the ring of all linear transformations of the n-dimensional* $\mathfrak{S}$*-vector module.*

M_n is said to be a *matrix ring of n-th degree over* $\mathfrak{S}$. M_n is an $\mathfrak{S}$-ring.

The basis elements are n^2 *matrix unities* e_{ik} $(i, k=1, 2, \ldots, n)$, where e_{ik} is the matrix which has a 1 at the intersection of the i-th column and k-th row, and otherwise all zeros. The multiplication rules of the matrix unities are

(23) $$e_{ik}e_{rs} = \delta_{kr}e_{is}$$
$$\left(i, k, r, s = 1, 2, \ldots, n; \ \delta_{kr} = \begin{cases} 1, & \text{if } k = r \\ 0, & \text{if } k \neq r \end{cases}\right).$$

The unit element of M_n is the identity matrix $E = \begin{pmatrix} 1 & & & 0 \\ & 1 & & \\ & & \ddots & \\ 0 & & & 1 \end{pmatrix}$

The zero element is the matrix of all zeros.

If the ring $\mathfrak{S}$ of matrix coefficients is commutative, then one usually applies the linear transformation σ of the vector module $\mathfrak{M}$ on the left, so that σ is a single-valued mapping of $\mathfrak{M}$ into itself for which

1. $\sigma(u + v) = \sigma u + \sigma v,$
2. $\sigma(\alpha u) = \alpha \sigma u.$

Moreover we set, differing from (19) above,

(24). $$\sigma u_k = \sum_1^n \alpha_{ik} u_i, \qquad (k = 1, 2, \ldots, n)$$

but define the associated matrix again as

$$A_\sigma = (\alpha_{ik}).$$

Since $\mathfrak{S}$ is now a commutative ring, the mapping $\sigma \rightarrow A_\sigma$ is again an isomorphism between the ring of linear transformations of $\mathfrak{M}$ and the matrix ring M_n.

DEFINITION: A quasi-ring over a field as coefficient ring is called an *algebra*. An algebra always has a basis over its coefficient field. An *associative algebra* is an algebra satisfying the associative law of multiplication. An example is the *semi-group ring* of a given semi-group over a field k. As a basis of the k-ring, we take the elements of the given semi-group, and as multiplication rule we take the multiplication table of the given semi-group.

An algebra which is a Lie-ring, is called a *Lie-algebra*. The Lie-ring belonging to an associative algebra, for example, is a Lie-algebra. An algebra with a finite number of basis elements over its coefficient field is called *finite dimensional* or simply *finite*. A finite associative algebra also is called a *hypercomplex system*. For example, the group ring of a given finite group over a field is a hypercomplex system. Another example is provided by the *finite extensions* of a given field of reference k. The *extensions* of k are defined simply as the fields containing k as subfield. They are finite if and only if they contain a finite basis over k.

6. *Galois Fields.*

A field with a finite number of elements is called a *Galois field.*

The number of elements of the prime field k contained in a Galois field K is finite, and is therefore a natural prime p. Since K contains only a finite number of elements, K is a finite extension of k. *The number of elements in a Galois field is thus a prime power* p^n. *The exponent* n *is equal to the degree of* K *over the prime field consisting of* p *elements.*

In order to investigate the multiplicative group of a Galois field, we need the

LEMMA: *A finite group must be cyclic if, for every natural number* n, *it has at most* n *elements whose* n*-th power is* e.[1]

Proof: Let $\mathfrak{G}$ have the order N; let $\mathfrak{Z}$ be the cyclic group with order N. An element in $\mathfrak{G}$ generates a cyclic subgroup $\mathfrak{U}$ whose order d is a divisor of N. It was shown earlier that $\mathfrak{Z}$ contains exactly one cyclic subgroup $\mathfrak{V}$ of order d. The d-th power of each of the d elements of $\mathfrak{U}$ is e; therefore by hypothesis $\mathfrak{U}$ contains all the elements of $\mathfrak{G}$ whose d-th power is e. Since $\mathfrak{U}$ and $\mathfrak{V}$ have the same structure, $\mathfrak{G}$ contains at most as many elements of order d as $\mathfrak{Z}$ does. This holds for every divisor of N. Since $\mathfrak{G}:1 = \mathfrak{Z}:1$, $\mathfrak{G}$ and $\mathfrak{Z}$ contain the same number of elements of order d. $\mathfrak{Z}$ contains an element of order N and therefore $\mathfrak{G}$ contains one also, Q.E.D.

In a field, according to a familiar theorem, the equation $x^n = 1$ has at most n different solutions. Therefore by the preceding lemma, the multiplicative group of any finite field is cyclic.

The multiplicative group of a Galois field is cyclic.

In the proof of the following theorem some acquaintance with cyclotomic polynomials is assumed.

THEOREM 16: *A finite division ring is a field.*[2]

Witt's Proof: Let K be a division ring with a finite number of elements. If k is a division ring contained in K then K is a finite k-module, and by 4. the number of elements of K is a power of the number of elements of k.

---

[1] Equivalently: For each given index, or for each given order, there is only one subgroup.

[2] J. H. M. Wedderburn, *A theorem on finite algebras*, Trans. Amer. Math. Soc., Vol. 6, p. 349.

The center of K is (as was shown earlier) a field; say it has q elements. Then K consists of q^n different elements. All the elements of K which commute with an element a form a division ring k_a, which contains the center of K. Therefore k_a contains q^d elements, where d is a positive divisor of n. We decompose the multiplicative group of K into classes of conjugate elements and obtain as the class equation

$$q^n - 1 = (q-1) + \sum_{\substack{\text{some } d/n \\ 0<d<n}} \frac{q^n-1}{q^d-1}.$$

Each summand $\frac{q^n-1}{q^d-1}$ and the number q^n—1 are divisible by $\varphi_n(q)$ where $\varphi_n(x)$ is the n-th cyclotomic polynomial. Therefore q-1 is also divisible by $\varphi_n(q)$. If $n>1$, then in the decomposition $\varphi_n(q) = \prod_1^{\varphi(n)} (q-\zeta_i)$, where the ζ_i are the primitive *n-th* roots of unity, each factor is greater than $q-1$ in absolute value, and therefore $\varphi_n(q)$ is also greater than $q-1$ in absolute value. Therefore $n=1$ and K is identical with its center, as was to be shown.

7. *Near-Rings and Near-Fields.*

We wish to add the operators of a given group $\mathfrak{G}$ and investigate the rules of combination which will obtain.

Single-valued mappings $\binom{x}{x^\pi}$ of a group into itself are added in the following way:

$$x^{\pi+\varrho} = x^\pi \cdot x^\varrho.$$

In general this addition is not commutative. All the single-valued mappings of $\mathfrak{G}$ into itself form an additive group. Moreover we know that they form a multiplicative semi-group.

In the domain $\Pi_{\mathfrak{G}}$ of all single-valued mappings of $\mathfrak{G}$ into itself, we also have the right distributive law

$$(\pi+\varrho)\sigma = \pi\sigma + \varrho\sigma,$$

which can be verified immediately. The left distributive law holds for all π and ϱ in $\Pi_{\mathfrak{G}}$ if and only if σ is an operator belonging to $\mathfrak{G}$.

Under what conditions is the sum of two operators an operator?

When, therefore, is

$$(xy)^{\Theta_1+\Theta_2} = x^{\Theta_1+\Theta_2} \cdot y^{\Theta_1+\Theta_2}$$

for two operators Θ_1, Θ_2 ?

Since by definition

$$(xy)^{\Theta_1+\Theta_2} = (xy)^{\Theta_1}(xy)^{\Theta_2} = x^{\Theta_1}y^{\Theta_1}x^{\Theta_2}y^{\Theta_2},$$

we have

$$x^{\Theta_2}\cdot y^{\Theta_1} = y^{\Theta_1}\cdot x^{\Theta_2}$$

as a necessary and sufficient condition.

DEFINITION: Two operators Θ_1, Θ_2 are said to be *additive* if $\mathfrak{G}^{\Theta_1}$ commutes with $\mathfrak{G}^{\Theta_2}$ elementwise.

The sum of two operators is an operator if and only if the summands are additive. For additive operators Θ_1, Θ_2, addition is commutative: $\Theta_1 + \Theta_2 = \Theta_2 + \Theta_1$, as is immediately seen. The sum of n operators is certainly additive if the summands are pairwise additive. The sum of pairwise additive operators is independent of order or parenthesizing.

It is precisely the center operators that are additive with respect to any operator. (If Θ is additive with respect to $\underline{1}$, then x^{Θ} is in the center of $\mathfrak{G}$). Two automorphisms are additive if and only if $\mathfrak{G}$ is abelian.

In the domain $\Pi_{\mathfrak{G}}$ with operators of $\mathfrak{G}$ as multipliers, almost all the ring axioms are fulfilled. We call such a domain a *near-ring*.

The axioms which must be fulfilled in a (left-) near-ring are:

I. A near-ring F is an additive group, not necessarily commutative.

II. There is a multiplier domain $\mathfrak{M}$ in F, such that for every element μ in $\mathfrak{M}$ and α in F the product $\mu\alpha$ is defined uniquely as an element in F. The following rules hold:

$$(\mu\mu')\alpha = \mu(\mu'\alpha) \qquad (\mu, \mu' \in \mathfrak{M})$$

$$\mu(\alpha+\beta) = \mu\alpha + \mu\beta \qquad (\alpha, \beta \in F).$$

The *near-fields* are special near-rings. A near-field is a near-ring whose multiplier domain forms a group. If 1 is the identity element of $\mathfrak{M}$, then we should have $1\cdot\alpha = \alpha$ for all $\alpha \in F$, and we should have the cancellation law:

$$\mu\alpha = \mu'\alpha,\ \alpha \neq 0$$

implies

$$\mu = \mu'.$$

The multiplicative group $\mathfrak{M}$ of a near-field F is mapped isomorphically onto a group of automorphisms of the additive groups of F by the mapping $\mu \rightarrow \begin{pmatrix}\alpha\\ \mu\alpha\end{pmatrix}$. Because of the cancellation law we have a group of regular automorphisms.[1]

[1] An automorphism is said to be *regular* if it permutes regularly the group elements different from the identity. A *regular automorphism group* is a group consisting entirely of regular automorphisms.

If conversely a regular automorphism group $\mathfrak{M}$ of a group $\mathfrak{G}$, which contains at least two elements, is given, then we consider $\mathfrak{G}$ as an additive group, find a non-zero element, denote it by 1, and introduce the notation $\mu = \mu(1)$ for all μ in $\mathfrak{M}$. Since $\mathfrak{M}$ is regular, the notation is single-valued. Now we define multiplication by the statement $\mu\alpha = \mu(\alpha)$ for all μ in $\mathfrak{M}$. Then we have found a near-field F with the additive group $\mathfrak{G}$ and the multiplicative group $\mathfrak{M}$.

The holomorph of $\mathfrak{M}$ over $\mathfrak{G}$ is the group of all permutations $\begin{pmatrix} \alpha \\ \beta + \mu\alpha \end{pmatrix}$ of elements of F.

This permutation group can be formed for every near-field F and is denoted by $\mathfrak{P}_F$. $\mathfrak{P}_F$ is transitive and each permutation in $\mathfrak{P}_F$ either leaves all of the elements of F fixed or leaves at most one element of F fixed. In order to prove the last property we must show that

$$\beta + \mu\alpha = \alpha, \quad \beta + \mu\alpha' = \alpha', \; \alpha \neq \alpha',$$

implies $\mu = 1, \beta = 0$. In fact

$$\alpha - \alpha' = \mu\alpha - \mu\alpha' = \mu(\alpha - \alpha'),$$

and since $\alpha - \alpha' \neq 0$, we find $\mu = 1$. Since $\beta + \alpha = \alpha$, we have $\beta = 0$.

The permutations $\begin{pmatrix} \alpha \\ \beta + \alpha \end{pmatrix}$ form a regular normal subgroup of $\mathfrak{P}_F$ isomorphic to the additive group of F.

If conversely the permutation group $\mathfrak{P}$ contains a regular normal subgroup $\mathfrak{G}$ and each permutation leaves either all or at most one letter fixed, then we can consider the group $\mathfrak{P}$ as a holomorph of a certain automorphism group $\mathfrak{M}$ over $\mathfrak{G}$ because of a remark in § 4, 6. Because of the second assumption, $\mathfrak{M}$ is a group of regular automorphisms. Consequently we can construct a near-field F with additive group $\mathfrak{G}$ and with $\mathfrak{M}$ as multiplier group, so that $\mathfrak{P} = \mathfrak{P}_F$.

A near-ring in which every element is a multiplier is said to be a *complete near-ring* or in accordance with a suggestion of Mr. Wieland a *stem*. For example, the previously constructed near-ring $\Pi_{\mathfrak{G}}$ is a *right stem*.

A near-field is said to be a *complete near-field* if the group of multipliers consists of all non-zero elements. For example every division ring is a complete near field.

The determination of all the types of complete near-fields which contain a finite number of elements, is an interesting problem which will be solved later.

Exercises

1. We have lost the first row and first column of a group table. Show that the associated abstract group is still uniquely determined by the incomplete table.

2a. All rational integers a, for which $X^a = e$ for all elements X of a group $\mathfrak{G}$, form a module—the *exponential module.*

2b. The non-negative rational integer generating the exponential module is called the *exponent.* The exponent is the least common multiple of the orders of all the elements of $\mathfrak{G}$.

3. The exponent is the smallest natural number a such that $X^a = e$ for all X, if there are any rational integers different from zero with this property.

The exponent is a divisor of the group order. The exponent of a cyclic group is equal to its order. Is the converse true for finite groups?

A finite abelian group whose exponent is equal to its order, is cyclic. More generally, show that in a finite abelian group every divisor of the exponent occurs as the order of an element.

Hint: Prove and use the fact that the product of two elements which have relatively prime orders and which commute has an order equal to the product of the orders of the factors.

4. The exponent of a subgroup is a divisor of the exponent of the whole group. The same holds true for a homomorphic image of $\mathfrak{G}$; in particular, the exponent of a factor group is a divisor of the exponent of $\mathfrak{G}$.

5. The greatest common divisor of all rational integers n with the property $X^n = e$ implies $X^{p^m} = e$, is called the p-exponent (p is a natural prime). Set up and prove statements analogous to those made in Exercises 2 - 4.

6. An automorphism α of a group $\mathfrak{G}$ which leaves both the normal subgroup $\mathfrak{N}$ and the factor group $\mathfrak{G}/\mathfrak{N}$ elementwise fixed multiplies each element in $\mathfrak{G}$ by an element of the center $\mathfrak{z}$ of $\mathfrak{N}$. Its order is a divisor of the exponent of $\mathfrak{z}$. All such automorphisms α form an abelian group.

7. A finite group has a non-zero central operator precisely when the order of the factor commutator group and the order of the center have a common prime factor. (Use Exercise 3 of Chap. 1).

8. A normal subgroup of a finite group contains every subgroup whose order is relatively prime to the index of the normal subgroup.

9. Prove the simplicity of $\mathfrak{A}_n$ for $n>4$ by the following method.

a) If the permutation π moves (does not leave fixed) more than 3 letters, then there is a three-cycle ϱ, such that $\varrho\pi\varrho^{-1}\pi^{-1}$ leaves more letters fixed than π but is not the identity permutation.

b) In any normal subgroup of $\mathfrak{A}_n$ which is different from $\underline{1}$ there is a three-cycle.

c) Apply Exercise 7 of Chapter I (following v.d. Waerden, *Modern Algebra I.*)

III. THE STRUCTURE AND CONSTRUCTION OF COMPOSITE GROUPS

§ 1. Direct Products

By the second isomorphy theorem in a group $\mathfrak{G}$ which is the product of two normal subgroups $\mathfrak{N}_1$ and $\mathfrak{N}_2$, the factor groups are isomorphic to the factor groups of the $\mathfrak{N}_i$ with respect to their intersection $\mathfrak{D}$; that is,

$$\mathfrak{G}/\mathfrak{N}_1 \simeq \mathfrak{N}_2/\mathfrak{D}, \quad \mathfrak{G}/\mathfrak{N}_2 \simeq \mathfrak{N}_1/\mathfrak{D}.$$

We ask to what extent the structure of $\mathfrak{G}/\mathfrak{D}$ is uniquely determined by the structures of $\mathfrak{N}_1/\mathfrak{D}$ and $\mathfrak{N}_2/\mathfrak{D}$.

To do this we can and will assume that $\mathfrak{G}$ is the product of the two normal subgroups $\mathfrak{N}_1$ and $\mathfrak{N}_2$ with e as their intersection:

$$\mathfrak{G} = \mathfrak{N}_1 \mathfrak{N}_2, \ \mathfrak{N}_1 \cap \mathfrak{N}_2 = e.$$

THEOREM 1. *Every element in $\mathfrak{G}$ can be represented as the product of an element from $\mathfrak{N}_1$ and one from $\mathfrak{N}_2$ in one and only one way.*

The multiplication rule is

$$(a_1 a_2) \cdot (b_1 b_2) = (a_1 b_1) \cdot (a_2 b_2),$$

where a_i, b_i are in $\mathfrak{N}_i$.

Proof: Since $\mathfrak{G} = \mathfrak{N}_1 . \mathfrak{N}_2$, every element in $\mathfrak{G}$ is of the form $a_1 . a_2$ with a_i in $\mathfrak{N}_i$.

From $a_1 a_2 = b_1 b_2$ with b_i in $\mathfrak{N}_i$, it follows that

$$b_1^{-1} a_1 = b_2 a_2^{-1} = d.$$

Since $\mathfrak{N}_1 \cap \mathfrak{N}_2 = e$,

$$d = e, \ b_1 = a_1, \ b_2 = a_2.$$

Next,

$$\begin{aligned} a_1 a_2 a_1^{-1} a_2^{-1} &= a_1 (a_2 a_1^{-1} a_2^{-1}) \in \mathfrak{N}_1 \\ &= (a_1 a_2 a_1^{-1}) a_2^{-1} \in \mathfrak{N}_2 , \end{aligned}$$

because $\mathfrak{N}_1$ and $\mathfrak{N}_2$ are normal subgroups, and therefore it follows from $\mathfrak{N}_1 \cap \mathfrak{N}_2 = e$ that $a_1 . a_2 = a_2 . a_1$. It follows from this that we have the multiplication rule, as was to be proved.

Conversely we now form a group $\mathfrak{G}$ with given normal subgroups $\mathfrak{N}_1$ and $\mathfrak{N}_2$ by defining it as follows: $\mathfrak{G}$ consists of all ordered pairs (a_1, a_2) with a_i in $\mathfrak{N}_i$.

We define multiplication by

$$(a_1, a_2) \cdot (b_1, b_2) = (a_1 b_1, a_2 b_2).$$

We immediately verify the validity of the group axioms. The identity element is $e = (e_1, e_2)$.

The mappings $a_1 \to (a_1, e_2)$ and $a_2 \to (e_1, a_2)$ are, respectively, isomorphisms of $\mathfrak{N}_1$ and $\mathfrak{N}_2$ with normal subgroups $\overline{\mathfrak{N}}_1$ and $\overline{\mathfrak{N}}_2$ of $\mathfrak{G}$ and

$$\overline{\mathfrak{N}}_1 \overline{\mathfrak{N}}_2 = \mathfrak{G}; \; \overline{\mathfrak{N}}_1 \cap \overline{\mathfrak{N}}_2 = e.$$

The group $\mathfrak{G}$ just constructed is called the *direct product* of the groups $\mathfrak{N}_1$ and $\mathfrak{N}_2$; in symbols:

$$\mathfrak{G} = \mathfrak{N}_1 \times \mathfrak{N}_2.$$

We restate the previous theorem as follows:

If $\mathfrak{G}$ is the product of the normal subgroups $\mathfrak{N}_1$ and $\mathfrak{N}_2$ with intersection $\mathfrak{D}$, then the factor group $\mathfrak{G}/\mathfrak{D}$ is the direct product of the factor groups $\mathfrak{N}_1/\mathfrak{D}$ and $\mathfrak{N}_2/\mathfrak{D}$.

As the direct product of three groups $\mathfrak{H}_i$, $i = 1, 2, 3$, we define

$$\mathfrak{H}_1 \times \mathfrak{H}_2 \times \mathfrak{H}_3 = (\mathfrak{H}_1 \times \mathfrak{H}_2) \times \mathfrak{H}_3.$$

It it obvious that there is a simple isomorphism between

$$(\mathfrak{H}_1 \times \mathfrak{H}_2) \times \mathfrak{H}_3 \text{ and } \mathfrak{H}_1 \times (\mathfrak{H}_2 \times \mathfrak{H}_3),$$

and likewise between

$$\mathfrak{H}_1 \times \mathfrak{H}_2 \text{ and } \mathfrak{H}_2 \times \mathfrak{H}_1.$$

Accordingly $\mathfrak{H}_1 \times \mathfrak{H}_2 \times \cdots \times \mathfrak{H}_n$ is *uniquely* defined as the direct product of the $\mathfrak{H}_i$ in any order and with any parenthesizing; and indeed we may define $\mathfrak{G} = \mathfrak{H}_1 \times \mathfrak{H}_2 \times \cdots \times \mathfrak{H}_n$ as the set of all ordered n-tuples $x = (x_1, \ldots, x_n)$ of elements x_i in $\mathfrak{H}_i$ with the multiplication rule

$$(x_1, x_2, \ldots, x_n) \cdot (y_1, y_2, \ldots, y_n) = (x_1 y_1, x_2 y_2, \ldots, x_n y_n).$$

After we have identified x_i with the element $(e_1, \ldots, e_{i-1}, x_i, e_{i+1}, \ldots, e_n)$, $\mathfrak{G}$ becomes the direct product of its normal subgroups $\mathfrak{H}_i$.

Necessary and sufficient conditions that a group $\mathfrak{G}$ with normal subgroups $\mathfrak{H}_1, \ldots, \mathfrak{H}_n$ be the direct product of these normal subgroups, are:

1. $$\mathfrak{G} = \mathfrak{H}_1 \cdot \mathfrak{H}_2 \cdot \ldots \cdot \mathfrak{H}_n,$$

2. $$\mathfrak{H}_i \cap (\mathfrak{H}_1, \ldots, \mathfrak{H}_{i-1}, \mathfrak{H}_{i+1}, \ldots, \mathfrak{H}_n) = e \qquad (i = 1, 2, \ldots, n).$$

Condition 2. can be replaced by

2 a. $$\mathfrak{H}_i \cap (\mathfrak{H}_{i+1}, \mathfrak{H}_{i+2}, \ldots, \mathfrak{H}_n) = e \qquad (i = 1, 2, \ldots, n-1)$$

or

2b. The representation $e = e_1 \cdot e_2 \cdot \ldots \cdot e_n$ with e_i in $\mathfrak{H}_i$, of the identity element of $\mathfrak{G}$ is unique.

We say that the x_i are the $\mathfrak{H}_i$-components of an element x of $\mathfrak{G}$. The mapping $x \to x_i$ is an operator H_i of $\mathfrak{G}$. H_i is said to be the i-th *decomposition operator* of $\mathfrak{G} = \mathfrak{H}_1 \times \mathfrak{H}_2 \times \cdots \times \mathfrak{H}_n$. The H_i are additive operators and $x = x^{H_1} \cdot x^{H_2} \cdot \ldots \cdot x^{H_n}$; hence

$$H_1 + H_2 + \cdots + H_n = \underline{1}$$

Moreover $H_i H_k = \underline{0}$ if $i \neq k$ and $H_i^2 = H_i$.

If, conversely, we are given additive operators H_i with the properties $\sum_1^n H_i = \underline{1}$, $H_i H_k = \underline{0}$ for $i \neq k$, then they are associated with the direct decomposition

$$\mathfrak{G} = \mathfrak{G}^{H_1} \times \mathfrak{G}^{H_2} \times \cdots \times \mathfrak{G}^{H_n}.$$

H_i is a normal operator over $\mathfrak{G}$ since for all b in $\mathfrak{G}^{H_i}$, x in $\mathfrak{G}$, we have $b^x = b^{xH_i} = b^{x^{H_i}}$, and therefore for all a in G, $a^{xH_i} = a^{x^{H_i}H_i} = a^{H_i x}$. The order of a direct product is equal to the product of the orders of its factors, as the component representation shows.

The center, the commutator group and the comutator form of a direct product are the direct products of the centers, the commutator groups and the commutator forms of the factors, respectively.

If $\mathfrak{G}$ is decomposed into the direct product of characteristic factors, then the automorphism group of $\mathfrak{G}$ is the direct product of the automorphism groups of the factors.

Every group is the direct product of the identity element and itself. A group which has only this direct decomposition is said to be *directly indecomposable.*

Direct products occur in the investigation of factors of a principal series of a given group. These factors are simple over a certain automorphism domain, and therefore are characteristically simple.

THEOREM 2: *If the group $\mathfrak{G} \neq e$ is characteristically simple and the double chain law holds for normal subgroups, then it is the direct product of (merely) simple groups which are isomorphic to each other.*

Proof: Since the minimal chain law holds, there is a minimal normal subgroup $\mathfrak{N}$ in $\mathfrak{G}$. The automorphisms α of $\mathfrak{G}$ map $\mathfrak{N}$ onto the minimal normal subgroups $\mathfrak{N}^\alpha$ of $\mathfrak{G}$, all isomorphic to $\mathfrak{N}$. We wish to form the largest possible direct product of these. By the maximal chain law, there is certainly a largest direct product $\mathfrak{M} = \mathfrak{N}^{\alpha_1} \times \mathfrak{N}^{\alpha_2} \times \cdots \times \mathfrak{N}^{\alpha_r}$. If $\mathfrak{M}$ were not equal to $\mathfrak{G}$, then by hypothesis $\mathfrak{M}$ would not be mapped into itself under all automorphisms of $\mathfrak{G}$; therefore there is an automorphism α of $\mathfrak{G}$ such that $\mathfrak{N}^\alpha$ does not lie in $\mathfrak{M}$. However we would then have

$\mathfrak{N}^{\alpha} \cap \mathfrak{M} = e$, since $\mathfrak{N}^{\alpha}$ is a smallest normal subgroup of $\mathfrak{G}$, and therefore $\mathfrak{N}^{\alpha} \times \mathfrak{N}^{\alpha_1} \times \cdots \times \mathfrak{N}^{\alpha_r}$ is greater than $\mathfrak{M}$. Consequently

$$\mathfrak{G} = \mathfrak{N}^{\alpha_1} \times \mathfrak{N}^{\alpha_2} \times \cdots \times \mathfrak{N}^{\alpha_r}.$$

The factors of this direct decomposition are the minimal normal subgroups of $\mathfrak{G}$ and therefore simple, as can easily be seen.

Remark: Every factor of a principal series (or of a characteristic series) of a finite solvable group is the direct product of cyclic groups of equal prime order.

§ 2. Theorems on Direct Products[1]

The following theorems also hold for groups with an operator domain Ω.

From the first isomorphism theorem we derive:

THEOREM 3: *If a homomorphism of a group $\mathfrak{G}$ onto a multiplicative system $\mathfrak{H}$ induces an isomorphism with $\mathfrak{H}$ of a normal subgroup $\mathfrak{N}_2$ of $\mathfrak{G}$, then $\mathfrak{G}$ is the direct product of the normal subgroup $\mathfrak{N}_1$ of $\mathfrak{G}$, consisting of all elements of $\mathfrak{G}$ which map onto $\bar{e}$, with the normal subgroup $\mathfrak{N}_2$.*

Proof: From the hypothesis, $\mathfrak{N}_1 \mathfrak{N}_2 = \mathfrak{G}$, $\mathfrak{N}_1 \cap \mathfrak{N}_2 = e$.

THEOREM 4: *If σ is a homomorphism of the group $\mathfrak{G}$ which is different from e, onto the normal subgroup $\overline{\mathfrak{G}}$ of the indecomposable group $\mathfrak{H}$, τ a homomorphy of $\mathfrak{H}$ onto $\mathfrak{J}$, $\tau\sigma$ a $\mathfrak{G}\mathfrak{J}$-isomorphism, then σ is a $\mathfrak{G}\mathfrak{H}$-isomorphism and τ an $\mathfrak{H}\mathfrak{J}$-isomorphism.*

Proof: One can show easily that σ is a $\mathfrak{G}\mathfrak{H}$-isomorphy and that τ is a $\overline{\mathfrak{G}}\mathfrak{J}$-isomorphy. Since $\tau\overline{\mathfrak{G}} = \tau\sigma\mathfrak{G} = \mathfrak{J}$, τ is a $\overline{\mathfrak{G}}\mathfrak{J}$-isomorphism. Since $\mathfrak{G} \neq e$, $\overline{\mathfrak{G}}$ is a normal subgroup of $\mathfrak{H}$ that is distinct from e. From the indecomposability of $\mathfrak{H}$ it follows, by Theorem 3, that $\overline{\mathfrak{G}} = \mathfrak{H}$, and therefore σ is a $\mathfrak{G}\mathfrak{H}$-isomorphism and τ is an $\mathfrak{H}\mathfrak{J}$-isomorphism, Q.E.D.

If ω is an operator of $\mathfrak{G}$, then all the elements of $\mathfrak{G}$ for which $x^{\omega} = e$ form a normal subgroup $\mathfrak{n}_{\omega}$ of $\mathfrak{G}$. All the elements of $\mathfrak{G}$ for which $x^{\omega^m} = e$ is solvable likewise form a normal subgroup of $\mathfrak{G}$, which is denoted by $\mathfrak{N}_{\omega}$. $\mathfrak{N}_{\omega}$ is the union of all $\mathfrak{n}_{\omega^m}$. ω is a meromorphism of $\mathfrak{G}/\mathfrak{N}_{\omega}$, and every normal subgroup of $\mathfrak{G}$ for which ω induces a meromorphism in its factor group, contains $\mathfrak{N}_{\omega}$.

If the minimal chain condition holds in $\mathfrak{G}$, then ω is actually an automorphism of $\mathfrak{G}/\mathfrak{N}_{\omega}$. It follows from this that $\mathfrak{G} = \mathfrak{N}_{\omega} \cdot \mathfrak{G}^{\omega^m}$ for all m. If the maximal chain condition also holds in $\mathfrak{G}$, then the chain

[1] Following Fitting, *Math. Zeitschr. 39 (1934)*; one will find further bibliographical material there.

$\mathfrak{n}_\omega, \mathfrak{n}_{\omega^2}, \mathfrak{n}_{\omega^3}, \ldots$ terminates, and $\mathfrak{N}_\omega = \mathfrak{n}_{\omega^n}$ is solvable. The intersection of $\mathfrak{G}^{\omega^n}$ with $\mathfrak{N}_\omega$ is e, since an element x of this intersection satisfies the equation $x^{\omega^n} = e$ and is of the form $x = y^{\omega^n}$ but $y^{\omega^{2n}} = x^{\omega^n} = e$ implies $y^{\omega^n} = x = e$. If, conversely, $\mathfrak{G} = \mathfrak{N}_\omega \cdot \mathfrak{U}$, where $\mathfrak{U}$ is a subgroup of $\mathfrak{G}$ with the properties $\mathfrak{N}_\omega \cap \mathfrak{U} = e$, $\mathfrak{U}^\omega \subseteq \mathfrak{U}$, then ω induces a meromorphism of $\mathfrak{U}$, and from the minimal chain condition of $\mathfrak{G}/\mathfrak{N}_\omega$ it follows that $\mathfrak{U}^\omega = \mathfrak{U}$, and therefore $\mathfrak{U} = \mathfrak{U}^{\omega^n} = \mathfrak{N}^{\omega^n}\mathfrak{U}^{\omega^n} = \mathfrak{G}^{\omega^n}$. We specialize to the case for which the operator domain Ω of $\mathfrak{G}$ contains the inner automorphisms of $\mathfrak{G}$ whenever it contains ω, and we thus obtain

THEOREM 5.[1] *With a normal operator ω of a group in which the double chain theorem for normal subgroups is fulfilled is associated a direct decomposition $\mathfrak{G} = \mathfrak{N}_\omega \times \mathfrak{G}^{\omega^n}$. The second factor of the decomposition is uniquely determined by having $\mathfrak{N}_\omega$ as the first factor.*

Hereafter we shall assume that the double chain condition for the normal subgroups holds in $\mathfrak{G}$.

THEOREM 6: *If the sum of additive normal operators of a directly indecomposable $\mathfrak{G}$ is an automorphism, then the same is true of one of the summands.*

Proof: We can assume immediately that the sum contains only two summands.

If $\omega_1 + \omega_2 = \omega$ is an automorphism then $\omega^{-1}\omega_1 + \omega^{-1}\omega_2 = \underline{1}$, and if we prove the theorem for this sum, it follows for the other. Therefore let $\omega_1 + \omega_2 = \underline{1}$ be the sum of additive normal operators ω_1 and ω_2. From $\omega_1 = \omega_1(\omega_1 + \omega_2) = (\omega_1 + \omega_2)\omega_1$ it follows that $\omega_1\omega_2 = \omega_2\omega_1$. If we had both $\omega_1{}^n = \underline{0}$ and $\omega_2{}^n = \underline{0}$, then

$$\underline{1} = (\omega_1 + \omega_2)^{2n} = \sum_0^{2n} c_i \omega_1{}^i \omega_2{}^{2n-i} = \underline{0}$$

i.e., $\mathfrak{G} = e$. In this case the theorem is trivial. If $\mathfrak{G} \neq e$ then for at least one of the two operators ω_i, let us say ω_1, every power is different from zero. From Theorem 5 and the indecomposability of $\mathfrak{G}$, it follows that $\mathfrak{N}_{\omega_1} = e$. ω_1 is a meromorphism, and, because of the minimal chain condition for normal subgroups, it is an automorphism, Q.E.D.

DEFINITION: A direct decomposition of a group into directly indecomposable factors not equal to e is said to be a *Remak decomposition*. If the group is directly indecomposable, then it is itself the only factor of its Remak decomposition.

[1] This is known as Fittings Lemma. (See Jacobson, *Theory of Rings*.) (*Ed.*)

THEOREM 7: a) *Every group (satisfying the double chain condition for normal subgroups) has a Remak decomposition.*

b) *If*
$$\mathfrak{G} = \mathfrak{H}_1 \times \mathfrak{H}_2 \times \cdots \times \mathfrak{H}_n$$
and
$$\mathfrak{G} = \mathfrak{J}_1 \times \mathfrak{J}_2 \times \cdots \times \mathfrak{J}_m$$

are two Remak decompositions with decomposition operators $H_1, H_2, \ldots, H_n$ *and* $J_1, J_2, \ldots, J_m$, *respectively, then* $n=m$, *and the* $\mathfrak{J}_i$ *can be renumbered so that*
$$\omega = J_1H_1 + J_2H_2 + \cdots + J_nH_n$$

is a normal automorphism of $\mathfrak{G}$ *which maps the* $\mathfrak{H}$*-decomposition onto the* $\mathfrak{J}$*-decomposition.*

c) *For the appropriate ordering of the* $\mathfrak{J}_i$ *we have the exchange equations*
$$\mathfrak{G} = \mathfrak{J}_1 \times \mathfrak{J}_2 \times \cdots \times \mathfrak{J}_k \times \mathfrak{H}_{k+1} \times \mathfrak{H}_{k+2} \times \cdots \times \mathfrak{H}_n.$$

Proof: a) Among all the decompositions $\mathfrak{G} = \mathfrak{H}_1 \times \mathfrak{H}_2 \times \ldots \times \mathfrak{H}_n$ with indecomposable factors $\mathfrak{H}_1, \mathfrak{H}_2 \ldots, \mathfrak{H}_{n-1}$ different from e, $\mathfrak{H}_n \neq e$, and n any natural number, choose one with minimal $\mathfrak{H}_n$. Among all the decompositions $\mathfrak{H}_n = \mathfrak{A} \times \mathfrak{B}$ having $\mathfrak{A} \neq e$, choose one with minimal $\mathfrak{A}$. It follows that $\mathfrak{A}$ is indecomposable and not equal to e, $\mathfrak{B} \subset \mathfrak{H}_n$; hence $\mathfrak{B} = e$, $\mathfrak{H}_n = \mathfrak{A}$; and hence we have obtained a Remak decomposition.

b) We remark that the additivity of two operators ω_1 and ω_2 implies the additivity of $\omega_1 J_i$ and ω_2, and of $J_k\omega_1$ and ω_2. Therefore
$$\sum_1^m H_1 J_k = H_1 \cdot \sum_1^m J_k = H_1,$$

and, by Theorem 6, at least one of the operators H_1J_k induces an automorphism in $\mathfrak{H}_1$.

The J_k can be re-indexed so that it is H_1J_1 that induces an automorphism in $\mathfrak{H}_1$. By Theorem 4, J_1 induces an $\mathfrak{H}_1\mathfrak{J}_1$-isomorphism. By the remark made above,
$$\omega_1 = J_1H_1 + \sum_2^n H_i$$

is a normal operator. An equation $x^{\omega_1} = e$ implies $e = x^{H_1\omega_1} = x^{H_1J_1H_1}$, and since H_1J_1 induces an automorphism of $\mathfrak{H}_1$, we have $x^{H_1} = e$. But then $x^{\omega_1} = x^{\omega_1 H_2} \ldots x^{\omega_1 H_n} = x^{H_2} \cdot x^{H_3} \ldots x^{H_n} = e$, and therefore
$$x^{H_1} = e, \; x = e.$$

The normal operator ω_1 is a meromorphism, and by the double chain condition it is, in fact, an automorphism of $\mathfrak{G}$ which maps the $\mathfrak{H}$-decomposition onto the Remak decomposition

$$\mathfrak{G} = \mathfrak{J}_1 \times \mathfrak{H}_2 \times \mathfrak{H}_3 \times \cdots \times \mathfrak{H}_n .$$

If $n=1$, then the theorem is now complete. We apply induction with respect to n and assume $n>1$. $\mathfrak{G}/\mathfrak{J}_1$ has $\mathfrak{H}_2 \times \mathfrak{H}_3 \times \cdots \times \mathfrak{H}_n$ as well as $\mathfrak{J}_2 \times \mathfrak{J}_3 \times \cdots \times \mathfrak{J}_m$ as representative groups. Since the $\mathfrak{H}_i, \mathfrak{J}_i$, for $i>1$, remain indecomposable in $\mathfrak{G}/\mathfrak{J}_1$, by the induction hypothesis $n=m$, and the $\mathfrak{J}_i$ with $i>0$, can be reindexed so that $\overline{\omega} = \sum_2^n \overline{J}_i \overline{H}_i$ transforms the $\mathfrak{H}$-decomposition of $\mathfrak{G}/\mathfrak{J}_1$ into the $\mathfrak{J}$-decomposition. Here $\overline{H}_i$ and $\overline{J}_i$ are the decomposition operators in $\mathfrak{G}/\mathfrak{J}_1 \cong \mathfrak{H}_2 \times \mathfrak{H}_3 \times \cdots \times \mathfrak{H}_n$ and $\mathfrak{G}/\mathfrak{J}_1 \cong \mathfrak{J}_2 \times \mathfrak{J}_3 \times \cdots \times \mathfrak{J}_n$ respectively. From this it follows that $\sum_2^n J_i H_i$ maps the decomposition $\mathfrak{H}_2 \times \mathfrak{H}_3 \times \cdots \times \mathfrak{H}_n$ isomorphically onto the decomposition $\mathfrak{J}_2 \times \mathfrak{J}_3 \times \cdots \times \mathfrak{J}_n$, and hence $\omega = \sum_1^n J_i H_i$ is the normal automorphism of $\mathfrak{G}$ which was sought.

c) Moreover it also follows from the induction hypothesis that, after appropriate reindexing of the $\mathfrak{J}_i$ for $i>1$,

$$\mathfrak{J}_2 \times \mathfrak{J}_3 \times \cdots \times \mathfrak{J}_h \times \mathfrak{H}_{h+1} \times \cdots \times \mathfrak{H}_n$$

is a representative system of $\mathfrak{G}$ over $\mathfrak{J}_1$. Therefore the exchange equations

$$\mathfrak{G} = \mathfrak{J}_1 \times \mathfrak{J}_2 \times \cdots \times \mathfrak{J}_h \times \mathfrak{H}_{h+1} \times \cdots \times \mathfrak{H}_n .$$

follow.

THEOREM 8: *If* $\mathfrak{G} = \mathfrak{H}_1 \times \mathfrak{H}_2$, *then a homomorphism* σ *of* $\mathfrak{H}_1$ *onto* $\mathfrak{H}_2$ *is normal*[1] *if and only if* $\mathfrak{H}_1{}^\sigma$ *is in the center of* $\mathfrak{G}$.

Proof: 1. Let σ be normal, $a \in \mathfrak{H}_1$, $b \in \mathfrak{H}_2$. Then $a^{b\sigma} = a^{\sigma b} = a^\sigma$. a^σ is in the center of $\mathfrak{H}_2$ and therefore of $\mathfrak{G}$.

2. Let $\mathfrak{H}_1{}^\sigma$ is in the center of $\mathfrak{H}_2$. Then it follows that for all b in $\mathfrak{H}_2$: $a^{b\sigma} = a^{\sigma b}$. If b is in $\mathfrak{H}_1$, then $a^\sigma \cdot b^\sigma = b^\sigma \cdot a^\sigma$. Therefore $(a^b)^\sigma = a^\sigma$, $a^{\sigma b} = a^\sigma = (a^\sigma)^b = a^{b\sigma}$, and therefore σ is normal in $\mathfrak{G}$.

From Theorem 8, with the notation of Theorem 7, there follows:

THEOREM 9: *A non-abelian factor of the* $\mathfrak{H}$*-decomposition is normally isomorphic in* $\mathfrak{G}$ *to one and only one factor of the* $\mathfrak{J}$*-decomposition.*

Proof: If $\mathfrak{H}_1$ is normally isomorphic to $\mathfrak{J}_1, \mathfrak{J}_2$, then it is also normally isomorphic to $\mathfrak{H}_2$, and by Theorem 8, $\mathfrak{H}_1$ is then abelian.

---

[1] The mapping σ of $\mathfrak{H}_1$ onto $\mathfrak{H}_2$ is said to be normal in $\mathfrak{G}$ if $a^{b\sigma} = a^{\sigma b}$ for all $a \in \mathfrak{H}_1$, $b \in \mathfrak{G}$.

THEOREM 10: *The Remak decomposition is uniquely determined if and only if all of its factors are invariant under normal operators of* $\mathfrak{G}$.

Proof: We need only prove the second part of the statement. Therefore, let ω be a normal operator of $\mathfrak{G}$ which does not transform the factor $\mathfrak{H}_1$ of the Remak decomposition $\mathfrak{G} = \mathfrak{H}_1 \times \mathfrak{H}_2 \times \cdots \times \mathfrak{H}_n$ into itself. ω is the sum of the normal operators $\omega_{ik} = H_i \omega H_k$. By Theorem 8 the operators ω_{ik} with $i \neq k$ are central operators.

If, for all $i > 1$, $\omega_{i1} = \underline{0}$, then we would have

$$\mathfrak{H}_1^{\omega} = \mathfrak{H}_1^{\Sigma \omega_{ik}} = \mathfrak{H}_1^{\Sigma \omega_{i1}} = \mathfrak{H}_1^{\omega_{11}} \subseteq \mathfrak{H}_1,$$

and therefore there would be an $i > 1$, say $i = 2$, such that $\omega_{21} \neq \underline{0}$. Since the operator ω_{21} is central, $\pi = \omega_{21} + \underline{1}$ is an operator. It does not map $\mathfrak{H}_1$ into itself.

Since $\omega_{21}{}^2 = \underline{0}$, then

$$(\omega_{21} + \underline{1}) \cdot (-\omega_{21} + \underline{1}) = \underline{1}, \quad (-\omega_{21} + \underline{1}) \cdot (\omega_{21} + \underline{1}) = \underline{1},$$

and therefore π is an automorphism of $\mathfrak{G}$. π maps the original Remak decomposition onto a different Remak decomposition

$$\mathfrak{G} = \mathfrak{H}_1^{\pi} \times \mathfrak{H}_2 \times \cdots \times \mathfrak{H}_n \quad ; \qquad \text{Q.E.D.}$$

THEOREM 11: *If* $\mathfrak{G} = \mathfrak{H}_1 \times \mathfrak{H}_2$, *then* $\mathfrak{H}_1$ *is invariant under normal operators of* $\mathfrak{G}$ *if and only if there exists no* $\mathfrak{H}_1\mathfrak{H}_2$*-homomorphy normal in* $\mathfrak{G}$ *other than the trivial one*: $\mathfrak{H}_1 \to e$.

Proof: 1. Let ω be a non-trivial normal $\mathfrak{H}_1\mathfrak{H}_2$-homomorphism. Then ωH_1 is a normal operator of $\mathfrak{G}$ which does not map $\mathfrak{H}_1$ onto itself.

2. If $\bar{\omega}$ is a normal operator of $\mathfrak{G}$ which does not map $\mathfrak{H}_1$ onto itself, then $H_2 \bar{\omega}$ induces a non-trivial normal $\mathfrak{H}_1\mathfrak{H}_2$-homomorphy, Q.E.D.

By Theorem 8, a normal $\mathfrak{H}_1\mathfrak{H}_2$-homomorphy is characterized by an abelian factor group of $\mathfrak{H}_1$ and a subgroup of the center of $\mathfrak{H}_2$ isomorphic to it.

From the previous theorems we derive

THEOREM 12: *The Remak decomposition* $\mathfrak{G} = \mathfrak{H}_1 \times \mathfrak{H}_2 \times \cdots \times \mathfrak{H}_n$ *is uniquely determined if and only if an abelian factor group of* $\mathfrak{H}_i$ *is isomorphic to no subgroup of the center of* $\mathfrak{H}_k$, $i \neq k$, *different from e.*

THEOREM 13 (Speiser): *A group whose factor commutator group or center is of order* 1 *has exactly one Remak decomposition.*

THEOREM 14: *The Remak decomposition of a finite group*

$$\mathfrak{G} = \mathfrak{H}_1 \times \mathfrak{H}_2 \times \cdots \times \mathfrak{H}_n$$

is uniquely determined if and only if the order of the factor commutator group of each of its factors is relatively prime to the order of the center of each of its other factors.

Proof: By Theorem 8 and Theorem 11 the condition is sufficient. If, however, the prime number p divides $\mathfrak{H}_1 : \mathfrak{H}_1'$ and $\mathfrak{z}(\mathfrak{H}_2) : 1$ then it follows from the basis theorem for abelian groups (which is proven in § 4) that $\mathfrak{H}_1$ has a normal subgroup of index p, and that $\mathfrak{z}(\mathfrak{H}_2)$ contains a subgroup of order p. Since the factor group and the subgroup are isomorphic, there is a non-trivial normal $\mathfrak{H}_1\mathfrak{H}_2$-homomorphy; by Theorem 11 and Theorem 10 the Remak decomposition of $\mathfrak{G}$ is therefore not uniquely determined.

§ 3. Abelian Groups

Let P be a ring with unity element and let $\mathfrak{A}$ be a finite P-module. Thus there are a finite number of elements $v_1, v_2, \ldots, v_n$ in $\mathfrak{A}$ such that every element v in $\mathfrak{A}$ is of the form.

$$v = a_1 v_1 + a_2 v_2 + \cdots + a_n v_n$$

with the a_i in P. Let $\mathfrak{M}_n$ be the P-vector module $(u_1, \ldots, u_n)$.

The mapping

$$a_1 u_1 + a_2 u_2 + \cdots + a_n u_n \rightarrow a_1 v_1 + a_2 v_2 + \cdots + a_n v_n$$

is an operator homomorphy of $\mathfrak{M}_n$ onto $\mathfrak{A}$, and the abelian group $\mathfrak{A}$ with the operator domain P is completely determined by the P-module $\mathfrak{R}$ consisting of all vectors which are mapped onto the zero element of $\mathfrak{A}$:

$$\mathfrak{A} \simeq \mathfrak{M}_n / \mathfrak{R}.$$

With every system of generators $R_1, R_2, \ldots, R_\omega$ of $\mathfrak{R}$ over P we associate the matrix $A_{\mathfrak{R}}$ whose row vectors are precisely the vectors R_i. A matrix $A_{\mathfrak{R}} = (a_{ik})$ is characterized by the properties

1. $$\sum_{k=1}^{n} a_{ik} v_k = 0 \qquad (i = 1, 2, \ldots),$$

2. if $$\sum b_k v_k = 0,$$

then $b_k = \sum_i a_{ik} x_i$ is solvable in P where the summation is over only a finite number of i.

Since each row of $A_{\mathfrak{R}}$ corresponds to a relation valid in $\mathfrak{A}$, we say that $A_{\mathfrak{R}}$ is a *relation matrix* belonging to $\mathfrak{A}$.

Conversely the row vectors of a matrix A with n columns generate a P-module $\mathfrak{R}$ in $\mathfrak{M}_n$ such that A is a relation matrix belonging to $\mathfrak{M}_n / \mathfrak{R}$.

When do two relation matrices belong to the same finite P-module $\mathfrak{A}$?

We say that two matrices with coefficients in P are *equivalent* if they are relation matrices of the same finite P-module $\mathfrak{A}$.

This equivalence has the three familiar properties.

A rule which uniquely assigns to each matrix an equivalent matrix is called an *elementary transformation.*

We want to find the simplest elementary transformations which by repeated application will transform any two equivalent matrices into one-another.

The zero vector can be adjoined to the generators of $\mathfrak{R}$:

N_0: The elementary transformation N_0 adjoins a row of zeros to the matrix A:

$$A = (a_{ik}) \to \begin{pmatrix} 0, & \ldots, & 0 \\ a_{11}, & \ldots, & a_{1n} \\ \ldots & \ldots & \ldots \end{pmatrix}.$$

N_0': Delete the first row if at least two rows occur and if the first row is a series of zeros.

The generator R_i may be replaced by $R_i + aR_j$ where $i \neq j$:

$T^a_{i,j}$: The elementary transformation $T^a_{i,j}$ replaces the i-th row of (a_{ik}) by $(a_{i1} + a a_{j1},\ a_{i2} + a a_{j2}, \ldots,\ a_{in} + a a_{jn})$, where $i \neq j$.

We have
$$T^a_{i,j}\, T^b_{i,j}\, A = T^{a+b}_{i,j}\, A.$$

T: By repeated application of $T^a_{i,j}$ an arbitrary linear combination of the other rows can be added to the i-th row.

Moreover we can obtain, by composition of the $T^a_{i,j}$:

$(i \neq j)$ $V_{i,j}$: Exchanges the i-th with the j-th row and changes the sign in the j-th row:

$$V_{i,j}\, A = T_{i,j}\, T^{-1}_{j,i}\, T_{i,j}\, A.$$

N: By applying N_0 and T an arbitrary linear combination of rows can be adjoined to a matrix.

$M_{i,\varepsilon}$: Moreover by repeated application of $T^a_{i,j}$ and N_0, N_0', the i-th row of A can be multiplied by a unit ε of P:

$$M_{i,\varepsilon}\, A = N_0'\, T^{-\varepsilon^{-1}}_{1,i+1}\, V_{1,i+1}\, T^{-\varepsilon}_{1,i+1}\, N_0\, A.$$

$M_{j,-1}\, V_{i,j}$ exchanges the i-th and j-th rows.

If we adjoin $v_0 = a_1 v_1 + a_2 v_2 + \cdots + a_n v_n$ to the generators $v_1, \ldots, v_n$ then we obtain a single new relation:

$$v_0 - a_1 v_1 - a_2 v_2 - \cdots - a_n v_n = 0.$$

S: The elementary transformation S adjoins to the matrix $A = (a_{ik})$

the row $(1, -a_1, -a_2, \ldots, -a_n)$ and the column $\begin{pmatrix} 1 \\ 0 \\ 0 \\ \vdots \end{pmatrix}$:

$$A \rightarrow \begin{pmatrix} 1 & -a_1, & \ldots, & & -a_n \\ 0 & a_{11}, & a_{12}, & \ldots, & a_{1n} \\ \vdots & a_{21}, & \ldots & & \end{pmatrix}$$ (bordering of a matrix).

S': Changes to an unbordered matrix if the first column is of the form $\begin{pmatrix} 1 \\ 0 \\ 0 \\ \vdots \end{pmatrix}$.

$R_{i,j}$: Exchange of the i-th with the j-th generator of $\mathfrak{A}$ induces an exchange of the i-th with the j-th column in A.

From the elementary transformations found up to now we can form

$(k \neq j)$ $S^a_{k,j}$: Adds the j-th column multiplied by a to the k-th column:

Let S denote bordering with the vector which has 1 at the zero-th, –1 at the j-th and a at the k-th place, and has zeros elsewhere. In SA the first row is multiplied by a_{ij} and added to the $(i+1)$-st row, $(i=1, 2, \ldots)$. After this the first and $(j+1)$-st columns are interchanged, the first line is multiplied by –1 and, finally, the border removed.

THEOREM 15: *By repeated application of the elementary transformations*

$$N_0, N_0', T^a_{i,k}, S, S', R_{i,k},$$

a matrix can be transformed into any matrix equivalent to it.

Proof: Let the matrices A, B be equivalent. Then there is a finite P-module $\mathfrak{A}$ with two systems of generators

$$S_1, S_2, \ldots, S_n$$

and

$$T_1, T_2, \ldots, T_{n'},$$

such that A is the relation matrix of the S_i and B is the relation matrix of the T_i.

After n' applications of S, A goes into a relation matrix belonging to the system of generators $T_1, T_2, \ldots, T_{n'}, S_1, S_2, \ldots, S_n$. After n applications of S and repeated application of $R_{i,k}$, B goes into a relation matrix belonging to $T_1, T_2, \ldots, T_{n'}, S_1, S_2, \ldots, S_n$.

Since the inverse of every transformation in the set of the six types of elementary transformations is also in the set, we need only carry out the proof for $S_i = T_i$, $n = n'$. By repeated application of N, A goes into the matrix $\binom{B}{A}$ and B goes into the matrix $\binom{A}{B}$. The new matrices can be obtained from each other by row permutations and thus the theorem is proven.

Are there easily calculated invariants under elementary transformation?

DEFINITION: The ideal generated by all the $(n-r)$-rowed subdeterminants of a matrix A with n columns is called the *r-th elementary ideal* $\mathfrak{E}_r(A)$. From the expansion theorem for determinants it follows that $\mathfrak{E}_r(A) \subseteq \mathfrak{E}_{r-1}(A)$. We set $\mathfrak{E}_n(A) = \mathfrak{E}_{n+1}(A) = \ldots = \mathrm{P}$. $\mathfrak{E}_0, \mathfrak{E}_1$, is called the *sequence of elementary ideals of A.*

THEOREM 17: *The sequence of elementary ideals is invariant under elementary transformation.*

By Theorem 15, it suffices to show the equality of the sequence of elementary ideals in the four cases

a) $B = N_0 A$, b) $B = T^a_{i,k} A$, c) $B = SA$, d) $B = R_{i,k} A$.

In case a): except for the elements of $\mathfrak{E}_r(A)$, $\mathfrak{E}_r(B)$ contains only determinants with a zero row, and therefore $\mathfrak{E}_r(B) = \mathfrak{E}_r(A)$.

In case b), the subdeterminants of B in which the i-th row takes no part have the same value as the corresponding subdeterminants in A. The same holds true, by the properties of determinants, for the subdeterminants in which both the i-th and the k-th rows take part.

Let $D(\bar{\mathfrak{b}}_{\nu_1}, \bar{\mathfrak{b}}_{\nu_2}, \ldots, \bar{\mathfrak{b}}_{\nu_{n-r}})$ be a subdeterminant in which the row vectors $\mathfrak{b}_{\nu_1}, \mathfrak{b}_{\nu_2}, \ldots, \mathfrak{b}_{\nu_{n-r}}$ are involved, among them $\mathfrak{b}_i = \mathfrak{a}_i + a\,\mathfrak{a}_k$, while the index k does not occur among the ν_j. Then

$$D(\bar{\mathfrak{b}}_{\nu_1}, \bar{\mathfrak{b}}_{\nu_2}, \ldots, \bar{\mathfrak{b}}_{\nu_{n-r}}) = D(\bar{\mathfrak{a}}_{\nu_1}, \bar{\mathfrak{a}}_{\nu_2}, \ldots, \bar{\mathfrak{a}}_{\nu_{n-r}}) + a D(\bar{\mathfrak{a}}_{\nu_1}, \ldots, \overset{i}{\bar{\mathfrak{a}}_k}, \ldots, \bar{\mathfrak{a}}_{\nu_{n-r}}).$$

From this we conclude $\mathfrak{E}_r(B) \subseteq \mathfrak{E}_r(A)$. Since conversely $A = T^{-a}_{i,k} B$, it follows that $\mathfrak{E}_r(A) \subseteq \mathfrak{E}_r(B)$, therefore $\mathfrak{E}_r(B) = \mathfrak{E}_r(A)$.

In case c), $\mathfrak{E}_r(B) = \mathfrak{E}_r(A)$.

In case d), $\mathfrak{E}_r(B) = \mathfrak{E}_r(A)$.

A matrix is said to be a *diagonal matrix* if all the elements not on the principal diagonal are zero. Then a relation matrix A can be transformed into a diagonal matrix

$$\begin{pmatrix} d_1 & & & \\ & d_2 & & \\ & & \ddots & \\ & & & d_r \end{pmatrix}$$

by means of elementary transformations precisely when $\mathfrak{A}$ is operator isomorphic to the direct sum of residue class rings

$$\mathrm{P}/(d_1) + \mathrm{P}/(d_2) + \ldots + \mathrm{P}/(d_r).$$

The diagonal matrix A is in *elementary divisor form* if the diagonal elements form a chain of divisors such that $d_1/d_2/d_3/ \ldots /d_r$ and d_1 is not a unit of P. Making use of the elementary divisor form, we may write the sequence of elementary ideals as follows:

$$\mathfrak{E}_0(A) = (d_1 d_2 \ldots d_r), \ldots, \mathfrak{E}_{r-1}(A) = (d_1),\ \mathfrak{E}_r(A) = \mathfrak{E}_{r+1}(A) = \ldots = \mathrm{P}.$$

If the d_i are not zero or divisors of zero, then A is equivalent to a second elementary divisor form with the diagonal elements $d_1', d_2', \ldots, d_s'$ if and only if $r=s$ and $(d_i) = (d_i')$, i.e., if and only if the diagonal elements in the same place differ only by a unit of P.

§ 4. Basis Theorem for Abelian Groups

THEOREM 17: *A cyclic group $\mathfrak{G}$ whose order N is the product of pairwise relatively prime numbers $n_1, n_2, \ldots, n_r$, can be decomposed in one and only one way into a direct product of cyclic groups of orders $n_1, n_2, \ldots, n_r$.*

Proof: The uniqueness is clear since in a cyclic group only one subgroup exists having a given order.

We set $m_i = \frac{N}{n_i}$ and seek the i-th decomposition operator. By hypothesis the congruence $x_i m_i \equiv 1 \pmod{n_i}$ is solvable. We define the operator H_i by means of the condition: $a^{H_i} = a^{x_i m_i}$ for all a in $\mathfrak{G}$. We obtain directly the equations:

$$\sum H_i = \underline{1},\ H_i H_k = 0\ (i \neq k),\ H_i^2 = H_i.$$

Therefore $\mathfrak{G} = \mathfrak{G}^{H_1} \times \mathfrak{G}^{H_2} \times \cdots \times \mathfrak{G}^{H_r}$. Since $n_i H_i = 0$ and $\mathfrak{G}$ is cyclic, the order of $\mathfrak{G}^{H_i}$ is a divisor of n_i, and so, because of the order relation, $\mathfrak{G}^{H_i}$ has the order n_i, Q.E.D.

THEOREM 18: *An abelian group $\neq e$ with a finite number of generators is the direct product of cyclic groups having prime power order or having order zero. The multiplicity of each basis order is uniquely determined.*

Proof: In order to prove the first part of the theorem, it suffices by Theorem 17 to prove decomposability into cyclic factors. Our operator ring P is the ring of integers, and our aim is to put the relation matrix A belonging to the abelian group $\mathfrak{A}$ into diagonal form by means of elementary transformations.

We apply the following reductions:

I. If A is the zero matrix, then A is already in diagonal form.

If $A \neq 0$, then among those integers a_{ik} which are different from zero there is one which is smallest in absolute value. By appropriate row and column interchange we may take a_{11} to be this number.

a) If $a_{1k} \neq 0$, $k > 1$, then $a_{1k} = qa_{11} + r$ where q and r belong to P and

$$0 \leqq r < |a_{11}|.$$

Multiplying the first column by q and subtracting it from the k-th column we obtain $b_{1k} = a_{1k} - qa_{11} = r$ which is smaller than a_{11} in absolute value.

b) If $a_{i1} \neq 0$, $i > 1$, $a_{i1} = qa_{11} + r$, where q and r belong to P, $0 \leqq r < |a_{11}|$, then subtract q times the first row from i-th row.

After a finite number of reductions of type I.a) or I.b), we find that all the elements of the first row and first column are divisible by a_{11}.

If A has n columns, then after at most $(n-1)$ reductions by I.a) we find that the first row is of the form $(a_{11}, 0, \ldots, 0)$. The first column remains unchanged by this reduction, and we may now replace it by $\begin{pmatrix} a_{11} \\ 0 \\ 0 \\ \vdots \end{pmatrix}$. (Here reduction I.b) may have to be applied infinitely often).

The unbordered matrix which is obtained from A by deleting the first row and first column has only $n-1$ columns. We apply the reduction process described above to it; here the bordered matrix is transformed without changing the first row and column. After at most n such reductions A goes into diagonal form.

In order to prove the uniqueness we assume that $S_1, S_2, \ldots, S_n$ is a basis of $\mathfrak{A} \neq e$ such that $S_1, S_2, \ldots, S_r$ are of prime power order, and that the remaining basis elements are of order zero.

Let $S_1, S_2, \ldots, S_s$ be all the basis elements whose order is a power of the natural prime p, so that S_i is of order p^{n_i} with $0 < n_1 \leqq n_2 \cdots \leqq n_s$. All elements of $\mathfrak{A}$ whose order is a power of p form a subgroup $\mathfrak{S}_p$ with $S_1, S_2, \ldots, S_s$ as basis ($p^k \cdot \sum a_i S_i = \sum p^k a_i S_i = 0$ implies $a_i = 0$, if

$i>s$). The relation matrix of $\mathfrak{S}_p$ with respect to this system of generators has the elementary divisor form

$$\begin{pmatrix} p^{n_1} & & & \\ & p^{n_2} & & \\ & & \ddots & \\ & & & p^{n_s} \end{pmatrix}.$$

By § 3 the numbers in the diagonal are uniquely determined to within sign.

The n_i are uniquely determined.

$S_1, S_2, \ldots, S_r$ generate the subgroup $\mathfrak{U}$ of all elements with positive order in $\mathfrak{A}$. The factor group $\mathfrak{A}/\mathfrak{U}$ has $S_{r+1}, \ldots, S_n$ as basis with the basis orders 0, 0, . . . , 0. Since the associated relation matrix, being a null matrix, is in the elementary divisor form, the number $n-r$ is uniquely determined, Q.E.D.

§ 5. The Order Ideal

The following investigations are closely related to § 3.

DEFINITION: The first member in the sequence of elementary ideals of a matrix A is said to be the *order ideal* $\mathfrak{O}_A$ of A. The order ideal is generated by all the subdeterminants of greatest order of A.

The order ideal is invariant under elementary transformations.

It follows from the basis theorem that the order ideal of an ordinary abelian group with a finite number of generators is generated by the group order (where P is the ring of rational integers).

THEOREM 19 (Analogous to the Fermat Theorem of group theory):
For all operators D of the order ideal $\mathfrak{O}_A$ and all elements v in $\mathfrak{A}$,

$$Dv=0.$$

Proof: It suffices to assume that D is an n-rowed subdeterminant of a relation matrix A with n columns, and that v is one of the corresponding generators v_i of $\mathfrak{A}$. After appropriate renumbering of the rows of A, let $D=|a_{ik}|$ $(i, k=1, \ldots, n)$.

Then

$$\sum_{k=1}^{n} a_{ik} v_k = 0,$$

and therefore

$$\sum_{k=1}^{n} a_{ik} A_{ih} v_k = 0 \qquad (k=1,2\ldots n),$$

where A_{ih} is the algebraic complement of a_{ih} in D. After summation over these n equations, we get

$$\sum_{k=1}^{n}\left(\sum_{i=1}^{n} a_{ik} A_{ih}\right) v_k = 0 .$$

As is well known,the inner sum has the value $\delta_{kh} D$, and therefore

$$Dv_k = 0, \qquad \text{Q.E.D.}$$

§ 6. Extension Theory

The extension problem posed and solved by Otto Schreier reads: *Given two abstract groups* $\mathfrak{N}$ and $\mathfrak{F}$, *find all groups* $\mathfrak{G}$ *which contain* $\mathfrak{N}$ *as a normal subgroup, such that*

(1) $$\mathfrak{G}/\mathfrak{N} \simeq \mathfrak{F}.$$

First we shall investigate the groups $\mathfrak{G}$ which contain $\mathfrak{N}$ as a normal subgroup with factor group isomorphic to $\mathfrak{F}$. The elements in $\mathfrak{F}$ are designated by $1, \sigma, \tau \ldots$.

There is a decomposition $\mathfrak{G} = \sum_{\sigma \in \mathfrak{F}} \mathfrak{N} S_\sigma$ of $\mathfrak{G}$ into cosets with respect to $\mathfrak{N}$ such that $\mathfrak{N}S_\sigma \cdot \mathfrak{N}S_\tau = \mathfrak{N}S_{\sigma\tau}$, and therefore

(2) $$S_\sigma S_\tau = C_{\sigma,\tau} S_{\sigma\tau}, \quad C_{\sigma,\tau} \in \mathfrak{N}.$$

Since $\mathfrak{N}$ is a normal subgroup of $\mathfrak{G}$, for the elements $E, A, B, C, \ldots$ in $\mathfrak{N}$:

(3) $$S_\sigma A S_\sigma^{-1} = A^{S_\sigma} \in \mathfrak{N},$$

(4) $$(AB)^{S_\sigma} = A^{S_\sigma} B^{S_\sigma},$$

(5) $$(A^{S_\tau})^{S_\sigma} = A^{S_\sigma S_\tau} = A^{C_{\sigma,\tau} S_{\sigma\tau}} = (A^{S_{\sigma\tau}})^{C_{\sigma,\tau}},$$

(6) $$S_1 S_1 = C_{1,1} S_1, \text{ hence } S_1 = C_{1,1},$$

(7) $$A^{S_1} = A^{C_{1,1}},$$

(8) $$AS_\sigma \cdot BS_\tau = AB^{S_\sigma} C_{\sigma,\tau} S_{\sigma\tau}.$$

From the associative law in $\mathfrak{G}$ it follows that

$$\begin{aligned}(S_\sigma S_\tau) S_\varrho = (C_{\sigma,\tau} S_{\sigma\tau}) S_\varrho &= C_{\sigma,\tau} C_{\sigma\tau,\varrho} S_{\sigma\tau\varrho} \\ &= S_\sigma (S_\tau S_\varrho) = S_\sigma C_{\tau,\varrho} S_{\tau\varrho} = C_{\tau,\varrho}^{S_\sigma} C_{\sigma,\tau\varrho} S_{\sigma\tau\varrho},\end{aligned}$$

and from this follow the *associativity relations*

(9) $$C_{\sigma,\tau} C_{\sigma\tau,\varrho} = C_{\tau,\varrho}^{S_\sigma} C_{\sigma,\tau\varrho}.$$

Conversely, let single-valued mappings S_σ of $\mathfrak{N}$ onto itself be given and let a system of elements $C_{\sigma,\tau}$ in $\mathfrak{N}$ be given also, so that

I. $(AB)^{S_\sigma} = A^{S_\sigma} B^{S_\sigma}$,

II. $(A^{S_\tau})^{S_\sigma} = A^{S_\sigma S_\tau} = (A^{S_{\sigma\tau}})^{C_{\sigma,\tau}} = A^{C_{\sigma,\tau} S_{\sigma\tau}}, \quad A^{S_1} = A^{C_{1,1}}$,

III. $C_{\sigma,\tau} C_{\sigma\tau,\varrho} = C_{\tau,\varrho}^{S_\sigma} C_{\sigma,\tau\varrho}$.

A system of elements $C_{\sigma,\tau}$ which occurs in a solution of these three equations is said to be a *factor system in* $\mathfrak{N}$ *belonging to* $\mathfrak{F}$.

A group will be constructed which contains $\mathfrak{N}$ as a normal subgroup and whose factor group is isomorphic to $\mathfrak{F}$ such that the multiplication of the representatives of $\mathfrak{G}$ over $\mathfrak{N}$ follows rule (8).

Let $\mathfrak{G}$ be the set of all symbols AS_σ with A in $\mathfrak{N}$; $AS_\sigma = BS_\tau$ if and only if $A = B$, $\sigma = \tau$. We define multiplication in $\mathfrak{G}$ by

$$AS_\sigma \cdot BS_\tau = AB^{S_\sigma} C_{\sigma,\tau} S_{\sigma\tau}.$$

In $\mathfrak{G}$ the associative law is valid:

$$\begin{aligned}(AS_\sigma \cdot BS_\tau)\cdot CS_\varrho &= AB^{S_\sigma} C_{\sigma,\tau} S_{\sigma\tau} \cdot CS_\varrho\\ &= AB^{S_\sigma} C_{\sigma,\tau} C^{S_{\sigma\tau}} C_{\sigma\tau,\varrho} S_{\sigma\tau\varrho}\\ &= AB^{S_\sigma} C^{C_{\sigma,\tau} S_{\sigma\tau}} C_{\sigma,\tau} C_{\sigma\tau,\varrho} S_{\sigma\tau\varrho}\\ &= AB^{S_\sigma} C^{S_\sigma S_\tau} C_{\tau,\varrho}^{S_\sigma} C_{\sigma,\tau\varrho} S_{\sigma\tau\varrho} && \text{(by II. and III.)}\\ &= A(BC^{S_\tau} C_{\tau,\varrho})^{S_\sigma} C_{\sigma,\tau\varrho} S_{\sigma\tau\varrho} && \text{(by I.)}\\ &= AS_\sigma \cdot BC^{S_\tau} C_{\tau,\varrho} S_{\tau\varrho}\\ &= AS_\sigma \cdot (BS_\tau \cdot CS_\varrho).\end{aligned}$$

If we set $\sigma = \tau = 1$ in III, then

$$C_{1,1} C_{1,\varrho} = C_{1,\varrho}^{S_1} C_{1,\varrho}, \quad C_{1,1} = C_{1,\varrho}^{C_{1,1}}, \text{ therefore}$$

(10)
$$C_{1,\varrho} = C_{1,1}.$$

If on the other hand we set $\tau = \varrho = 1$ in III, then

$$C_{\sigma,1} C_{\sigma,1} = C_{1,1}^{S_\sigma} C_{\sigma,1}, \quad \text{therefore}$$

(11)
$$C_{\sigma,1} = C_{1,1}^{S_\sigma}.$$

$e = C_{1,1}^{-1} S_1$ is left identity of $\mathfrak{G}$, since

$$e \cdot AS_\varrho = C_{1,1}^{-1} A^{C_{1,1}} C_{1,\varrho} S_\varrho = AS_\varrho .$$

As solution of $$XS_\sigma \cdot BS_\tau = e,$$

that is, of
$$\begin{cases} XB^{S_\sigma} C_{\sigma,\tau} = C_{1,1}^{-1} \\ \sigma\tau = 1, \end{cases}$$

we find
$$X = C_{1,1}^{-1} C_{\sigma,\tau}^{-1} B^{-S_\sigma}$$
$$\sigma = \tau^{-1}.$$

$\mathfrak{G}$ is a group.

We set $$\bar{A} = A C_{1,1}^{-1} S_1 .$$

Then
$$\begin{aligned} \bar{A}\bar{B} &= A C_{1,1}^{-1} S_1 \cdot B C_{1,1}^{-1} S_1 = A C_{1,1}^{-1} (B C_{1,1}^{-1})^{S_1} \cdot C_{1,1} S_1 \\ &= A B C_{1,1}^{-1} S_1 = \overline{AB}. \end{aligned}$$

Now $\bar{A} = e$ implies $A = 1$.

Therefore $A \to \bar{A}$ is an isomorphic mapping of $\mathfrak{N}$ onto a subgroup $\bar{\mathfrak{N}}$ of $\mathfrak{G}$.

Set $\bar{S}_\sigma = 1 S_\sigma$. Then

$$\begin{aligned} \bar{A} \cdot \bar{S}_\sigma &= A C_{1,1}^{-1} S_1 \cdot 1 S_\sigma = A C_{1,1}^{-1} C_{1,\sigma} S_\sigma = AS_\sigma \qquad \text{by (10)}, \\ \bar{S}_\sigma \cdot \bar{S}_\tau &= 1 S_\sigma \cdot 1 S_\tau = C_{\sigma,\tau} S_{\sigma\tau} = \bar{C}_{\sigma,\tau} \cdot \bar{S}_{\sigma\tau}, \\ \bar{S}_\sigma \cdot \bar{A} &= 1 S_\sigma A C_{1,1}^{-1} S_1 = A^{S_\sigma} C_{1,1}^{-S_\sigma} C_{\sigma,1} S_\sigma = A^{S_\sigma} S_\sigma \\ &= \overline{A^{S_\sigma}} \cdot \bar{S}_\sigma , \end{aligned}$$

therefore $$\bar{A}^{\bar{S}_\sigma} = \overline{A^{S_\sigma}}.$$

$\bar{\mathfrak{N}}$ is a normal subgroup of $\mathfrak{G}$ with the $\bar{S}_\sigma$ as a system of representatives, the $\bar{C}_{\sigma,\tau}$ as factor system, and the $\bar{S}_\sigma$ as automorphisms, and the mapping $A \to \bar{A}$ is an operator isomorphism between $\mathfrak{N}$ and $\bar{\mathfrak{N}}$. Thus the problem stated above is completely solved.

Let $\mathfrak{G}$ and $\bar{\mathfrak{G}}$ be two extensions of $\mathfrak{N}$ which belong to the same factor system $C_{\sigma,\tau}$ and the same automorphism set S_σ. The elements in $\mathfrak{G}$ are uniquely of the form AS_σ, those in $\bar{\mathfrak{G}}$ uniquely of the form $A\bar{S}_\sigma$, where

$$A S_\sigma B S_\tau = A B^{S_\sigma} C_{\sigma,\tau} S_{\sigma\tau}$$
$$A \bar{S}_\sigma B \bar{S}_\tau = A B^{S_\sigma} C_{\sigma,\tau} \bar{S}_{\sigma\tau}.$$

Then the identity automorphism of $\mathfrak{N}$ can be extended, by means of the mapping $A S_\sigma \rightarrow A \bar{S}_\sigma$, to an isomorphism between $\mathfrak{G}$ and $\bar{\mathfrak{G}}$. We say more briefly: $\mathfrak{G}$ and $\bar{\mathfrak{G}}$ are $\mathfrak{N}$-isomorphic.

From these investigations we conclude:

THEOREM 20: *To each extension $\mathfrak{G}$ of a normal subgroup $\mathfrak{N}$ with given factor group $\mathfrak{F}$ there belongs a factor system and a set of automorphisms of $\mathfrak{N}$ such that conditions I, II, III are fulfilled.*

Conversely, to a given factor system and a given set of automorphisms of $\mathfrak{N}$, which fulfill I, II, III, there belongs an extension of $\mathfrak{N}$, unique to within isomorphy over $\mathfrak{N}$.

If instead of choosing S_σ as a representative of $S_\sigma \mathfrak{N}$ we choose $T_\sigma = A_\sigma S_\sigma$ with A_σ in $\mathfrak{N}$, then the automorphism S_σ of $\mathfrak{N}$ is replaced by $A_\sigma S_\sigma = T_\sigma$ and $C_{\sigma,\tau}$ is replaced by $A_\sigma A_\tau^{S_\sigma} C_{\sigma,\tau} A_{\sigma\tau}^{-1}$, since

$$T_\sigma T_\tau = A_\sigma S_\sigma A_\tau S_\tau = A_\sigma A_\tau^{S_\sigma} C_{\sigma,\tau} S_{\sigma\tau} = A_\sigma A_\tau^{S_\sigma} C_{\sigma,\tau} A_{\sigma\tau}^{-1} T_{\sigma\tau}.$$

The converse is clear.

DEFINITION: Two factor systems $(S_\sigma, C_{\sigma,\tau})$, $(T_\sigma, D_{\sigma,\tau})$ are said to be equivalent if there are elements A_σ such that

$$A^{T_\sigma} = A^{A_\sigma S_\sigma} \text{ for all } A \subseteq \mathfrak{N}$$

$$D_{\sigma,\tau} = A_\sigma A_\tau^{S_\sigma} C_{\sigma,\tau} A_{\sigma\tau}^{-1} .$$

We then write $$(T_\sigma, D_{\sigma,\tau}) \sim (S_\sigma, C_{\sigma,\tau}).$$

For this equivalence the three rules are valid.

Two factor systems with sets of automorphisms induce extensions which are isomorphic over $\mathfrak{N}$ and for which the coset R_σ maps onto the coset $\bar{R}_\sigma$, if and only if the factor systems are equivalent.

There always exists at least one factor system, namely that belonging to the direct product $\mathfrak{F} \times \mathfrak{N}$:

$$C_{\sigma,\tau} = 1,$$
$$A^{S_\sigma} = A.$$

A factor system is equivalent to $(T_\sigma, 1)$ if and only if

$$1 = A_\sigma A_\tau^{S_\sigma} C_{\sigma,\tau} A_{\sigma\tau}^{-1}$$

is solvable.

An equivalent condition is: In some associated extension, a subgroup can be found which is a system of representatives with respect to $\mathfrak{N}$.

Then $\mathfrak{G}$ decomposes into a product of $\mathfrak{F}$ and $\mathfrak{N}$ where $\mathfrak{F} \cap \mathfrak{N} = e$. Therefore we say that a factor system equivalent to $(T_\sigma, 1)$ is a *retracting factor system.*

If the given normal subgroup $\mathfrak{N}$ of $\mathfrak{G}$ is abelian, then the automorphism $A \to A^{S_\sigma}$ is independent of the particular choice of the representative S_σ and therefore we simply set $A^{S_\sigma} = A^\sigma$.

Three necessary and sufficient conditions in terms of $(\sigma, C_{\sigma,\tau})$ are then

$$\text{I. } (A\,B)^\sigma = A^\sigma B^\sigma,$$

$$\text{II. } (A^\tau)^\sigma = A^{\sigma\tau},\ A^1 = A.$$

$$\text{III. } C_{\sigma,\tau} C_{\sigma\tau,\varrho} = C^\sigma_{\tau,\varrho} C_{\sigma,\tau\varrho}.$$

The factor systems belonging to the same group $(\sigma, \tau, \ldots)$ of automorphisms of $\mathfrak{N}$ form a group $(C_{\sigma,\tau})$. The number of non-equivalent factor systems is equal to the index

$$(C_{\sigma,\tau}) : (A_\sigma A^\sigma_\tau A^{-1}_{\sigma\tau}).$$

The number of different retracting factor systems is equal to the index

$$(A_\sigma) : (\delta_\sigma), \quad \text{where} \quad \delta_\sigma \delta^\sigma_\tau = \delta_{\sigma\tau}.$$

If $\mathfrak{N}$ is of order m and $\mathfrak{F}$ is of order n, then the last index is equal to

$$m^n/((\delta_\sigma) : 1).$$

§ 7. Extensions with Cyclic Factor Group

Let the factor group of $\mathfrak{G}$ over the normal subgroup $\mathfrak{N}$ be isomorphic to the cyclic group $\mathfrak{F} = (\sigma)$.

Let S be a representative of the coset associated with σ.

If $\mathfrak{F}:1 = 0$, then $1, S^{\pm 1} S^{\pm 2}, \ldots$ is a system of representatives of $\mathfrak{G}$ over $\mathfrak{N}$ and $C_{\sigma^i, \sigma^k} = 1$.

If $\mathfrak{F}:1 = n > 0$ then $1, S, S^2, \ldots, S^{n-1}$ is a system of representatives of $\mathfrak{G}$ over $\mathfrak{N}$. Then $S^n = N$ is an element in $\mathfrak{N}$ for which $N^S = N$ holds. Moreover $A^{S^n} = A^N$ for all A in $\mathfrak{N}$ and

$$C_{\sigma^i, \sigma^k} = \begin{cases} 1, & i + k < n \\ N, & i + k \geqq n \end{cases} \qquad (0 \leqq i,\ k < n).$$

Conversely if $\mathfrak{F}:1 = 0$, and $A \to A^S$ is any automorphism of $\mathfrak{N}$, then we set $C_{\sigma^i,\sigma^k}=1$ and see that one and only one extension group $\mathfrak{G}$ of $\mathfrak{N}$ exists such that $\mathfrak{G}/\mathfrak{N}=(S\mathfrak{N})\simeq\mathfrak{F}$ and $SAS^{-1}=A^S$ for all A in $\mathfrak{N}$.

If $\mathfrak{F}:1=n>0$ and $A\to A^\sigma$ is a single-valued mapping of $\mathfrak{N}$ onto itself with the properties Ia. $(AB)^\sigma=A^\sigma B^\sigma$,

IIa. $A^{\sigma^n}=A^N,\ N\in\mathfrak{N}$,

IIIa. $N^\sigma=N$, then there exists one and only one extension group $\mathfrak{G}$ of $\mathfrak{N}$ such that $\mathfrak{G}/\mathfrak{N}=(S\mathfrak{N})\simeq\mathfrak{F}$ and $SAS^{-1}=A^\sigma$ for all A in $\mathfrak{N}$, $S^n=N$.

Proof: In order to see that $\mathfrak{G}$ exists, we set

$$C_{\sigma^i,\sigma^k}=\begin{cases}1, & \text{if}\quad i+k<n\\ N, & \phantom{\text{if}}\quad i+k\geqq n\end{cases}\qquad (0\leqq i,\ k<n).$$

Then I and II hold.

The validity of III means, since $N^\sigma=N$, that certain identities with the factors 1, N are fulfilled, no matter what the structure of $\mathfrak{N}$. Now since the infinite cyclic group (S) is a cyclic extension of index n over (S^n), and $N=S^n$ generates an infinite cyclic group, the identities are valid in all groups.

Since $\mathfrak{G}$ is uniquely determined by $\mathfrak{N}$, σ and N, we denote $\mathfrak{G}$ by $(\mathfrak{N},\sigma,N)$.

When is $(\mathfrak{N},\sigma,N)$ isomorphic to $(\mathfrak{N},\sigma^*,N^*)$ over $\mathfrak{N}$?

We can assume that the two groups are identical, and then $S_{\sigma^*}=AS_\sigma^\nu$ where $0<\nu\leqq n$ and $(\nu,n)=1$:

$$N^*=(AS_\sigma^\nu)^n=A^{1+\sigma^\nu+\cdots+\sigma^{\nu(n-1)}}N^\nu.$$

Conversely if $x^{\sigma^*}=x^{A\sigma^\nu}$ for all x in N, where $(\nu,n)=1$, and if $N^*=A^{1+\sigma^\nu+\cdots+\sigma^{\nu(n-1)}}N^\nu$, then $(\mathfrak{N},\sigma^*,N^*)$ is a cyclic extension of index n of N, which is isomorphic to $(\mathfrak{N},\sigma,N)$ over $\mathfrak{N}$.

Example: Let $\mathfrak{N}$ be a finite cyclic group with order m. If $\mathfrak{N}=(A)$, then $N=A^t$ and $A^\sigma=A^r$. $\mathfrak{G}$ is uniquely described by the four numbers n, m, t, r. IIa. implies $r^n\equiv 1\,(m)$, and conversely. IIIa. implies $rt\equiv t\,(m)$, and conversely. We obtain:

THEOREM 21 (Hölder): *A group $\mathfrak{G}$ of finite order $n\cdot m$ with cyclic normal subgroup (A) and with cyclic factor group $(B\ (A))$ of finite order n has the two generators A, B with the defining relations*:

$$A^m=e,\ B^n=A^t,\ BAB^{-1}=A^r,$$

and with the numerical conditions

a) $0 < n, m.$

b) $r^n \equiv 1 (m),$

c) $t(r-1) \equiv 0 (m).$

Conversely if the numerical conditions are fulfilled, then a group with the previously given properties is defined by the three relations. For fixed n and m the replacement of r, t by

$$r^* = r^\nu, \quad (\nu, n) = 1,$$

$$t^* = \nu t + (1 + r^\nu + r^{2\nu} + \cdots + r^{(n-1)\nu})$$

leads to $\mathfrak{N}$-isomorphic extensions.

§ 8. Extensions with Abelian Factor Group

Let the factor group of the group $\mathfrak{G}$ over the normal subgroup $\mathfrak{N}$ be isomorphic to the direct product

$$\mathfrak{F} = (\sigma_1) \times (\sigma_2) \times \cdots \times (\sigma_r).$$

of cyclic groups (σ_i) of orders n_i. Let $(S_i \mathfrak{N})$ be cosets associated with σ_i. The following relations hold in $\mathfrak{G}$:

$$S_i A S_i^{-1} = A^{S_i} \in \mathfrak{N} \qquad \text{if} \quad A \in \mathfrak{N},$$

$$S_i^{n_i} = A_i \in \mathfrak{N},$$

$$S_i S_k S_i^{-1} S_k^{-1} = A_{i,k} \in \mathfrak{N}.$$

For the mappings $A \to A^{S_i}$ of $\mathfrak{N}$ onto itself the rules

$$(A B)^{S_i} = A^{S_i} B^{S_i},$$

$$A^{S_i^{n_i}} = A^{A_i},$$

$$A^{S_i S_k} = A^{A_{i,k} S_k S_i}$$

hold.

Moreover $A_{i,k}\,A_{k,i}=1$, $n_i \geqq 0$, and if $n_i=0$ then $A_i=1$.

$$\begin{aligned} A_k^{S_i} &= S_i A_k S_i^{-1} = S_i S_k^{n_k} S_i^{-1} = (S_i S_k S_i^{-1})^{n_k} \\ &= (A_{i,k} S_k)^{n_k} = A_{i,k}(S_k A_{i,k} S_k^{-1})\, S_k^2 A_{i,k} \ldots \\ &= A_{i,k} A_{i,k}^{S_k} \ldots A_{i,k}^{S_k^{n_k-1}} A_k \end{aligned}$$

$$\begin{aligned} S_i A_{k,l} S_i^{-1} &= A_{k,l}^{S_i} \\ &= (S_i S_k S_i^{-1}) \cdot (S_i S_l S_i^{-1})\,(S_i S_k S_i^{-1})^{-1}\,(S_i S_l S_i^{-1})^{-1} \\ &= A_{i,k} S_k \cdot A_{i,l} S_l \cdot (A_{i,k} S_k)^{-1} \cdot (A_{i,l} S_l)^{-1} \\ &= A_{i,k} A_{i,l}^{S_k} \cdot S_k S_l S_k^{-1} S_l^{-1} \cdot A_{i,k}^{-S_l} A_{i,l}^{-1} \\ &= A_{i,k} A_{i,l}^{S_k} A_{k,l} A_{i,k}^{-S_l} A_{i,l}^{-1}. \end{aligned}$$

Now conversely let a group $\mathfrak{N}$ be given which contains the elements $A_i, A_{i,k}$ $(i, k = 1, 2, \ldots, r;\ i \neq k)$ and let single valued mappings $A \rightarrow A^{S_i}$ of $\mathfrak{N}$ onto itself be defined with the following properties:

1. $(A\,B)^{S_i} = A^{S_i}\,B^{S_i}$,
2. $A^{S_i^{n_i}} = A^{A_i}$ (if $n_i > 0$), $A_i^{S_i} = A_i$,

2a. $n_i \geqq 0$, and if $n_i = 0$, then $A_i = 1$,

3. $A^{S_i S_k} = A^{A_{i,k} S_k S_i}$ $(i > k)$,

3a. $A_{i,k} A_{k,i} = 1$ $(i > k)$,

4. $A_k^{S_i} = A_{i,k}^{1+S_k+\cdots+S_k^{n_k-1}} A_k$ (if $n_k > 0$, $i \neq k$),
5. $A_{i,k}^{S_l} A_{k,l}^{-1} A_{l,i}^{S_k} A_{i,k}^{-1} A_{k,l}^{S_i} A_{l,i}^{-1} = 1$ $(i < k < l)$.

THEOREM 22: *The group $\mathfrak{G}$ with generators $\bar{A}\,(A \in N), S_1, S_2, \ldots, S_r$ and with the defining relations*

a) $\overline{AB} = \bar{A}\,\bar{B}$,

b) $S_i \bar{A} S_i^{-1} = \overline{A^{S_i}}$,

c) $S_i^{n_i} = \bar{A}_i$,

d) $S_i S_k S_i^{-1} S_k^{-1} = \bar{A}_{i,k}$ $(i > k)$

contains the normal subgroup $\bar{\mathfrak{N}}$ of all $\bar{A}$ such that the factor group is the direct product of the cyclic groups $(S_i\mathfrak{N})$ of order n_i and the mapping $A \rightarrow \bar{A}$ is an isomorphism of $\mathfrak{N}$ onto $\bar{\mathfrak{N}}$.

Proof: If $r=1$ then the theorem follows from § 6. Let $r>1$ and assume that the theorem has been proven when there are only r–1 generators S_i.

Let $\mathfrak{G}_1$ be the group generated by $\bar{A}$ and $S_1, \ldots, S_{r-1}$ which satisfy conditions 1. to 5. and relations a) to d) where the indices i, k, l run from 1 to $r-1$.

We define $\quad \bar{A}^{S_r} = \overline{A^{S_r}},\ S_k^{S_r} = \bar{A}_{r,k} S_r \qquad (k = 1, 2, \ldots, r-1).$

Conditions 1., 3., 3*a*. and 4. with $i=r$, 5. with $l=r$ merely state that the mapping S_r just defined can be extended (uniquely) to an operator S_r of $\mathfrak{G}_1$.

The conditions $A^{A_r} = A^{S_r^{n_r}}$, $A_r^{S_r} = A_r$, state, by § 6, that the relations

b) $S_r \bar{A} S_r^{-1} = \bar{A}^{S_r}$,

c) $S_r^{n_r} = \bar{A}_r$,

d) $S_r S_i S_r^{-1} S_i^{-1} = \bar{A}_{r,i} \quad (i < r)$

define a cyclic extension $\mathfrak{G}$ of index n_r over $\mathfrak{G}_1$.

By the induction hypothesis, $A \to \bar{A}$ is an isomorphism of $\mathfrak{N}$ onto $\bar{\mathfrak{N}}$. It follows from b) that $\bar{\mathfrak{N}}$ is a normal subgroup of $\mathfrak{G}$. It follows from d) that $\mathfrak{G}/\bar{\mathfrak{N}}$ is abelian and is generated by $S_1\bar{\mathfrak{N}}, S_2\bar{\mathfrak{N}}, \ldots, S_r\bar{\mathfrak{N}}$.

Now
$$\prod_1^r S_i^{\nu_i} \equiv 1\ (\bar{\mathfrak{N}})$$
implies
$$S_r^{\nu_r} \equiv 1\ (\mathfrak{G}_1),$$
therefore
$$\nu_r \equiv 0\ (n_r),$$
hence $\prod_1^{r-1} S_i^{\nu_i} \equiv 1\ (\bar{\mathfrak{N}})$, and the induction hypothesis applied to $\mathfrak{G}_1$ gives
$$\nu_i \equiv 0\ (n) \qquad (i = 1, 2, \ldots, r-1).$$
$\mathfrak{G}/\bar{\mathfrak{N}}$ is the direct product of the cyclic groups $(S_i\bar{\mathfrak{N}})$ with orders n_i $(i=1, 2, \ldots, r)$, Q.E.D. If the normal subgroup $\mathfrak{N}$ is abelian then the conditions can be stated more simply:

1. $(A B)^{\sigma_i} = A^{\sigma_i} B^{\sigma_i}$,
2. $A^{\sigma_i^{n_i}} = A$,

2a. $n_i \geqq 0$, and if $n_i = 0$, then $A_i = 1$,

3. $A^{\sigma_i \sigma_k} = A^{\sigma_k \sigma_i}$,

3a. $A_{i,k} A_{k,i} = 1 \quad (i < k)$,

4. $A_k^{\sigma_k - 1} = A_{i,k}^{1+\sigma_k+\sigma_k^2+\cdots+\sigma_k^{n_k-1}} \quad (\text{if } n_k > 0,\ i \neq k)$,
5. $A_{i,k}^{\sigma_l - 1} A_{k,l}^{\sigma_i - 1} A_{l,i}^{\sigma_k - 1} = 1 \quad (i < k < l)$

and the relations

b) $S_i A S_i^{-1} = A^{\sigma_i}$,

c) $S_i^{n_i} = A_i$,

d) $S_i S_k S_i^{-1} S_k^{-1} = A_{i,k} \quad (i < k)$.

§ 9. Splitting Groups

Definition: A group $\overline{\mathfrak{G}}$ which contains the extension $\mathfrak{G}$ of $\mathfrak{N}$ by $\mathfrak{F}$ with the system of representatives S_σ is said to be a *splitting group* of $\mathfrak{G}$ over $\mathfrak{N}$, if $\overline{\mathfrak{G}}$ has a normal subgroup $\overline{\mathfrak{N}}$, containing $\mathfrak{N}$, with S_σ as system of representatives, such that $\overline{\mathfrak{G}}$ splits over $\overline{\mathfrak{N}}$.

THEOREM 23 (Artin): *Every group with abelian normal subgroup $\mathfrak{N}$ has a splitting group.*

Proof: We set $A_1 = C_{1,1}^{-1}$, but let (A_σ) be the infinite cyclic group generated by the new element A_σ, if $\sigma \neq 1$. Let $\overline{\mathfrak{N}}$ be the direct product of $\mathfrak{N}$ with the (A_σ), $(\sigma \neq 1)$. Then an operator S_σ of $\overline{\mathfrak{N}}$ is defined by

$$A_\tau^{S_\sigma} = A_\sigma^{-1} A_{\sigma\tau} C_{\sigma,\tau}^{-1} \qquad (\tau \neq 1)$$

and $$\overline{N}^{S_\sigma} = \Big(N \cdot \prod_\tau A_\tau^{m_\tau}\Big)^{S_\sigma} = N^{S_\sigma} \cdot \prod_\tau (A_\tau^{S_\sigma})^{m_\tau}.$$

The same formula holds for $A_1^{S_\sigma}$, since

$$A_1^{S_\sigma} = (C_{1,1}^{-1})^{S_\sigma} = C_{\sigma,1}^{-1} = A_\sigma^{-1} A_\sigma C_{\sigma,1}^{-1}.$$

The factor system $C_{\sigma,\tau}$ and the mapping S_σ satisfy conditions I and III. Now II must be verified.

$A^{S_1} = A^{C_{1,1}}$ holds for all A in $\mathfrak{N}$:

$$A_\tau^{S_1} = A_1^{-1} A_\tau C_{1,1}^{-1} = C_{1,1} A_\tau C_{1,1}^{-1} = A_\tau^{C_{1,1}},$$

and therefore $\overline{N}^{S_1} = \overline{N}^{C_{1,1}}$ for all $\overline{N}$ in $\overline{\mathfrak{N}}$.

$$\begin{aligned}(A_\varrho^{S_\tau})^{S_\sigma} &= (A_\tau^{-1} A_{\tau\varrho} C_{\tau,\varrho}^{-1})^{S_\sigma} = (A_\tau^{S_\sigma})^{-1} A_{\tau\varrho}^{S_\sigma} (C_{\tau,\varrho}^{S_\sigma})^{-1} \\ &= (A_\sigma^{-1} A_{\sigma\tau} C_{\sigma,\tau}^{-1})^{-1} A_\sigma^{-1} A_{\sigma\tau\varrho} C_{\sigma,\tau\varrho}^{-1} (C_{\tau,\varrho}^{S_\sigma})^{-1} = C_{\sigma,\tau} A_{\sigma\tau}^{-1} A_{\sigma\tau\varrho} C_{\sigma,\tau\varrho}^{-1} (C_{\tau,\varrho}^{S_\sigma})^{-1} \\ &= (A_{\sigma\tau}^{-1} A_{\sigma\tau\varrho})^{C_{\sigma,\tau}} C_{\sigma,\tau} (C_{\sigma,\tau} C_{\sigma\tau,\varrho})^{-1} = (A_{\sigma\tau}^{-1} A_{\sigma\tau\varrho} C_{\sigma\tau,\varrho}^{-1})^{C_{\sigma,\tau}} \text{ (by III.!)} \\ &= A_\varrho^{C_{\sigma,\tau} S_{\sigma\tau}}.\end{aligned}$$

Since II is valid in $\mathfrak{N}$ it holds also in $\overline{\mathfrak{N}}$. Therefore there exists an extension group $\overline{\mathfrak{G}}$ of $\overline{\mathfrak{N}}$ with the elements S_σ as system of representatives, the $C_{\sigma,\tau}$ as factor system, and the S_σ as automorphisms of $\overline{\mathfrak{N}}$.

This extension naturally contains the extension $\mathfrak{G}$ of $\mathfrak{N}$ with the factor group $\mathfrak{F}$ and system of representatives S_σ. $\bar{\mathfrak{G}}$ is a splitting group of $\mathfrak{G}$ since

$$e = A_\sigma A_\tau^{S_\sigma} C_{\sigma,\tau} A_{\sigma\tau}^{-1}.$$ [1]

[1] If we take $\bar{\mathfrak{G}}$ as the free product (which will be defined later) of $\mathfrak{G}$ with the infinite cyclic groups (A_σ) then it follows in exactly the same way that every group has a splitting group.

IV. SYLOW p-GROUPS AND p-GROUPS

§ 1. The Sylow Theorems

In a finite group $\mathfrak{G}$ of order N, the order of every subgroup is a divisor of N. On the other hand there need not be a subgroup with order d for every divisor d of N. For example, in the tetrahedral group, as one can see easily, there is no subgroup of order 6. We shall now prove, however, that for every power p^a of a prime dividing N there is a subgroup with the order p^a.

DEFINITION: A group is said to be a *p-group* if the order of each of its elements is a power of the prime p.

We determine the largest possible p-groups in the finite group $\mathfrak{G}$.

DEFINITION: A subgroup of $\mathfrak{G}$ is said to be a *Sylow p-group,* if its order is equal to the greatest power of the natural prime p dividing N.

For example, the four group is a Sylow 2-group of the tetrahedral group. A Sylow p-group of $\mathfrak{G}$ is denoted by S_p or by $\mathfrak{P}$. The normalizer of S_p in $\mathfrak{G}$ is denoted by N_p, the center of S_p by z_p.

THEOREM 1. *For every natural prime p, every finite group contains a Sylow p-group.*

Proof: If the order N of $\mathfrak{G}$ is 1, then the theorem is clear. Now let $N>1$ and assume the theorem proven for groups of order smaller than N.

If in the center $\mathfrak{z}$ of $\mathfrak{G}$ there is an element a of order $m.p$, then the factor group $\mathfrak{G}/(a^m)$ is of order $\frac{N}{p}$ and contains by the induction assumption a Sylow p-group $\mathfrak{P}/(a^m)$ of order p^{n-1}, where $\frac{N}{p^n}$ is not divisible by p.

$\mathfrak{P}$ is of order p^n and therefore is a Sylow p-group of $\mathfrak{G}$.

Now let there be no element of order divisible by p in the center $\mathfrak{z}$ of $\mathfrak{G}$. If the order of $\mathfrak{z}$ were divisible by p, then the factor group of $\mathfrak{z}$ with respect to a cyclic normal subgroup $(a) \neq 1$ is of order divisible by p. But then by the induction hypothesis $\mathfrak{z}/(a)$ would contain a Sylow p-group $\neq 1$, and therefore would contain elements $b(a)$ of order divisible by p. Then the order of b in $\mathfrak{z}$ would be divisible by p. Therefore the order of $\mathfrak{z}$ is not divisible by p. If $p \nmid N$ then e is the Sylow p-group sought. If p/N then it follows from the class equation

$$N = (\mathfrak{z} : 1) + \sum_{h_i > 1} h_i$$

and from $p \nmid (\mathfrak{G}:1)$, p/N, that at least one $h_i > 1$ is not divisible by p. $\mathfrak{G}$ contains a normalizer N_i of index $h_i > 1$ and therefore N_i contains, by the induction hypothesis, a Sylow p-group $\mathfrak{P}$. Since $p \nmid h_i$, $\mathfrak{P}$ is also a Sylow p-group of $\mathfrak{G}$.

COROLLARY: For every prime divisor p of the order of a finite group there is an element of order p (Cauchy).

The order and exponent of a finite group have the same prime divisors.

It is a p-group if and only if its order is a power of p.

THEOREM 2: *If $\mathfrak{P}$ is a Sylow p-group of $\mathfrak{G}$ and $\mathfrak{N}$ a normal subgroup of $\mathfrak{G}$, then $\mathfrak{N} \cap \mathfrak{P}$ is a Sylow p-group of $\mathfrak{N}$; $\mathfrak{P}\mathfrak{N}/\mathfrak{N}$ is a Sylow p-group of $\mathfrak{G}/\mathfrak{N}$.*

Proof:[1] A subgroup $\mathfrak{U}$ of $\mathfrak{G}$ is a Sylow p-group if and only if

1. The order of $\mathfrak{U}$ is a power of p (written: *pp*),
2. The index of $\mathfrak{U}$ is prime to p (written: *prime*).

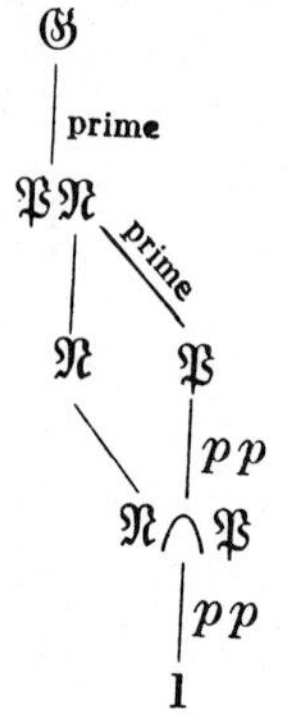

Now we may construct the diagram to the left and, observe first that $\mathfrak{P}\mathfrak{N}:\mathfrak{P}$ is prime to p, and $\mathfrak{P}:\mathfrak{N} \cap \mathfrak{P}$ is a p-power. From the second isomorphism theorem it follows that $\mathfrak{P}\mathfrak{N}:\mathfrak{N}$ is a p-power, and $\mathfrak{N} : \mathfrak{N} \cap \mathfrak{P}$ is prime to p, from which the theorem follows.

If a Sylow p-group $\mathfrak{P}$ is a normal subgroup of $\mathfrak{G}$ then it is the only Sylow p-group, since for every other Sylow p-group $\mathfrak{P}_1$ it follows that $\mathfrak{P}_1\mathfrak{P}$ is of p-power order, but $\mathfrak{G}:\mathfrak{P}_1\mathfrak{P}$ is prime to p, and therefore $\mathfrak{P}_1\mathfrak{P} = \mathfrak{P} = \mathfrak{P}_1$. Consequently a Sylow p-group S_p of a finite group $\mathfrak{G}$ is the only Sylow p-group of its normalizer N_p.

THEOREM 3: *All Sylow p-groups of a finite group $\mathfrak{G}$ are conjugate under $\mathfrak{G}$. Their number when divided by p leaves a remainder 1.*

Proof: Let the Sylow p-groups of $\mathfrak{G}$ be $\mathfrak{P} = \mathfrak{P}_1, \ldots, \mathfrak{P}_r$.

Under the mapping $x \rightarrow \begin{pmatrix} a \\ x\,a\,x^{-1} \end{pmatrix}$ of $\mathfrak{G}$ onto the group of inner automorphisms, $\mathfrak{P}$ is represented as a permutation group. Since conjugate subgroups have the same order, the $\mathfrak{P}_i$ are transformed into each other by $\mathfrak{P}$, so that we obtain a representation Δ of $\mathfrak{P}$ as a permutation group of degree r. By a remark above, $\mathfrak{P}$ transforms only $\mathfrak{P}_1$ and no other $\mathfrak{P}_i$ into itself. Consequently there is only one system of transitivity of first degree. The other systems of transitivity of Δ have a degree > 1 which is a divisor of $\mathfrak{P}:1$ and which, therefore, is a p-power. Consequently $r \equiv 1 (p)$.

$\mathfrak{P}$ transforms the $s = \mathfrak{G}:N_p$ Sylow p-groups conjugate to $\mathfrak{P}$ under $\mathfrak{G}$

[1] According to a communication from E. Witt.

among themselves, and as above it follows that $s \equiv 1\,(p)$. If there were another system of conjugate Sylow p-groups, then its members would be transformed into each other by $\mathfrak{P}$ in systems of transitivity whose degree would be divisible by p. The system would therefore contain a number s_1, divisible by p, of Sylow p-groups; on the other hand we conclude for s_1, just as we did for s, that $s_1 \equiv 1\ (p)$. Consequently all Sylow p-groups are conjugate to $\mathfrak{P}$, Q.E.D.

THEOREM 4: *Every p-group $\mathfrak{U}$ in $\mathfrak{G}$ is contained in a Sylow p-group.*

Proof: We replace $\mathfrak{P}$ by $\mathfrak{U}$ in the proof of the previous theorem. Let the transformed objects again be $\mathfrak{P}_1, \ldots, \mathfrak{P}_r$. The degree of a system of transitivity of $\varDelta$ is either 1 or a p-power. Since $r \equiv 1\,(p)$, there is certainly a system of transitivity of degree 1. Therefore there is a $\mathfrak{P}_i$ which is transformed into itself by all the elements of $\mathfrak{U}$. Since $\mathfrak{U}\mathfrak{P}_i$ is a p-group which contains $\mathfrak{P}_i$, we have $\mathfrak{U}\mathfrak{P}_i = \mathfrak{P}_i$, $\mathfrak{U} \subseteq \mathfrak{P}_i$, Q.E.D.

THEOREM 5: *Every subgroup $\mathfrak{U}$ of $\mathfrak{G}$ which contains the normalizer N_p of a Sylow p-group S_p, is its own normalizer.*

Proof: We must show that $x\,\mathfrak{U}\,x^{-1} \subseteq \mathfrak{U}$ implies

$$x \in \mathfrak{U}.$$

In any case S_p and xS_px^{-1} are Sylow p-groups of $\mathfrak{U}$, and by Theorem 3 there is a U in $\mathfrak{U}$ such that $U\,x\,S_p\,x^{-1}U^{-1} = S_p$;

therefore $\qquad\qquad U\,x \in N_p \subseteq \mathfrak{U},$

therefore $\qquad\qquad x \in \mathfrak{U},$ $\qquad\qquad$ Q.E.D.

THEOREM 6: *If the p-group $\mathfrak{U}$ contained in the finite group $\mathfrak{G}$ is not a Sylow p-group, then the normalizer $N_\mathfrak{U}$ of $\mathfrak{U}$ is larger than $\mathfrak{U}$.*

Proof: If $p \nmid \mathfrak{G} : N_\mathfrak{U}$ then the theorem is clear; if, however $\mathfrak{G} : N_\mathfrak{U} = pr$, then $\mathfrak{U}$ transforms the pr subgroups conjugate to $\mathfrak{U}$ in systems of transitivity whose degrees are 1 or numbers divisible by p. Since $\mathfrak{U}$ is transformed into itself, there are at least p subgroups $\mathfrak{U}_1 = \mathfrak{U}, \mathfrak{U}_2, \ldots, \mathfrak{U}_p$, conjugate to $\mathfrak{U}$ which are transformed into themselves by $\mathfrak{U}$. Consequently $N_{\mathfrak{U}_2}$ is greater than $\mathfrak{U}_2$, and therefore $N_\mathfrak{U}$ is greater than $\mathfrak{U}$, Q.E.D.

COROLLARIES:

1. Every maximal subgroup of a p-group is a normal subgroup; therefore it is of index p.

2. If a p-group is simple then it is of order p.

3. The composition factors of a p-group are of order p and therefore every p-group is solvable.

§ 2. Theorems on Sylow p-Groups

Information on the intersection of different Sylow p-groups is given by

THEOREM 7: *In the normalizer of a maximal*[1] *intersection* $\mathfrak{D}$ *of two different Sylow p-groups of* $\mathfrak{G}$ *we have:*

1. *Every Sylow p-group of* $N_{\mathfrak{D}}$ *contains* $\mathfrak{D}$ *properly.*
2. *The number of Sylow p-groups of* $N_{\mathfrak{D}}$ *is greater than* 1.
3. *The intersection of two distinct Sylow p-groups of* $N_{\mathfrak{D}}$ *is equal to* $\mathfrak{D}$.
4. *Every Sylow p-group of* $N_{\mathfrak{D}}$ *is the intersection of* $N_{\mathfrak{D}}$ *with exactly one Sylow p-group of* $\mathfrak{G}$.
5. *The intersection of* $N_{\mathfrak{D}}$ *with a Sylow p-group of* $\mathfrak{G}$ *which contains* $\mathfrak{D}$ *is a Sylow p-group of* $N_{\mathfrak{D}}$.
6. *The normalizer of a Sylow p-group of* $N_{\mathfrak{D}}$ *in* $N_{\mathfrak{D}}$ *is equal to the intersection of* $N_{\mathfrak{D}}$ *with the normalizer of a Sylow p-group of* $\mathfrak{G}$ *which contains* $\mathfrak{D}$.

Proof: $\mathfrak{D}$ is in a Sylow p-group $\mathfrak{P}$ of $\mathfrak{G}$ and by hypothesis $\mathfrak{D} \neq \mathfrak{P}$. Therefore by Theorem 6: $\mathfrak{D} \neq \mathfrak{p} = N_{\mathfrak{D}} \cap \mathfrak{P}$. $\mathfrak{p}$ is a p-group in $N_{\mathfrak{D}}$, thus by Theorem 4 it lies in a Sylow p-group $\bar{\mathfrak{p}}$ of $N_{\mathfrak{D}}$. By Theorem 4, $\bar{\mathfrak{p}}$ lies in a Sylow p-group $\bar{\mathfrak{P}}$ of $\mathfrak{G}$. Since $\mathfrak{P} \cap \bar{\mathfrak{P}}$ contains $\mathfrak{p}$ and thus is larger than $\mathfrak{D}$, $\mathfrak{P} = \bar{\mathfrak{P}}$. Therefore $\mathfrak{p} = \mathfrak{P} \cap N_{\mathfrak{D}} = \bar{\mathfrak{p}}$ is a Sylow p-group of $N_{\mathfrak{D}}$, and the $\mathfrak{P}$ in $\mathfrak{p} = \mathfrak{P} \cap N_{\mathfrak{D}}$ is uniquely determined by $\mathfrak{p}$. Since every p-group in $N_{\mathfrak{D}}$ is in a Sylow p-group of $\mathfrak{G}$, the intersection of two distinct Sylow p-groups of $N_{\mathfrak{D}}$ is equal to $\mathfrak{D}$. Since $\mathfrak{D}$ is the intersection of two different Sylow p-groups of $\mathfrak{G}$, $N_{\mathfrak{D}}$ contains several Sylow p-groups. $\mathfrak{p} = \mathfrak{P} \cap N_{\mathfrak{D}}$ is a normal subgroup of $n_{\mathfrak{p}} = N_{\mathfrak{P}} \cap N_{\mathfrak{D}}$. If we have

$$x \mathfrak{p} x^{-1} = \mathfrak{p},$$

for an x in $N_{\mathfrak{D}}$, then it follows that $\mathfrak{p} \subseteq x \mathfrak{P} x^{-1}$, and therefore by 4., $\mathfrak{P} = x \mathfrak{P} x^{-1}$, $x \subseteq N_{\mathfrak{P}}$, $x \in n_{\mathfrak{p}}$, consequently $n_{\mathfrak{p}} = N_{\mathfrak{P}} \cap N_{\mathfrak{D}}$ is the normalizer of $\mathfrak{p}$ in $N_{\mathfrak{D}}$, Q.E.D.

As an application of this theorem, we shall show that *every group* $\mathfrak{G}$ of order $p^n q$ *is solvable* (p, q *are two distinct primes*).

If a Sylow p-group $\mathfrak{P}$ is a normal subgroup, then $\mathfrak{G}/\mathfrak{P}$ is cyclic and by Theorem 6, Corollary 3, $\mathfrak{P}$ is solvable. Hence $\mathfrak{G}$ is solvable. Now suppose $\mathfrak{P}$ is not a normal subgroup of $\mathfrak{G}$; then $\mathfrak{G} : N_{\mathfrak{P}} = q$, $N_{\mathfrak{P}} = \mathfrak{P}$.

---

[1] If $\mathfrak{D}$ is the intersection of two Sylow p-groups and no group containing $\mathfrak{D}$ properly is contained in the intersection of *any* two Sylow p-groups, $\mathfrak{D}$ is called a *maximal* intersection of two Sylow p-groups.

If the intersection of any two different Sylow p-groups is 1, then there are $1+q.(p^n-1)$ elements of p-power order, and therefore there is at most one subgroup with order q. Consequently a Sylow q-group $\mathfrak{Q}$ is a normal subgroup of $\mathfrak{G}$, and $\mathfrak{G}/\mathfrak{Q}$ is isomorphic to $\mathfrak{P}$. Since $\mathfrak{G}/\mathfrak{Q}$ and $\mathfrak{Q}$ are solvable, $\mathfrak{G}$ is also solvable. Finally let $\mathfrak{D}$ be a maximal intersection of different Sylow p-groups greater than 1. The number of Sylow p-groups of $N_{\mathfrak{D}}$ is >1, is not divisible by p, is a divisor of $p^n q$ and therefore is equal to q. Also it follows from the previous theorem that $\mathfrak{D}$ lies in q different Sylow p-groups of $\mathfrak{G}$. Therefore $\mathfrak{D}$ is the intersection of all the Sylow p-groups of $\mathfrak{G}$. $\mathfrak{D}$ is a normal subgroup of $\mathfrak{G}$, and the factor group $\mathfrak{G}/\mathfrak{D}$ has as the maximal intersection of different Sylow p-groups the element 1. By what has already been proven, $\mathfrak{G}/\mathfrak{D}$ is solvable. Moreover the p-group $\mathfrak{D}$ is solvable. Consequently $\mathfrak{G}$ is solvable, Q.E.D.

For many applications the following theorem is useful:

THEOREM 8 (Burnside): *If the p-group $\mathfrak{h}$ in the finite group $\mathfrak{G}$ is a normal subgroup of one Sylow p-group but is not a normal subgroup of another Sylow p-group, then there is a number r, relatively prime to p, of subgroups $\mathfrak{h}_1, \mathfrak{h}_2, \ldots, \mathfrak{h}_r$ $(r>1)$ conjugate to $\mathfrak{h}$ which are all normal subgroups of $\mathfrak{H} = \mathfrak{h}_1\mathfrak{h}_2\ldots\mathfrak{h}_r$ but which are not all normal subgroups of the same Sylow p-group of $\mathfrak{G}$, so that the normalizer of $\mathfrak{H}$ transforms the $\mathfrak{h}_i$ transitively among themselves.*

Proof: Among the Sylow p-groups which contain $\mathfrak{h}$ as a non-normal subgroup, $\mathfrak{Q}$ is chosen so that the intersection $\mathfrak{D}$ of $\mathfrak{Q}$ with the normalizer $N_{\mathfrak{h}}$ of $\mathfrak{h}$ is as large as possible. Let $\mathfrak{h} = \mathfrak{h}_1, \mathfrak{h}_2, \ldots, \mathfrak{h}_s$ be the subgroups conjugate to $\mathfrak{h}$ in the normalizer $N_{\mathfrak{D}}$ of $\mathfrak{D}$. Along with $\mathfrak{h}$, all the $\mathfrak{h}_i$ are also normal subgroups of $\mathfrak{D}$. The normalizer $N_{\mathfrak{H}}$ of $\mathfrak{H} = \mathfrak{h}_1\mathfrak{h}_2\ldots\mathfrak{h}_s$ contains $N_{\mathfrak{D}}$. Let $\mathfrak{h}_1, \mathfrak{h}_2, \ldots, \mathfrak{h}_s, \ldots, \mathfrak{h}_r$ be all the groups conjugate to $\mathfrak{h}$ in $N_{\mathfrak{H}}$. Along with $\mathfrak{h}$, all the $\mathfrak{h}_i$ are normal subgroups of $\mathfrak{H}$. $\mathfrak{D}$ is contained in a Sylow p-group $\mathfrak{p}^*$ of $N_{\mathfrak{D}} \cap N_{\mathfrak{h}}$. Since $\mathfrak{D}$ is not a Sylow p-group of $\mathfrak{G}$, while, by hypothesis, a Sylow p-group of $N_{\mathfrak{h}}$ is also a Sylow p-group of $\mathfrak{G}$, then $\mathfrak{p}^*$ is larger than $\mathfrak{D}$. $\mathfrak{p}^*$ is in a Sylow p-group $\bar{\mathfrak{p}}$ of $N_{\mathfrak{H}} \cap N_{\mathfrak{h}}$, $\bar{\mathfrak{p}}$ is in a Sylow p group $\mathfrak{p}$ of $N_{\mathfrak{H}}$, and $\mathfrak{p}$ in a Sylow p-group $\mathfrak{P}$ of $\mathfrak{G}$. Since the intersection of $\mathfrak{P}$ with $N_{\mathfrak{h}}$ contains $\mathfrak{p}^*$, and therefore is larger than $\mathfrak{D}$, then by the construction of $\mathfrak{D}$ the Sylow p-group $\mathfrak{P}$ of $\mathfrak{G}$ is contained in $N_{\mathfrak{h}}$, and therefore a fortiori $\mathfrak{p}$ is contained in $N_{\mathfrak{h}}$.

Since $\mathfrak{p}$ contains the Sylow p-group $\bar{\mathfrak{p}}$ of $N_{\mathfrak{H}} \cap N_{\mathfrak{h}}$, we have $\mathfrak{p} = \bar{\mathfrak{p}}$. Since therefore a Sylow p-group of $N_{\mathfrak{H}} \cap N_{\mathfrak{h}}$ is already a Sylow p-group of $N_{\mathfrak{H}}$, $r = N_{\mathfrak{H}} : N_{\mathfrak{H}} \cap N_{\mathfrak{h}}$ is relatively prime to p. If all the $\mathfrak{h}_i$ were

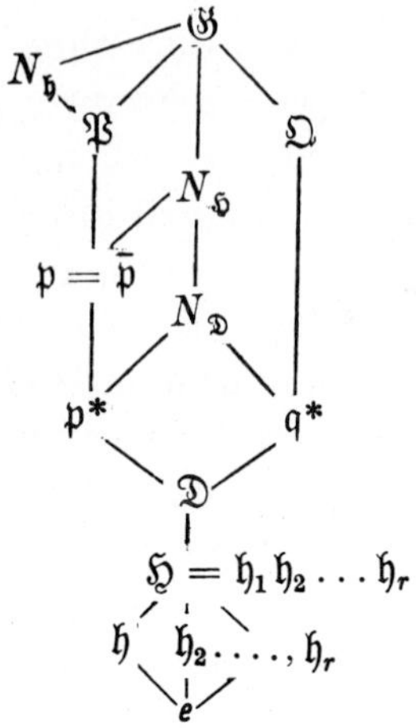

normal subgroups of the same Sylow p-group of $\mathfrak{G}$, then the latter would be contained in $N_{\mathfrak{H}}$. But then the groups $\mathfrak{h}_i$ conjugate to each other in $N_{\mathfrak{H}}$ would be normal subgroups in all the Sylow p-groups of $N_{\mathfrak{H}}$. Then $\mathfrak{h}$ would be a normal subgroup of the Sylow p-group $\mathfrak{q}^*$ of the intersection of $N_{\mathfrak{D}}$ with $\mathfrak{Q}$. But $\mathfrak{q}^*$ is larger than $\mathfrak{D}$, and this contradicts the definition of $\mathfrak{D}$ as the intersection of $\mathfrak{Q}$ with $N_{\mathfrak{h}}$; therefore the $\mathfrak{h}_i$ are not all normal subgroups of the same Sylow p-group of $\mathfrak{G}$, Q.E.D.

The positional relationships of the subgroups of $\mathfrak{G}$ constructed in this proof can be seen from the diagram on the left.

§ 3. On p-Groups

1. *Nilpotent Groups.*

Fundamental for the theory of p-groups is the following statement:

THEOREM 9: *The center of a p-group different from e is itself different from e.*

Proof: From the class equation for a group of order $p^n > 1$:

$$p^n = \mathfrak{z} : 1 + \sum_{i>0} p^i,$$

where the summands p^i run through indices > 1 of certain normalizers. Therefore $\mathfrak{z} : 1$ is divisible by p, and consequently $\mathfrak{z} \neq e$.

COROLLARY: The $(n+1)$-th member of the ascending central series of a group $\mathfrak{G}$ of order p^n is equal to the whole group.

The members of the ascending central series are defined as the normal subgroups $\mathfrak{z}_i$ of $\mathfrak{G}$ such that $\mathfrak{z}_0 = e$, $\mathfrak{z}_{i+1}/\mathfrak{z}_i$ is the center of $\mathfrak{G}/\mathfrak{z}_i$. Now either $\mathfrak{z}_i = \mathfrak{G}$ or, as just proven, $\mathfrak{z}_{i+1}$ is larger than $\mathfrak{z}_i$, and therefore certainly $\mathfrak{z}_n = \mathfrak{G}$.

By refinement of the ascending central series of a p-group we obtain a principal series in which every factor is of order p. It follows from the Jordan - Hölder - Schreier theorem that:

Every principal series of a p-group has steps of prime order.

The index of the center of a non-abelian p-group is divisible by p^2. This follows from the useful lemma: *If a normal subgroup $\mathfrak{N}$ of a group $\mathfrak{G}$ is contained in the center and has a cyclic factor group, then $\mathfrak{G}$ is abelian.*

Since $\mathfrak{G}/\mathfrak{N}$ is generated by a coset $A\mathfrak{N}$, all the elements of $\mathfrak{G}$ are of the form A^iZ where Z is in the center.
Therefore

$$A^iZ \cdot A^kZ' = A^{i+k}ZZ' = A^kZ' \cdot A^iZ$$

and $\mathfrak{G}$ is abelian.

If we apply the result found above to a p-group in which $\mathfrak{G} = \mathfrak{z}_c \neq \mathfrak{z}_{c-1} \neq e$, then: $p^2/\mathfrak{G} : \mathfrak{z}_{c-1}$, and since $\mathfrak{G}/\mathfrak{z}_{c-1}$ is abelian, it follows that:

The factor commutator group of a non-abelian p-group has an order divisible by p^2.

A group of order p or p^2 is abelian. In a non-abelian group of order p^3, the center and the commutator group are identical and are of order p.

DEFINITION: A group $\mathfrak{G}$ is said to be *nilpotent*[1] if the ascending central series contains the whole group as a member, i.e., if

$$e = \mathfrak{z}_0 \subset \mathfrak{z}_1 \subset \mathfrak{z}_2 \subset \cdots \subset \mathfrak{z}_c = \mathfrak{G}.$$

The uniquely determined number c is called, following Hall, the *class of the group*. Therefore "nilpotent of class 1" is the same as "abelian $\neq e$."

THEOREM 10: *In a nilpotent group of class c it is possible to ascend to the whole group from any subgroup by forming normalizers at most c times.*

Proof: Let $\mathfrak{G}$ be nilpotent of class c; let $\mathfrak{U}$ be a subgroup. Certainly $\mathfrak{z}_0$ is contained in $\mathfrak{U}$. If $\mathfrak{z}_i$ is already contained in $\mathfrak{U}$, then by the definition of $\mathfrak{z}_{i+1}$, it follows that $\mathfrak{z}_{i+1}$ is contained in the normalizer of $\mathfrak{U}$. By at most c repetitions of this procedure we obtain the result.

COROLLARY: *Every maximal subgroup of a nilpotent group is a normal subgroup and therefore is of prime index.*

Therefore in a p-group $\mathfrak{G}$ the intersection of all the normal subgroups of index p is equal to the Φ-subgroup defined earlier. The factor group $\mathfrak{G}/\Phi$ is an abelian group of exponent p. By its order p^d the important invariant $d = d(\mathfrak{G})$ is defined. The significance of d is made clear by the following BURNSIDE BASIS THEOREM: *From every system of generators of $\mathfrak{G}$ exactly d can be selected so that these alone generate $\mathfrak{G}$.* By the general basis theorem this theorem need only be proven for $\mathfrak{G}/\Phi$.

---

[1] This name is used because for finite continuous groups the associated Lie ring of infinitesimal transformations is nilpotent precisely when the ascending central series terminates with the full group.

2. *Elementary Abelian Groups.*

An abelian group $\mathfrak{A}$ with prime exponent p is called an *elementary abelian group.* If $\mathfrak{A}$ is of order p^d then it is possible to generate $\mathfrak{A}$ by d elements: Let S_1 be an element $\neq e$ in $\mathfrak{A}$; let S_2 be an element in $\mathfrak{A}$ not in (S_1) ; let S_3 be an element of $\mathfrak{A}$ not in $\{S_1, S_2\}$; let $S_{d'}$ be an element in $\mathfrak{A}$ and not in $\{S_1, S_2, \ldots, S_{d'-1}\}$ and $\{S_1, S_2, \ldots, S_{d'}\} = \mathfrak{A}$. Then (S_i) is of prime order p so that we must have

$$(S_i) \cap \{S_1, S_2, \ldots, S_{i-1}\} = e$$

It follows from this that

$$\{S_1, S_2, \ldots, S_i\} = \{S_1, S_2, \ldots, S_{i-1}\} \times (S_i)$$
$$\mathfrak{A} = (S_1) \times (S_2) \times \cdots \times (S_{d'}),$$

and since $\mathfrak{A} : 1 = p^d$, , we see that $d = d'$. Therefore a finite elementary abelian group is the direct product of a finite number of cyclic groups of prime order. Conversely a direct product of a finite number of cyclic groups $(S_1), (S_2), \ldots, (S_d)$ of order p is an elementary abelian group. The elements $S_1, S_2, \ldots, S_d$ in the direct product representation are said to be a basis of $\mathfrak{A}$. The above method of construction shows that every generating system of $\mathfrak{A}$ contains a basis. Therefore d is the minimal number of generators. Consequently every system of d generators is a basis of $\mathfrak{A}$. The number of basis systems of $\mathfrak{A}$ can be calculated easily:

In the above construction there are p^d-1 possibilities for S_1; after choosing S_1, there are p^d-p possibilities for S_2 and so forth, so that we obtain the number $(p^d - 1)(p^d - p) \cdot \ldots \cdot (p^d - p^{d-1})$ as the number of basis systems of $\mathfrak{A}$. If $S_1, S_2, \ldots, S_d$ is a fixed basis and $T_1, T_2, \ldots, T_d$ is an arbitrary basis then the mapping

$$S_1^{\alpha_1} S_2^{\alpha_2}, \ldots, S_d^{\alpha_d} \rightarrow T_1^{\alpha_1} T_2^{\alpha_2}, \ldots, T_d^{\alpha_d}$$

defines an automorphism of $\mathfrak{A}$ and conversely. Therefore it follows that:

The number of automorphisms of an elementary abelian group of order p^d is equal to $(p^d - 1)(p^d - p) \cdot \ldots \cdot (p^d - p^{d-1})$. If we set

$$k_d = (p^d - 1) \ (p^{d-1} - 1) \cdot \ldots \cdot (p - 1),$$

then the number is equal to $p^{\frac{1}{2}d(d-1)} k_d$. From the general basis theorem in II, § 4, it follows that:

The number of automorphisms of a p-group of order p^n $(n > 0)$ and d generators is a divisor of

$$p^{d(n-d)} \cdot (p^d - p)(p^d - p) \cdot \ldots \cdot (p^d - p^{d-1}).$$

Remark: The highest power of p which divides this number is $p^{\frac{1}{2}d(2n-1-d)}$, and since $0 < d \leq n$, this number is a divisor of $p^{\frac{1}{2}n(n-1)}$, as can easily be seen. Therefore the number of automorphisms of an arbitrary group of order p^n is a divisor of the number of automorphisms of the elementary abelian group of order p^n.

For later theorems it is important to obtain several formulae about the number $\varphi_{d,\alpha}$ of subgroups of order p^α in the elementary abelian group of order p^d. Let $0 < \alpha \leq d$. Every subgroup of order p^α is elementary.

If $S_1, S_2, \ldots, S_\alpha$ are the first α elements of a basis of the whole group, then these α elements are a basis of a subgroup of order p^α. Conversely, as we have seen previously, every basis of a subgroup of order p^α can be extended to a basis of the whole group. Since the elements $S_1, S_2, \ldots, S_\alpha$ can be chosen in

$$(p^d - 1)(p^d - p) \cdot \ldots \cdot (p^d - p^{\alpha-1})$$

different ways, and every subgroup of order p^α has $(p^\alpha - 1) \cdot \ldots \cdot (p^\alpha - p^{\alpha-1})$ different basis systems, then

(1) $$\varphi_{d,\alpha} = \frac{(p^d - 1) \cdot \ldots \cdot (p^d - p^{\alpha-1})}{(p^\alpha - 1) \cdot \ldots \cdot (p^\alpha - p^{\alpha-1})} = \frac{k_d}{k_\alpha k_{d-\alpha}} = \varphi_{d,d-\alpha},$$

where $k_0 = 1$. From this the reader can derive the recursion formula

(2) $$\varphi_{d+1,\alpha} = \varphi_{d,\alpha} + p^{d+1-\alpha} \varphi_{d,\alpha-1}$$

for $0 < \alpha \leqq d$, where $\varphi_{d,0} = 1 = \frac{k_d}{k_0 k_d}$. If we set $\varphi_{d,\alpha} = 0$ for rational integers α which are larger than d or smaller than 0, then the formula is valid generally. From this formula we derive the congruence

(3) $$\varphi_{d+i,\alpha} \equiv \varphi_{d,\alpha} \, (p^{d+1-\alpha}) \qquad (i \geqq 1)$$

and the polynomial identity

(4) $$\prod_{\nu=1}^{d} (x - p^{\nu-1}) = \sum_{0}^{d} (-1)^\alpha p^{\frac{1}{2}\alpha(\alpha-1)} \varphi_{d,\alpha} x^{d-\alpha}$$

by induction. If we set $x = 1$, then

(5) $$0 = 1 - \varphi_{d,1} + p\varphi_{d,2} + \cdots + (-1)^d p^{\frac{1}{2}d(d-1)}.$$

An abelian group of order p^n can be decomposed, by the basis theorem, into the direct product of cyclic groups of orders $p^{n_1}, p^{n_2}, \ldots, p^{n_r}$. Here the exponents $n_1, n_2, \ldots, n_r$ are determined uniquely to within order. Therefore we say: *The group is of type* $(p^{n_1}, p^{n_2}, \ldots, p^{n_r})$. If we order the n_i by size so that p^j occurs a_j times as the order of a basis element, then we say: *The group is of type*

$$a_1 \cdot 1 + a_2 \cdot 2 + \cdots + a_n \cdot n.$$

Here the non-negative integers a_i are bound only by the relation $\sum_{1}^{n} a_i \cdot i = n$.

3. *Finite Nilpotent Groups.*

The direct product of a finite number of nilpotent groups is nilpotent, as is easily seen. For example, the direct product of a finite number of p-groups is nilpotent. The following converse is important:

THEOREM 11: *Every finite nilpotent group is the direct product of its Sylow groups.*

Proof: The normalizer of a Sylow group is its own normalizer by Theorem 5, and therefore, by Theorem 10, it is equal to the whole group; consequently every Sylow group is a normal subgroup.

Let $p_1, p_2, \ldots, p_r$ be the various prime divisors of the group order, and assume we have already shown that

$$S_{p_1} S_{p_2} \cdot \ldots \cdot S_{p_i} = S_{p_1} \times S_{p_2} \times \cdots \times S_{p_i}, \quad i < r.$$

Then the normal subgroups $S_{p_1} \cdot S_{p_2} \cdot \ldots \cdot S_{p_i}$ and $S_{p_{i+1}}$ have relatively prime orders so that their intersection is e; and therefore

$$S_{p_1} \cdot S_{p_2} \cdot \ldots \cdot S_{p_{i+1}} = S_{p_1} \times S_{p_2} \times \cdots \times S_{p_{i+1}}.$$

But from the equation $S_{p_1} \cdot S_{p_2} \cdot \ldots \cdot S_{p_r} = S_{p_1} \times S_{p_2} \times \cdots \times S_{p_r}$ it follows, by comparing the orders, that the whole group is the direct product of its Sylow groups.

THEOREM 12: *The Φ-subgroup of a nilpotent group contains the commutator group.*

Proof: As we saw earlier, the Φ-subgroup is equal to the intersection of the whole group with its maximal subgroups. By the Corollary to Theorem 10, every maximal subgroup of a nilpotent group is a normal subgroup of prime index, and therefore every maximal subgroup of a nilpotent group contains the commutator group. Consequently the Φ-subgroup of a nilpotent group contains the commutator group.

Remark: We have further that $\Phi(\mathfrak{G})/\mathfrak{G}' = \Phi(\mathfrak{G}/\mathfrak{G}')$, which can be derived from the definition of the Φ-subgroup as the intersection of the whole group with its maximal subgroups.

For finite groups we have the converse:

THEOREM 13 (Wieland): *If the Φ-subgroup of a finite group contains the commutator group, then the group is nilpotent.*

Proof: As in the proof of Theorem 11 it suffices to prove that every Sylow group is a normal subgroup. If the normalizer of a Sylow group were not the whole group, then it would be contained in a maximal subgroup which on the one hand would contain the Φ-subgroup and therefore the commutator group; and on the other hand, by Theorem 5, must be its own normalizer. Since this is not possible, every Sylow group must be a normal subgroup of the whole group.

THEOREM 14 (Hall): *If the normal subgroup $\mathfrak{N}$ is not contained in $\mathfrak{z}_i$, but is contained in $\mathfrak{z}_{i+1}$, then the following is a normal subgroup chain without repetitions*: $\mathfrak{N} \supset \mathfrak{N} \cap \mathfrak{z}_i \supset \mathfrak{N} \cap \mathfrak{z}_{i-1} \supset \cdots \supset e$.

Proof: We have $(\mathfrak{G}, \mathfrak{N}) \subseteq \mathfrak{N} \cap (\mathfrak{G}, \mathfrak{z}_{i+1}) \subseteq \mathfrak{N} \cap \mathfrak{z}_i$. Since $\mathfrak{N}$ is not contained in $\mathfrak{z}_i$, $(\mathfrak{G}, \mathfrak{N})$ is not contained in $\mathfrak{z}_{i-1}$, and therefore $\mathfrak{N} \cap \mathfrak{z}_i$ is not contained in $\mathfrak{N} \cap \mathfrak{z}_{i-1}$. We apply the same argument to $\mathfrak{N} \cap \mathfrak{z}_i$, etc., **Q.E.D.**

4. *Maximal abelian normal subgroups.*

It is natural to consider the maximal abelian normal subgroups as well as the maximal abelian factor group. In general abelian normal subgroups which are contained in no other abelian normal subgroup are neither uniquely determined nor isomorphic to each other, as is easily seen in the example of the dihedral group of eight elements. The center seems to be more appropriate as a counterpart of the factor commutator group, as we already have seen in the theorems on direct products.

In any case, there is, in every group whose elements $e=a_1, a_2, \ldots$, are well ordered, a maximal abelian normal subgroup. We can construct an abelian normal subgroup $\mathfrak{A}_\omega$ for any index ω in the following way: $\mathfrak{A}_1 = e$; let $\mathfrak{B}_\omega$ be the union of all $\mathfrak{A}_\nu$ with $\nu < \omega$; let $\mathfrak{A}_\omega$ be equal to the normal subgroup generated by $\mathfrak{B}_\omega$ and a_ω if this normal subgroup is abelian. Otherwise let $\mathfrak{A}_\omega = \mathfrak{B}_\omega$. The union of all the $\mathfrak{A}_\omega$ is a maximal abelian normal subgroup.

A maximal abelian normal subgroup $\mathfrak{A}$ *of a nilpotent group is its own centralizer.*

Proof: The centralizer $Z_\mathfrak{A}$ is a normal subgroup of $\mathfrak{G}$. If $Z_\mathfrak{A}$ contained $\mathfrak{A}$ properly, then by Theorem 14, a center element $X\mathfrak{A}$ in $Z_\mathfrak{A}/\mathfrak{A}$ would be contained in $\mathfrak{G}/\mathfrak{A}$ [1] so that the subgroup generated by X and $\mathfrak{A}$ would be larger than $\mathfrak{A}$. But since this subgroup containing $\mathfrak{A}$ would also be an abelian normal subgroup, we must have $Z_\mathfrak{A} = \mathfrak{A}$.

If $\mathfrak{G}$ and $\mathfrak{A}$ are of orders p^n and p^m, respectively then the index p^{n-m} is a divisor of the number of automorphisms of $\mathfrak{A}$, whereupon, by Part 2, it follows that

$$p^{n-m}/p^{\frac{1}{2}m(m-1)},$$

$$2n \leqq m(m+1). \tag{6}$$

5. *The automorphism group of* Z_N.

We wish to determine the automorphism group of the cyclic group Z_N for $N>1$. For this purpose we consider Z_N as the residue class module (quotient module) $\mathfrak{o}(N)$ of the additive group of integers with respect to the submodule of integers divisible by N. The operators of Z_N are given by the multiplications $\underline{t} = \begin{pmatrix} x \\ tx \end{pmatrix}$ by the rational integers t; $\underline{t_1}$ and $\underline{t_2}$ are equal if and only if t_1 and t_2 are congruent mod N. $\underline{t}$ is an automorphism if and only if t is relatively prime to N. The number $\varphi(N)$ of automorphisms of Z_N is equal to the number of residue classes (cosets) mod N which contain numbers relatively prime to N (*prime residue classes*).

The automorphism group of Z_N *(cyclic group of order* N*) is isomorphic to the group of prime residue classes mod* N. If N is the product of relatively prime numbers m_1, m_2, then Z_N is the direct product of two characteristic cyclic groups of orders m_1, m_2. For the automorphism group we have the corresponding situation; in particular

$$\varphi(N) = \varphi(m_1)\varphi(m_2).$$

---

[1] Here we must anticipate the result of § 5 which is trivial for p-groups; namely, that every factor group of a nilpoint group is itself nilpotent.

If N is the n-th power of a prime p then a residue class is prime if and only if it consists of numbers relatively prime to p; the number of these residue classes is $p^n - p^{n-1}$. If $N = p_1^{n_1} p_2^{n_2} \cdot \ldots \cdot p_r^{n_r}$ is the prime power decomposition, then

$$(7) \qquad \varphi(N) = \prod_1^r \varphi(p_i^{n_i}) = \prod_1^r (p_i^{n_i} - p_i^{n_i - 1}) = N \cdot \prod_1^r \left(1 - \frac{1}{p_i}\right).$$

The residue class ring $\mathfrak{o}(p)$ is a field and therefore, by II, § 7, the automorphism group of Z_p is cyclic of order p–1. A rational number g whose order mod p is p–1 is said to be a *primitive congruence root* mod p. g has an order which is divisible by p–1 mod p^n; say therefore, it has the order $(p-1) \cdot p^\nu$. The order of $g_1 = g^{p^\nu}$ is then equal to p-1, mod p^n. If $a = 1 + kp^m$, then it follows from the binomial theorem that $a^p \equiv 1 + kp^{m+1} + \frac{p(p-1)}{2} k^2 p^{2m} (p^{m+2})$. Therefore $a \equiv 1 (p^m)$ implies that $a^p \equiv 1 (p^{m+1})$. However if $m > 1$ or if p is odd then $a \not\equiv 1 (p^{m+1})$ implies that $a^p \not\equiv 1 (p^{m+2})$. If p is odd, then $1 + p$ is of order p^{n-1} mod p^n, $(1 + p) \cdot g_1$ is of order $(p-1) \cdot p^{n-1}$. If $p=2$, then $1 + 2^2$ is of order 2^{n-2} mod 2^n $(n > 2)$. Since –1 is congruent to no power of 5 mod 4, there are, mod 2^n, the 2^{n-1} different prime residue classes $\pm 5^\nu$ $(0 \leqq \nu < 2^{n-2})$. As a result we obtain:

If $n < 3$ or p is odd, then the automorphism group of Z_{p^n} is cyclic of order $(p-1)p^{n-1}$. The automorphism group of Z_{2^n}, for $n > 2$, is abelian of type $(2^{n-2}, 2)$ with the associated basis automorphisms 5 and –1.

6. *p-Groups with only one Subgroup of Order p.*

A non-cyclic abelian group of exponent p^n contains at least two different subgroups of order p.

Proof: Let A be an element of order p^n and let B not be a power of A. Then the order p^ν of B mod (A) is greater than 1, but at most p^n. We have

$$B^{p^\nu} = A^r, \; B^{p^n} = A^{r \cdot p^{n-\nu}} = e, \; r = s \cdot p^\nu, \; (B \cdot A^{-s})^{p^\nu} = e.$$

Therefore $(BA^{-s})^{p^\nu - 1}$ and $A^{p^{n-1}}$ generate two different subgroups of order p.

We wish to find non-abelian groups of order p^n which contain only one subgroup of order p.

An example is the *quaternion group*. By the theorem of Hölder it is defined by the relations $A^4 = 1$, $BAB^{-1} = A^{-1}$, $B^2 = A^2$ as a group of order 8 with generators A and B. Its eight elements are called quaternions; they are

$$1, A, \quad A^2, \quad A^3$$

$$B, AB, A^2 B, A^3 B.$$

If instead we write

$$1, \mathfrak{i}, -1, -\mathfrak{i}$$

$$\mathfrak{j}, \mathfrak{k}, -\mathfrak{j}, -\mathfrak{k}$$

and set

$$-(-1)=1,\ -(-\mathfrak{i})=\mathfrak{i},\ -(-\mathfrak{j})=\mathfrak{j},\ -(-\mathfrak{k})=\mathfrak{k},$$

then we have the following calculational rules:

$$1\cdot x = x\cdot 1 = x,\ -1\cdot x = x\cdot -1 = -x,\ (-1)^2=1,$$
$$\mathfrak{i}^2=\mathfrak{j}^2=\mathfrak{k}^2=-1,\ \mathfrak{i}\mathfrak{j}=-\mathfrak{j}\mathfrak{i}=\mathfrak{k},\ \mathfrak{j}\mathfrak{k}=-\mathfrak{k}\mathfrak{j}=\mathfrak{i},$$
$$\mathfrak{k}\mathfrak{i}=-\mathfrak{i}\mathfrak{k}=\mathfrak{j}.$$

From this we conclude that there is only one subgroup of order 2 and exactly three subgroups of order 4. The center is equal to the commutator group which is equal to (-1).

The *generalized quaternion group* is defined by the relations

$$A^{2^{n-1}}=1,\ BAB^{-1}=A^{-1},\ B^2=A^{2^{n-2}} \qquad (n>2) \tag{8}$$

as a group generated by A, B, and of order 2^n, by the Hölder Theorem. Since

$$(BA^\nu)^2 = BA^\nu B^{-1}\cdot B^2\cdot A^\nu = A^{-\nu}B^2A^\nu = B^2,$$

this group contains only one subgroup of order 2. The elements $A^{2^{n-3}}$ and B generate a quaternion group.

The relations above can be written more elegantly in the form

$$A^{2^{n-2}} = B^2 = (AB)^2\ . \tag{9}$$

The new relations follow from those above.

From the new relations, however, it follows that

$$BAB^{-1} = A^{-1}(AB)^2B^{-2} = A^{-1}$$
$$BA^{2^{n-2}}B^{-1} = A^{-2^{n-2}} = BB^2B^{-1} = B^2 = A^{2^{n-2}}$$
$$A^{2^{n-1}} = 1,$$

and therefore the old relations follow.

If A' is of order 2^{n-1}, B' of order 4, and if A' and B' generate the whole group, then

$$A'^{2^{n-2}} = B'^2 = (A'B')^2 = A^{2^{n-2}}.$$

Therefore all the calculational rules which are valid for power products of A and B also remain valid for the corresponding power products of A' and B'.

Since A' and B' generate the whole group, (A') is a normal subgroup of index 2, and every element can be written uniquely in the form

$$A'^\nu B'^\mu\ (0\leqq \nu < 2^{n-1},\ 0\leqq \mu < 2).$$

Therefore the mapping $A^\nu B^\mu \to A'^\nu B'^\mu$ is an automorphism of the group. The number of all the automorphisms is equal to the number of pairs A', B'. It follows by simple enumeration that:

The quaternion group has exactly 24 automorphisms. The generalized quaternion group of order 2^n *has exactly* 2^{2n-3} *automorphisms for* $n > 3$.

In the automorphism group A of the quaternion group, the inner automorphisms form an abelian normal subgroup J of order 4. An automorphism which commutes with all the inner automorphisms is itself an inner automorphism. Since it changes each generator by a factor in the center, there are at most $2 \cdot 2$ such automorphisms.

A group A having order 24, and containing a normal subgroup J of order 4 which is its own centralizer, must be isomorphic to $\mathfrak{S}_4$.

This is because a central element of A must be in J, and an element of order 3 must transform the three elements $\neq e$ in J in a cyclic manner. Then, since according to the results of Sylow there are elements of order 3, the center is e, and there is no normal subgroup of order 3. From these results also, the index in A of the normalizer N_3 of a Sylow 3-group is 4. A transitive representation of A in 4 letters is associated with N_3. The representation is faithful since the intersection of all 3–normalizers contains only center elements with orders 1 or 2 and therefore is e. Since A consists of 24 elements, A is isomorphic to $\mathfrak{S}_4$.

The automorphism group of the quaternion group is isomorphic to the symmetric permutation group of four letters.

The quaternion group is the only p-group which contains two different cyclic subgroups of index p but only one subgroup of order p.

Proof:[1] Let $\mathfrak{G}$ be of order p^n and let it contain two different cyclic subgroups $\mathfrak{U}_1$ and $\mathfrak{U}_2$ of index p. $\mathfrak{U}_1$ and $\mathfrak{U}_2$ are different normal subgroups of index p, and therefore their intersection $\mathfrak{D}$ is of index p^2. Moreover $\mathfrak{D}$ is in the center and contains the commutator group. It follows for any two elements x, y that x^p and y^p are in $\mathfrak{D}$, and that

$$(y, x)^p = (y^p, x) = e, \quad (xy)^p = (y, x)^{\frac{1}{2}p(p-1)} x^p y^p.$$

If p is odd, then $(xy)^p = x^p y^p$, and therefore the operation of raising to power p is a homomorphy. Since the group of p-th powers is contained in $\mathfrak{D}$, by the first isomorphism theorem the elements whose p-th power is e form a subgroup whose order is at least p^2. There are at least two different subgroups of order p in this subgroup.

If $p = 2$, then $(xy)^4 = (y, x)^6 x^4 y^4 = x^4 y^4$. Now we conclude just as above that either $\mathfrak{D} = 1$ and $\mathfrak{U}_1$, $\mathfrak{U}_2$ are two different subgroups of order 2, or there are two subgroups $\mathfrak{u}_1 \neq \mathfrak{u}_2$ of order 4 by the first isomorphism theorem. We may assume that $\mathfrak{u}_1$ is in $\mathfrak{U}_1$. If $\mathfrak{u}_1$ is different from $\mathfrak{U}_1$, then $\mathfrak{u}_1$ is in $\mathfrak{D}$ and $\mathfrak{u}_1 \cdot \mathfrak{u}_2$ is an abelian group of order 8. Since it contains two different subgroups of index 2, it is not cyclic, and therefore it also contains two different subgroups of order 2. If, in conclusion, $\mathfrak{u}_1 = \mathfrak{U}_1$, then the whole group is of order 8. Let $\mathfrak{U}_1 = (A)$ and $\mathfrak{U}_2 = (B)$. If there is only one subgroup of order 2 then $B^2 = (AB)^2 = A^2$, and therefore the group is the quaternion group.

THEOREM 15: *A p-group which contains only one subgroup of order p is either cyclic or a generalized quaternion group.*

Proof: Let $\mathfrak{G}$ be of order p^n and let it contain only one subgroup of order p. First

[1] In accordance with a communication from Herr Maass, Hamburg.

let p be odd. If $n=0, 1$, then the theorem is clearly true. We now apply induction to n.

Every subgroup of index p is cyclic by the induction hypothesis, and therefore by what was proven previously there is only one subgroup of index p in $\mathfrak{G}$, and therefore $\mathfrak{G}$ itself is cyclic.[1]

Now let $p=2$ and let $\mathfrak{A}$ be a maximal abelian normal subgroup. $\mathfrak{A}$ is cyclic and its own centralizer. Therefore $\mathfrak{G}/\mathfrak{A}$ is isomorphic to a group of automorphisms of $\mathfrak{A}$. We shall show that only one automorphism of order 2 can occur, namely, the operation of a raising the elements of $\mathfrak{A}$ to the power -1. Since this automorphism is not the square of any other automorphism of $\mathfrak{A}$, it follows that $\mathfrak{G}:\mathfrak{A}$ is either 1 or 2. If we set $\mathfrak{A}=(A)$ and assume that $B \not\equiv e(\mathfrak{A})$, $B^2 \equiv e(\mathfrak{A})$, then as a preliminary BAB^{-1} must be shown to be equal to A^{-1}. In fact, we want to show further that the group generated by A and B is a generalized quaternion group with relations (8) and (9). Then the theorem will be proven.

Since B cannot commute with all the elements of A, $(B^2) \neq A$, and there is a subgroup $\mathfrak{A}_1$ of $\mathfrak{A}$ which contains (B^2) as a subgroup of index 2. The group $\mathfrak{A}_1(B)$ contains the two different cyclic subgroups $\mathfrak{A}_1$ and (B) of index 2; and therefore it is, as was previously shown, the quaternion group. If A is of order 2^m then: $B^2 = A^{2^{m-1}}$. We also conclude $(AB)^2 = A^{2^{m-1}}$. Therefore A and B generate the generalized quaternion group of order 2^{m+1}.

THEOREM 16: *A group of order p^n is cyclic if it contains only one subgroup of order p^m (where $1<m<n$).*

Proof: There is a subgroup $\mathfrak{U}$ of order p^m. $\mathfrak{U}$ is contained in a subgroup $\mathfrak{U}_1$ of order p^{m+1} and is the only subgroup of index p in $\mathfrak{U}_1$. Therefore $\mathfrak{U}_1$ is cyclic and consequently $\mathfrak{U}$ is cyclic. Since every subgroup of order p or p^2 is contained in a subgroup of order p^m, and since the only subgroup of order p^m is cyclic, there is only one subgroup of order p and one of order p^2. Since the generalized quaternion group contains some subgroups of order 4, we conclude from the previous theorem that the whole group is cyclic.

If in a p-group, every subgroup of order p^2 is cyclic, then there is only one subgroup of order p, and conversely.

If there were two different subgroups of order p then we can assume that one of them is contained in the center. But then the product of the two subgroups is a non-cyclic group of order p^2. Conversely, in a non-cyclic group of order p^2 there are certainly two different subgroups of order p. Now one can easily prove:

THEOREM 17: *A group of order p^n in which every subgroup of order p^m is cyclic, where $1<m<n$, is cyclic except in the case $p=2$, $m=2$ in which case the group can also be a generalized quaternion group.*

7. *p-Groups with a Cyclic Normal Subgroup of Index p.*

We shall determine all the p-groups which contain a cyclic normal subgroup of index p. This problem will now be solved for non-abelian p-groups, which contain some subgroups of order p. If $\mathfrak{G}$ is of order p^n, then in $\mathfrak{G}$ there is an element A

[1] This last by the basis theorem.

of order p^{n-1} and an element B of order p which is not a power of A. A and B generate $\mathfrak{G}$, and the subgroup (A) of index p is a normal subgroup. Therefore

$$A^{p^{n-1}} = B^p = 1, \qquad BAB^{-1} = A^r,$$
$$r \not\equiv 1\,(p^{n-1}), \qquad r^p \equiv 1\,(p^{n-1}).$$

If for odd p the element B is replaced by an appropriate power, then we can take $r = 1 + p^{n-2}$.

If $p = 2$, $n = 3$, then we must have $r \equiv -1\,(4)$. If $p=2$, $n>3$, then there are three possibilities for r,

$$r \equiv -1,\ 1 + 2^{n-2},\ -1 + 2^{n-2}(2^{n-1}).$$

The number r is not altered mod 2^{n-1} if B is replaced by BA^{μ}.

If $r \equiv 1 + 2^{n-2}$, then the commutator subgroup is of order 2; in the other two cases it is of order 2^{n-1}.

If $r \equiv -1$, then $(BA^{\nu})^2 = (BA^{\nu}B^{-1})B^2A^{\nu} = A^{-\nu}B^2A^{\nu} = 1$, and therefore there is only one cyclic subgroup of index 2. Thus r is uniquely determined by the group. As a result we obtain:

The groups $\mathfrak{G}$ of order p^n which contain an element A of order p^{n-1}, are of the following types:

a) *$\mathfrak{G}$ abelian:*

$n \geqq 1$ I $\quad Z_{p^n}: B^{p^n} = 1$

$n \geqq 2$ II $\quad A^{p^{n-1}} = 1,\quad B^p = 1,\quad AB = BA;$

b) *$\mathfrak{G}$ non-abelian, p odd:*

$n \geqq 3$ III $\quad A^{p^{n-1}} = 1,\quad B^p = 1,\quad BAB^{-1} = A^{1+p^{n-2}};$

c) *$\mathfrak{G}$ non-abelian, $p = 2$:*

$n \geqq 3$ III *generalized quaternion group:*

$$A^{2^{n-1}} = 1,\quad B^2 = A^{2^{n-2}},\quad BAB^{-1} = A^{-1}$$

$n \geqq 3$ IV *dihedral group D_{2^n}:*

$$A^{2^{n-1}} = 1,\quad B^2 = 1,\quad BAB^{-1} = A^{-1}$$

$n \geqq 4$ V $\quad A^{2^{n-1}} = 1,\quad B^2 = 1,\quad BAB^{-1} = A^{1+2^{n-2}},$

$n \geqq 4$ VI $\quad A^{2^{n-1}} = 1,\quad B^2 = 1,\quad BAB^{-1} = A^{-1+2^{n-2}}.$

Groups of different type are not isomorphic. From Hölder's theorem it follows that all types exist. For $n = 3$, V will coincide with IV, and VI with II.

Now it is simple to give all groups of order p^3. We must now investigate among such all those in which the p-th power of every element is e. A group in which all squares are equal to e is abelian since

$$x = x^{-1}, \text{ thus } xy = (xy)^{-1} = y^{-1}x^{-1} = yx.$$

If the group is non-abelian and p is odd, then it is generated by two elements

A and B such that the relations

$$A^p = B^p = (A, B)^p = 1, A(A, B) = (A, B)A, \quad B(A, B) = (A, B)B$$

hold. By III, Theorem 21, these relations define a non-abelian group with generators A, B and order p^3, in which, for any two elements x, y, we have:

$$(xy)^p = (x, y)^{-\frac{1}{2}p(p-1)} x^p y^p = x^p y^p.$$

Thus the p-th power of every element is equal to e. As a result we obtain:

There are, for every prime number p, five types of groups of order p^3, namely the three abelian types:

I. $Z_{p^3} : B^{p^3} = 1,$

II. $A^{p^2} = 1,\ B^p = 1,\ AB = BA,$

VII. $A^p = B^p = C^p = 1,\ AB = BA,\ AC = CA,\ BC = CB$,

and two non-abelian types, which are, for $p=2$: III, the quaternion group and IV, the dihedral group, and for odd p the types

III. $A^{p^2} = 1,\ B^p = 1,\ BAB^{-1} = A^{1+p},$

IV. $A^p = B^p = (A, B)^p = 1,\ A(A, B) = (A, B)A,\ B(A. B) = (A, B)B.$

Exercises

1. If a p-group contains a cyclic normal subgroup of index p, then every subgroup different from e has the same property.

2. For odd p, the following properties hold for abelian groups of type (p, p^{n-1}) and for non-abelian groups of order p^n having a cyclic subgroup of index p, where m is a number greater than zero and less than n:

a) The number of subgroups of order p^m is $1 + p$ in both cases.

b) The number of cyclic subgroups of order p^m is, in both cases, $1+p$ or p according to whether $m=1$ or $m>1$.

c) The number of elements whose p^m-th power$=e$ is p^{m+1} in both cases.

d) In both groups, every subgroup whose order is divisible by p^2 is a normal subgroup. Therefore for $m>1$ there are equally many normal subgroups of order p^m.

e) The number of automorphisms is $p^n(p-1)$.

3. The two types of non-abelian groups of order p^3 can be defined by the relations

III. $A^p = B^p = (A, B),$

IV. $A^p = B^p = (A, A, B) = (B, B, A) = 1$

for all p by an appropriate choice of generators A, B.

4. If a 2-group contains a cyclic subgroup of index 2 and is neither abelian of type (2, 2) nor the quaternion group, then the number of its automorphisms is a power of 2.

5. In a finite group, the index of the normalizer over the centralizer of a Sylow p-group with d generators is a divisor of k_d. If the order of the group is divisible neither by the third power of its smallest prime factor p, nor by 12, then every Sylow p-group is in the center of its normalizer.

6. In an abelian p-group $\mathfrak{G}$ with the exponent p^m, the characteristic chains

$$\mathfrak{G} \supset \mathfrak{G}^p \supset \mathfrak{G}^{p^2} \supset \cdots \supset \mathfrak{G}^{p^m} = e$$

and

$$\mathfrak{G} = \mathfrak{G}_{p^m} \supset \mathfrak{G}_{p^{m-1}} \supset \cdots \supset \mathfrak{G}_p \supset e$$

give rise to a characteristic series through the refinement process which was given in the proof of the Jordan-Hölder-Schreier Theorem. There is only this one characteristic series. (Here $\mathfrak{G}^{p^\nu}$ denotes the group of the p^ν-th powers and $\mathfrak{G}_{p^\nu}$ denotes the group of all elements whose p^ν-th power is e.)

7. Theorem 2 in § 1 admits the following corollaries: If $\mathfrak{P}$ is a Sylow p-group, in $\mathfrak{G}$, $N_{\mathfrak{P}}$ its normalizer, $\mathfrak{N}$ a normal subgroup of $\mathfrak{G}$, then

a) $N_{\mathfrak{P}}\mathfrak{N}/\mathfrak{N}$ is the normalizer of the Sylow p-group $\mathfrak{P}\mathfrak{N}/\mathfrak{N}$ of $\mathfrak{G}/\mathfrak{N}$;

b) $N_{\mathfrak{P}}$ is contained in the normalizer $N_{\mathfrak{p}}$ of the Sylow p-group $\mathfrak{p} = \mathfrak{P} \cap \mathfrak{N}$ of $\mathfrak{N}$;

c) $N_{\mathfrak{p}}\mathfrak{N} = \mathfrak{G}$; therefore by the Second Isomorphism Theorem

$$N_{\mathfrak{p}}/N_{\mathfrak{p}} \cap \mathfrak{N} \cong \mathfrak{G}/\mathfrak{N}.$$

(Hint for a): If $x\mathfrak{P}\mathfrak{N}x^{-1} = \mathfrak{P}\mathfrak{N}$, then by Theorem 3: $x\mathfrak{P}x^{-1} = \nu\mathfrak{P}\nu^{-1}$ is solvable for ν in $\mathfrak{N}$, therefore $\nu^{-1}x \in N_{\mathfrak{P}}$; (for c): for every x in $\mathfrak{G}$, $x\mathfrak{p}x^{-1} = \nu\mathfrak{p}\nu^{-1}$ is solvable for ν in $\mathfrak{N}$.)

With the help of c) it should be shown that the Φ-subgroup of a finite group is nilpotent.

§ 4. On the Enumeration Theorems of the Theory of p-Groups

In the study of finite groups the question arises naturally as to the number of elements or subgroups with some given property. The results obtained in connection with this question do not lie very deep.

The following systematic derivation of the enumeration theorems in p-groups is due to P. Hall.

THEOREM 18 (Counting Principle): Let $\mathfrak{G}$ be a finite p-group. $\mathfrak{M}_\alpha$ denotes any subgroup of index p^α which contains $\Phi(\mathfrak{G})$. Let $(\mathfrak{K})$ be a set of complexes such that each complex $\mathfrak{K}$ in $(\mathfrak{K})$ is contained in at least one subgroup of index p. Let $n(\mathfrak{M}_\alpha)$ be the number of complexes of $(\mathfrak{K})$ which are contained in $\mathfrak{M}_\alpha$. Then

$$n(\mathfrak{M}_0) - \sum_{(\mathfrak{M}_1)} n(\mathfrak{M}_1) + p\sum_{(\mathfrak{M}_2)} n(\mathfrak{M}_2) - p^3\sum_{(\mathfrak{M}_3)} n(\mathfrak{M}_3) + \cdots$$
$$+ (-1)^\alpha p^{\frac{1}{2}\alpha(\alpha-1)} \sum_{(\mathfrak{M}_\alpha)} n(\mathfrak{M}_\alpha) + \cdots + (-1)^d p^{\frac{1}{2}d(d-1)} n(\mathfrak{M}_d) = 0,$$

where the summation $\sum_{(\mathfrak{M}_\alpha)}$ is extended over the $\varphi_{d,\alpha}$ subgroups $\mathfrak{M}_\alpha$ of $\mathfrak{G}$.

Proof: We shall show that the number of times that an element $\mathfrak{K}$

in $(\mathfrak{K})$ is "counted" with the appropriate sign on the left of the equation above is equal to zero.

The intersection of all $\mathfrak{M}_\alpha$ which contain $\mathfrak{K}$, contains $\Phi(\mathfrak{G})$, and therefore is an $\mathfrak{M}_\varrho$. By hypothesis $\mathfrak{K}$ is contained in an $\mathfrak{M}_1$ and therefore $\varrho > 0$. The number of all $\mathfrak{M}_\alpha$'s which contain $\mathfrak{K}$ is equal to the number of all $\mathfrak{M}_\alpha$'s which contain $\mathfrak{M}_\varrho$, i.e., $\varphi_{\varrho,\alpha}$. Therefore the number of times that $\mathfrak{K}$ is "counted" is

$$1 - \varphi_{\varrho,1} + p\,\varphi_{\varrho,2} - p^3\varphi_{\varrho,3} + \cdots + (-1)^\alpha p^{\frac{1}{2}\alpha(\alpha-1)}\varphi_{\varrho,\alpha} + \cdots + (-1)^\varrho p^{\frac{1}{2}\varrho(\varrho-1)}.$$

But this number is zero, by § 3, Formula 5, Q.E.D.

THEOREM 19: *The number of subgroups of fixed order p^m $(0 \leq m \leq n)$ of a p-group $\mathfrak{G}$ of order p^n leaves* 1 *as a remainder when divided by p.*

Proof: If $n = 0$, then the theorem is clear. Now let $n > 0$ and assume that the theorem is proven for p-groups whose order is less than p^n. If $m = n$ then the theorem is trivial. Let $m < n$. For Theorem 19, let $(\mathfrak{K})$ denote the set of all subgroups of $\mathfrak{G}$ of order p^m. Then:

$$n(\mathfrak{M}_0) \equiv \sum_{\mathfrak{M}_1} n(\mathfrak{M}_1)\,(p),$$

and by the induction hypothesis

$$n(\mathfrak{M}_1) \equiv 1\,(p);$$

moreover the number of all $\mathfrak{M}_1$ is $\varphi_{d,d-1}$, and therefore by § 3 congruent to 1 mod p, so that

$$n(\mathfrak{M}_0) \equiv 1\,(p)$$

follows, Q.E.D.

THEOREM 20 (Kulakoff): *In a non-cyclic p-group of odd order p^n, the number of subgroups of order p^m $(0 < m < n)$ is congruent to $1+p$ modulo p^2.*

In the non-cyclic group of order p^2 there are $p+1$ subgroups of order p. We apply induction on n and assume $n > 2$.

The number of all $\mathfrak{M}_1$ is $\varphi_{d,d-1}$, and therefore, since $d > 1$, is congruent to $1+p$ mod p^2. Let $m < n$-1, $(\mathfrak{K})$ be the set of all subgroups of order p^m. By the Counting Principle it follows that

$$n(\mathfrak{M}_0) \equiv \sum_{\mathfrak{M}_1} n(\mathfrak{M}_1) - p\sum n(\mathfrak{M}_2).$$

By Theorem 19

$$n(\mathfrak{M}_2) \equiv 1\,(p)$$

and by § 3, 2., the number of all $\mathfrak{M}_2$ (namely $\varphi_{d,\,d-2}$) is congruent to 1 mod p. Consequently

$$n(\mathfrak{M}_0) \equiv \sum_{\mathfrak{M}_1} n(\mathfrak{M}_1) - p\,(p^2).$$

For the non-cyclic $\mathfrak{M}_1$, $n(\mathfrak{M}_1) \equiv 1 + p(p^2)$ by the induction hypothesis. As was shown, the number of all $\mathfrak{M}_1$ is congruent to $1 + p\ (p^2)$. If there is no cyclic subgroup of index p in $\mathfrak{G}$, then

$$\sum_{\mathfrak{M}_1} n(\mathfrak{M}_1) \equiv (1+p)^2 - p\,(\mathrm{mod}\,p) \equiv 1 + p\,(\mathrm{mod}\,p^2).$$

If $\mathfrak{G}$ contains a cyclic subgroup of index p, then the theorem follows from the solution of Exercise 2a at the end of § 3.

THEOREM 21 (Miller): *In a non-cyclic group of odd order p^n, the number of cyclic subgroups of order p^m $(1 < m < n)$ is divisible by p.*

Proof: If $\mathfrak{G}$ contains a cyclic subgroup of index p, then the theorem follows from the solution of Exercise 2b at the end of § 3. To continue, let every subgroup of index p in $\mathfrak{G}$ be non-cyclic, $m < n$-1 and assume the proof has been carried out already for smaller n. Let $(\mathfrak{K})$ be the set of cyclic subgroups of order p^m. We find the congruence:

$$n(\mathfrak{M}_0) \equiv \sum_{\mathfrak{M}_1} n(\mathfrak{M}_1)\,(p).$$

By the induction hypothesis each of the numbers $n(\mathfrak{M}_1)$ is divisible by p, and therefore the desired number $n(\mathfrak{M}_0)$ is also.

THEOREM 22 (Hall): *The number of subgroups of index p^α in $\mathfrak{G}$ $(0 \leq \alpha \leq d)$ is congruent to $\varphi_{d,\,\alpha}(\mathrm{mod}\,p^{d-\alpha+1})$. The number of those subgroups which do not contain $\Phi(\mathfrak{G})$ is consequently divisible by $p^{d-\alpha+1}$.*

Proof: If $d = n$, then the number in question is already known to be $\varphi_{d,\,\alpha}$. Let $n > 1$ and let the theorem be proved for smaller n. If $\alpha = 0$, then the theorem is clearly true. Let $\alpha > 0$, then $n(\mathfrak{M}_\beta)$ is equal to the number of all subgroups of index $p^{\alpha-\beta}$ in $\mathfrak{M}_\beta$. Therefore $n(\mathfrak{M}_\beta) = 0$ if $\beta > \alpha$; but otherwise by the induction hypothesis

$$n(\mathfrak{M}_\beta) \equiv \varphi_{d(\mathfrak{M}_\beta),\,\alpha-\beta}\,(p^{d(\mathfrak{M}_\beta)-(\alpha-\beta)+1}).$$

Since $d(\mathfrak{M}_\beta) \geq d - \beta$ and therefore by § 3

$$\varphi_{d(\mathfrak{M}_\beta),\,\alpha-\beta} \equiv \varphi_{d-\beta,\,\alpha-\beta}\,(p^{(d-\beta)-(\alpha-\beta)+1}),$$

we have
$$n(\mathfrak{M}_\beta) \equiv \varphi_{d-\beta,\,\alpha-\beta}\,(p^{d-\alpha+1}).$$

The Counting Principle now gives the congruence

$$n(\mathfrak{M}_0) \equiv \varphi_{d,1}\varphi_{d-1,\alpha-1} - p\,\varphi_{d,2}\varphi_{d-2,\alpha-2} + \cdots + (-1)^{\alpha-1}p^{\frac{1}{2}\alpha(\alpha-1)}\varphi_{d,\alpha}\varphi_{d-\alpha,0}(p^{d-\alpha+1}).$$

But by the Counting Principle, the right side of the congruence is exactly the number of subgroups of index p^α in an elementary abelian group of order p^d, so that

$$n(\mathfrak{M}_0) \equiv \varphi_{d,\alpha}(p^{d-\alpha+1}), \qquad \text{Q.E.D.}$$

Exercise (Kulakoff): In a non-cyclic p-group of odd order p^n, the number of solutions of $x^{p^m} = e\,(0 < m < n)$ is divisible by p^{m+1}.

Exercise: The number of normal subgroups of order p^m in a group of order $p^n\,(0 < m < n)$ is congruent to 1 (mod p).

If p is odd, $1 < m$, and $\mathfrak{G}$ is non-cyclic then, more precisely, the number is congruent to $1 + p$ (mod p^2).

§ 5. On the Descending Central Series

P. Hall has generalized the concept of a terminating ascending central series by defining:

A chain of normal subgroups of $\mathfrak{G}$

(1) $$\mathfrak{G} = \mathfrak{N}_1 \supseteqq \mathfrak{N}_2 \supseteqq \mathfrak{N}_3 \supseteqq \cdots \supseteqq \mathfrak{N}_{r+1} = e$$

is called a *central chain* if $\mathfrak{N}_i/\mathfrak{N}_{i+1}$ is contained in the center of $\mathfrak{G}/\mathfrak{N}_{i+1}$ $(i = 1, 2, \ldots, r)$.

If the ascending central series (See II § 4, 3.) terminates, then it is a central chain. The following definition is still more useful: A chain of subgroups

$$\mathfrak{G} = \mathfrak{N}_1 \supseteqq \mathfrak{N}_2 \supseteqq \cdots \supseteqq \mathfrak{N}_{r+1} = e$$

is said to be a central chain if the mutual commutator group $(\mathfrak{G}, \mathfrak{N}_i)$ is contained in $\mathfrak{N}_{i+1}$ $(i = 1, \ldots, r)$. Since for every x_i in $\mathfrak{N}_i$, x in $\mathfrak{G}$: $x x_i x^{-1} x_i^{-1} \equiv e(\mathfrak{N}_{i+1})$, and thus certainly $x x_i x^{-1} \in \mathfrak{N}_i$, it follows that $\mathfrak{N}_i$ is a normal subgroup of $\mathfrak{G}$ and that $\mathfrak{N}_i/\mathfrak{N}_{i+1}$ is contained in the center of $\mathfrak{G}/\mathfrak{N}_{i+1}$. The converse is clear. $\mathfrak{N}_{r+1}$ is contained in $\mathfrak{z}_0$; if it has already been shown that $\mathfrak{N}_{r+1-i}$ is in $\mathfrak{z}_i$ where $i < r$, then

$$(\mathfrak{G}, \mathfrak{N}_{r-i}) \subseteqq \mathfrak{N}_{r+1-i} \subseteqq \mathfrak{z}_i,$$

and therefore $\mathfrak{N}_{r-i} \subseteqq \mathfrak{z}_{i+1}$. Hence

(2) $$\mathfrak{N}_{r+1-i} \subseteqq \mathfrak{z}_i \qquad (i = 0, 1, 2, \ldots, r).$$

Consequently $\mathfrak{z}_r = \mathfrak{G}$.

If a group has a central chain, then it is nilpotent and the length of every central chain is at least equal to the class of the group.

Now it is natural to define the *descending central series* for an arbitrary group $\mathfrak{G}$ as $\mathfrak{Z}_1 \geqq \mathfrak{Z}_2 \geqq \mathfrak{Z}_3, \ldots,$ where $\mathfrak{Z}_1(\mathfrak{G}) = \mathfrak{Z}_1 = \mathfrak{G}$, $\mathfrak{Z}_2(\mathfrak{G}) = \mathfrak{Z}_2 = (\mathfrak{G}, \mathfrak{G}) = \mathfrak{G}', \ldots, \mathfrak{Z}_{n+1}(\mathfrak{G}) = \mathfrak{Z}_{n+1} = (\mathfrak{G}, \mathfrak{Z}_n) \ \ldots$

If $\mathfrak{G}$ has a central chain (1) then it follows by induction that: $\mathfrak{Z}_1 \leqq \mathfrak{N}_1, \mathfrak{Z}_2 \leqq \mathfrak{N}_2, \ldots, \mathfrak{Z}_{r+1} \leqq \mathfrak{N}_{r+1}$, and therefore $\mathfrak{Z}_{r+1} = e$. If, conversely, the descending central series is equal to e from the $(r+1)$-th place on, then $\mathfrak{G} = \mathfrak{Z}_1 \geqq \mathfrak{Z}_2 \geqq \cdots \geqq \mathfrak{Z}_{r+1} = e$ is a central chain. If c is the class of $\mathfrak{G}$, then $r \geqq c$, $\mathfrak{Z}_i \leqq \mathfrak{z}_{c+1-i}$, and therefore $\mathfrak{Z}_c \neq e$, $\mathfrak{Z}_{c+1} = e$.

In a nilpotent group, the class c can be found from the relation:

(3) $$\mathfrak{G} = \mathfrak{Z}_1 > \mathfrak{Z}_2 > \cdots > \mathfrak{Z}_{c+1} = e$$

By Chapter II, § 6, $\mathfrak{Z}_i$ is a commutator form of $\mathfrak{G}$ of weight[1] i and of degree 1 and is generated by the higher commutators $(G_1, G_2, \ldots, G_i)$ where $G_j \in \mathfrak{G}$. Therefore $\mathfrak{Z}_i$ is a fully invariant subgroup of $\mathfrak{G}$.

For every subgroup $\mathfrak{U}$ of $\mathfrak{G}$ it follows that

$$\mathfrak{Z}_i(\mathfrak{U}) \leqq \mathfrak{Z}_i(\mathfrak{G}).$$

If $\mathfrak{N}$ is a normal subgroup of $\mathfrak{G}$, then

$$\mathfrak{Z}_i(\mathfrak{G}/\mathfrak{N}) = \mathfrak{Z}_i(\mathfrak{G})\mathfrak{N}/\mathfrak{N}.$$

It follows from this that:

Every subgroup and every factor group of a nilpotent group is itself nilpotent, and the class of the subgroup or factor group is at most equal to the class of the whole group.

We wish to state something about the positional relationships, and the mutual commutator groups, of members of the descending central series and of an arbitrary central chain.

If $\mathfrak{N}_1 \geqq \mathfrak{N}_2 \geqq \mathfrak{N}_3 \ldots$ is a sequence of subgroups of an arbitrary group $\mathfrak{G}$, so that $(\mathfrak{G}, \mathfrak{N}_j) \leqq \mathfrak{N}_{j+1}$ $(j = 1, 2, \ldots)$, it follows immediately that $\mathfrak{N}_i$ is a normal subgroup of $\mathfrak{G}$. If moreover $\mathfrak{Z}_i \leqq \mathfrak{N}_j$, then it follows by induction that $\mathfrak{Z}_{i+k} \leqq \mathfrak{N}_{j+k}$ $(k = 0, 1, 2, \ldots)$. In nilpotent groups of class c we can conclude from this that:

(4) $\mathfrak{Z}_i$ is not contained in $\mathfrak{z}_{c-i}$ (since otherwise we would have $\mathfrak{Z}_c = e$).

---

[1] $\mathfrak{Z}_i$ is also called the i-th *Reidemeister commutator group.*

Now we claim that in the general case

(5) $$(\mathfrak{Z}_i, \mathfrak{N}_j) \subseteq \mathfrak{N}_{i+j}.$$

We carry out the proof by induction on i. By hypothesis

$$(\mathfrak{Z}_1, \mathfrak{N}_j) = (\mathfrak{G}, \mathfrak{N}_j) \subseteq \mathfrak{N}_{j+1}.$$

Let $i > 1$ and assume we have already proven that $(\mathfrak{Z}_{i-1}, \mathfrak{N}_k) \subseteq \mathfrak{N}_{i+k-1}$ for all k.

Then by II. Theorem 14:

$$(\mathfrak{Z}_i, \mathfrak{N}_j) = (\mathfrak{N}_j, \mathfrak{Z}_i) = (\mathfrak{N}_j, (\mathfrak{G}, \mathfrak{Z}_{i-1}))$$
$$= (\mathfrak{N}_j, \mathfrak{G}, \mathfrak{Z}_{i-1}) \subseteq (\mathfrak{G}, \mathfrak{Z}_{i-1}, \mathfrak{N}_j) \cdot (\mathfrak{Z}_{i-1}, \mathfrak{N}_j, \mathfrak{G})$$

and by the induction hypothesis:

$$(\mathfrak{G}, \mathfrak{Z}_{i-1}, \mathfrak{N}_j) = (\mathfrak{G}, (\mathfrak{Z}_{i-1}, \mathfrak{N}_j)) \subseteq (\mathfrak{G}, \mathfrak{N}_{i+j-1}) \subseteq \mathfrak{N}_{i+j}$$
$$(\mathfrak{Z}_{i-1}, \mathfrak{N}_j, \mathfrak{G}) = (\mathfrak{Z}_{i-1}, (\mathfrak{G}, \mathfrak{N}_j)) \subseteq (\mathfrak{Z}_{i-1}, \mathfrak{N}_{j+1}) \subseteq \mathfrak{N}_{i+j}.$$

Therefore $$(\mathfrak{Z}_i, \mathfrak{N}_j) \subseteq \mathfrak{N}_{i+j}.$$

If we set

$$\mathfrak{N}_1 = \mathfrak{Z}_1, \ \mathfrak{N}_2 = \mathfrak{Z}_2, \ldots, \mathfrak{N}_j = \mathfrak{Z}_j, \ldots$$

then

(6) $$(\mathfrak{Z}_i, \mathfrak{Z}_j) \subseteq \mathfrak{Z}_{i+j}.$$

We can now show by induction on the weight that:

An arbitrary commutator form $f(\mathfrak{G})$ of weight w is contained in $\mathfrak{Z}_w$.

This is true if $w = 1$. Now let $w > 1$, and assume that the statement is already known to be true for commutator forms with weight less than w. We have $f(\mathfrak{G}) = (f_1(\mathfrak{G}), f_2(\mathfrak{G}))$ where the f_i are commutator forms of weight w_i such that $w = w_1 + w_2$. By the induction hypothesis it follows that $f_1(\mathfrak{G}) \subseteq \mathfrak{Z}_{w_1}$, $f_2(\mathfrak{G}) \subseteq \mathfrak{Z}_{w_2}$; thus

$$f(\mathfrak{G}) \subseteq (\mathfrak{Z}_{w_1}, \mathfrak{Z}_{w_2}) \subseteq \mathfrak{Z}_{w_1+w_2} = \mathfrak{Z}_w.$$

In particular it follows that

(7) $$\mathfrak{Z}_i(\mathfrak{Z}_k(\mathfrak{G})) \subseteq \mathfrak{Z}_{ik}(\mathfrak{G}).$$

(8) $$D^k \mathfrak{G} \subseteq \mathfrak{Z}_{2^k}(\mathfrak{G}).$$

A nilpotent group of class c is always k-step metabelian, where k satisfies the inequality

$$2^{k-1} \leqq c \tag{9}$$

Moreover if we set $\mathfrak{z}_{-1} = \mathfrak{z}_{-2} = \cdots = e$, then in general

$$(\mathfrak{Z}_i, \mathfrak{z}_j) \subseteqq \mathfrak{z}_{j-i}. \tag{10}$$

In particular, $\mathfrak{Z}_i$ commutes with $\mathfrak{z}_i$ elementwise.

THEOREM 23 (Hall) : *If the non-abelian normal subgroup $\mathfrak{N}$ of the p-group $\mathfrak{G}$ is contained in $\mathfrak{Z}_i$, then its center is of order at least p^i, $\mathfrak{N}$ itself is at least of order p^{i+2}, its factor commutator group is at least of order p^{i+1}.*

Proof: Since $\mathfrak{z}_i$ commutes with $\mathfrak{Z}_i$ elementwise, $\mathfrak{N}$ is not contained in $\mathfrak{z}_i$. $\mathfrak{N} \cap \mathfrak{z}_i$ is in the center of $\mathfrak{N}$ and, by Theorem 14, is at least of order p^i, so that *a fortiori* the center of $\mathfrak{N}$ is of order divisible by p^i. Since $p^2/\mathfrak{N} : \mathfrak{z}(\mathfrak{N})$, the order of $\mathfrak{N}$ is divisible by p^{i+2}. Since $\mathfrak{N}$ is not abelian, we can find in the normal subgroup $\mathfrak{N}'$ of $\mathfrak{G}$ a normal subgroup $\mathfrak{N}_1$ of $\mathfrak{G}$ with $\mathfrak{N}_1$ of index p under $\mathfrak{N}'$. $\mathfrak{N}/\mathfrak{N}_1$ is a non-abelian normal subgroup of $\mathfrak{G}/\mathfrak{N}_1$ and so we conclude as above that $\mathfrak{N} : \mathfrak{N}_1$ is divisible by p^{i+2}. Consequently $\mathfrak{N}/\mathfrak{N}'$ has an order divisible by p^{i+1}, Q.E.D.

Now if in a p-group of order p^n, $D^i \mathfrak{G} \supset D^{i+1} \mathfrak{G} \supset e$, then $D^i \mathfrak{G} \subseteqq \mathfrak{Z}_{2^i}$ and therefore as was just shown, $D^i \mathfrak{G} : D^{i+1} \mathfrak{G} \geqq p^{2^i+1}$. If $\mathfrak{G}$ is now $(k+1)$–step metabelian, then

$$n \geqq 1 + \sum_{0}^{k-1} (2^i + 1) = 2^k + k.$$

The order of a $(k+1)$–step metabelian p-group is divisible by p^{2^k+k}.

Remark: Under the hypothesis of Theorem 23 it can be shown by the same methods that the factor groups of the ascending and descending central series of the normal subgroup $\mathfrak{N}$ have an order divisible by p^i, with the possible exception of the last factors different from 1. The proof is left to the reader.

Exercises

1. In a finite group $\mathfrak{G}$ the intersection of all the normal subgroups whose factor group is an abelian p-group is called the *p-commutator group* of $\mathfrak{G}$ and is denoted by $\mathfrak{G}'(p)$.

Prove: The *p-factor commutator group* $\mathfrak{G}/\mathfrak{G}'(p)$ is an abelian p-group. Moreover, the commutator group of $\mathfrak{G}$ is the intersection of the p-commutator groups, and the factor commutator group is isomorphic to the direct product of the p-factor commutator groups. Moreover, $\mathfrak{G}/\mathfrak{G}'(p) \simeq S_p/S_p \cap \mathfrak{G}'$.

2. For an arbitrary group $\mathfrak{G}$, the class may be defined by the following property: Let the class be equal to c, if $\mathfrak{Z}_{c+1}$ is a proper subgroup $\mathfrak{Z}_c$ and $\mathfrak{Z}_{c+1} = \mathfrak{Z}_{c+2} = \cdots$ Let the class be equal to zero, if the group coincides with its commutator group.[1] Let the class be infinite if $\mathfrak{Z}_{i+1}$ is a proper subgroup of $\mathfrak{Z}_i$ for all i.

For nilpotent groups the two definitions of class coincide.

Prove: If the class c is finite then $\mathfrak{Z}_{c+1}$ is the intersection of all normal subgroups with nilpotent factor group, and the factor group $\mathfrak{G}/\mathfrak{Z}_{c+1}$ is also nilpotent. Hence we shall call the factor group $\mathfrak{G}/\mathfrak{Z}_{c+1}$ the *maximal nilpotent factor group*. Its class is c. The class of every factor group is at most c.

If the class is infinite, then there are factor groups of any given class.

3. In finite groups of class c we can obtain $\mathfrak{Z}_{c+1}$ in the following way:

For every prime number p we form the intersection $\mathfrak{D}_p(\mathfrak{G})$ of all normal subgroups of p-power index.

Prove: $\mathfrak{D}_p$ itself is of p-power index. Hence we shall call the factor group $\mathfrak{G}/\mathfrak{D}_p$ the *maximal p-factor group* of $\mathfrak{G}$.

Prove: $\mathfrak{Z}_{c+1}$ is the intersection of all $\mathfrak{D}_p$, and the maximal nilpotent factor group is isomorphic to the direct product of the maximal p-factor groups over all prime divisors of the group order.

4. If p^a is divisible by the exponent of the maximal p-factor group of the finite group $\mathfrak{G}$ (see Exercise 3), then the subgroup generated by all p^a-th powers is equal to $\mathfrak{D}_p$. Therefore $\mathfrak{D}_p$ is a fully invariant subgroup of $\mathfrak{G}$. Moreover, prove that

$$\mathfrak{D}_p(\mathfrak{D}_p(\mathfrak{G})) = \mathfrak{D}_p(\mathfrak{G}).$$

5. a) An abelian group with a finite number of generators is finite if and only if the factor group over its Φ-subgroup is finite.

b) A nilpotent group with a finite number of generators is finite if and only if the factor group over its Φ-subgroup is finite. [Use a) and apply induction to the length of the descending central series!]

6. In a nilpotent group all the elements of finite order form a fully invariant subgroup. (Use Exercise 5.)

7. a) (Hilton.) In a nilpotent group any two elements with relatively prime orders commute.

(*Hint*: Show that the commutator of the two elements is in members of the descending central series with arbitrarily great subscript.)

b) Two elements with p-power order generate a p-group.

c) Prove the following generalization of Theorem 11: A nilpotent group in which every element is of finite order is the direct product of nilpotent groups in which every element is of prime power order.

§ 6. Hamiltonian Groups

In an abelian group every subgroup is a normal subgroup. What other groups also have this property?

DEFINITION: A non-abelian group in which every subgroup is a

[1] These groups are also said to be *perfect groups*.

normal subgroup is said to be a *Hamiltonian group*. For example, the quaternion group is a Hamiltonian group.

THEOREM: *A Hamiltonian group is the direct product of a quaternion group with an abelian group in which every element is of odd order and an abelian group of exponent 2, and conversely.*

Proof: In a Hamiltonian group $\mathfrak{H}$ there are two elements A, B which do not commute with each other. Since (A) and (B) are normal subgroups of $\mathfrak{H}$, the commutator $C = (A, B) = ABA^{-1}B^{-1}$ of A and B is contained in the intersection of (A) and (B), and therefore in the center of the subgroup $\mathfrak{Q} = \{A, B\}$ generated by A and B.

The commutator group $\mathfrak{Q}'$ of $\mathfrak{Q}$ is generated by C and is a proper subgroup of (A) and likewise of (B). Since $(C) \neq e$, $C = A^r = B^s$ where $r, s \neq 0$. By Chapter II § 6, $(A, B)^s = (A, B^s)$, and therefore $C^s = e$. Consequently A and B have finite orders m and n respectively. We choose A and B so that m and n are minimal. Then it follows for a prime divisor p of m that

$$(A^p, B) = e \text{ and therefore } C^p = (A, B)^p = (A^p, B) = e.$$

Similarly it follows for a prime divisor p of n that $C^p = e$. The orders of A, B are consequently powers of the same natural prime p; they are divisible by p^2 since (C) is a proper subgroup of both (A) and (B), while A^p, B^p are contained in the center of $\mathfrak{Q}$.

If, say $A^{p^a} = C^\nu$, $B^{p^b} = C^\mu$, where ν, μ are not divisible by p, then we replace A by A^μ, B by B^ν, and we may assume that

$$A^{p^a} = B^{p^b} = (A, B) = C \neq e.$$

where $a \geq b > 0$.

By chapter II § 6, in $\mathfrak{Q}$ we have the relation

$$(x\,y)^p = (x, y)^{-\frac{1}{2}p(p-1)}\, x^p y^p.$$

Now A, $A^{-p^{a-b}}B$ also generate $\mathfrak{Q}$, and therefore $B_1 = A^{-p^{a-b}}B$ must be of order at least equal to that of B. From this we conclude:

$$B_1^p = C^{p^{a-b+1} \cdot \frac{p-1}{2}} A^{-p^{a-b+1}} B^p,$$

$$B_1^{p^b} = C^{p^a \cdot \frac{p-1}{2}},$$

$$p = 2, \qquad a = b = 1.$$

Therefore $\mathfrak{Q}$ is a quaternion group with the relations $A^2 = B^2 = ABA^{-1}B^{-1} = C$, $C^2 = e$.[1]

[1] Instead of this process one can apply Theorem 15 of § 3 to the group $\mathfrak{Q}$!

We wish to show that $\mathfrak{H}$ is generated by $\mathfrak{Q}$ and the group $\mathfrak{B}$ of all elements of $\mathfrak{H}$ which commute with every element of $\mathfrak{Q}$.

If the element X does not commute with A, then $XAX^{-1} = A^{-1}$ and therefore BX commutes with A. If, now , BX does not commute with B, then ABX commutes with B. Consequently $\mathfrak{H} = \mathfrak{Q}\mathfrak{B}$.

Every element X in $\mathfrak{B}$ is of finite order, since BX does not commute with A, and therefore BX is of finite order. But B is of order 4 and commutes with X, therefore X is of finite order. Now, if $X^4 = e, X \in \mathfrak{B}$, then $(A, BX) \neq e$, $(A, BX) = A^2 = B^2$. Since $(BX)^4 = e$, we have $(A, BX) = (BX)^2 = B^2X^2$ and therefore $X^2 = e$.

In $\mathfrak{B}$ there is no element of order 4 and thus certainly no quaternion group. But since every subgroup in $\mathfrak{B}$ is a normal subgroup, $\mathfrak{B}$ is abelian. $\mathfrak{B}$ is the direct product of the subgroup $\mathfrak{U}$ of all elements of odd order, and the subgroup $\mathfrak{G}_1$ of all elements whose square is e. C is contained in $\mathfrak{G}_1$. Among all the subgroups of $\mathfrak{G}_1$ which do not contain C there is a largest $\mathfrak{G}$. For every element X in $\mathfrak{G}_1$ not contained in $\mathfrak{G}$, C must be contained in $\{\mathfrak{G}, X\}$. Since $X^2 = e$, we have $\{\mathfrak{G}, X\} : \mathfrak{G} = 2$ and likewise $\{\mathfrak{G}, C\} : \mathfrak{G} = 2$, and therefore $\{\mathfrak{G}, X\} = \{\mathfrak{G}, C\}$; it follows that $\{\mathfrak{G}, C\} = \mathfrak{G}_1$ and moreover $\mathfrak{G} \cap (C) = e$; therefore

$$\mathfrak{B} = \mathfrak{U} \times \mathfrak{G} \times (C).$$

Since $\mathfrak{Q} \cap \mathfrak{B} = (C)$, we have $\mathfrak{Q} \cap (\mathfrak{U} \times \mathfrak{G}) = e$, and moreover $\mathfrak{Q} \cdot (\mathfrak{U} \times \mathfrak{G}) = \mathfrak{H}$; therefore $\mathfrak{H} = \mathfrak{Q} \times \mathfrak{U} \times \mathfrak{G}$.

Conversely a group with this structure is Hamiltonian. For $\mathfrak{Q}$ is not abelian. We have yet to show that every cyclic subgroup (QUG) is a normal subgroup. Since $\mathfrak{Q}$ is the only non-abelian factor of the decomposition we only need show that the transform of QUG by A or B is in (QUG).

Now $A(QUG)A^{-1} = Q^iUG$ where i is either 1 or 3. The order of U is an odd number d. Therefore the congruences $r \equiv i\ (4)$, $r \equiv 1\ (d)$ can be solved, and $G^r = G$, $AQUGA^{-1} = (QUG)^r$, Q.E.D.

§ 7. Applications of Extension Theory

Let $\mathfrak{G}$ be an extension of the normal subgroup $\mathfrak{N}$ with the factor group $\mathfrak{F}$.

We say a factor system $(C_{\sigma,\tau})$ is an *abelian factor system* if all the $C_{\sigma,\tau}$ commute with each other.

THEOREM 24: *The* $(\mathfrak{F}:1)$*-th power of an abelian factor system is a retracting*[1] *factor system.*

[1] See end of § 6, Chapter III.

Proof: Let $(\mathfrak{F}:1) = n > 0$. We set $a_\sigma = \prod_\varrho C_{\sigma,\varrho}$ and form the product over ϱ of all the equations $C_{\sigma,\tau} C_{\sigma\tau,\varrho} = C^\sigma_{\tau,\varrho} C_{\sigma,\tau\varrho}$.

Then it follows that $C^n_{\sigma,\tau} = a^\sigma_\tau a_\sigma a^{-1}_{\sigma\tau}$, Q.E.D.

THEOREM 25 (Schur): *If the order n of the finite factor group $\mathfrak{F}$ is relatively prime to the order m of the finite normal subgroup $\mathfrak{N}$, then the extension $\mathfrak{G}$ splits over $\mathfrak{N}$.*

Proof: We need only show that $\mathfrak{G}$ contains a subgroup of order n.

If $m = 1$, this is clear. Let $m > 1$ and assume the statement proven when the order of the normal subgroup is less than m. For a prime divisor p of m, every Sylow p-group S_p of $\mathfrak{G}$ is contained in $\mathfrak{N}$. Since there are as many Sylow p-groups in $\mathfrak{N}$ as in $\mathfrak{G}$, $N_p : \mathfrak{N} \cap N_p = n$. Now $N_p \cap \mathfrak{N}/S_p$ is a normal subgroup of N_p/S_p with index n. By the induction hypothesis there is a subgroup $\mathfrak{H}/S_p$ of order n in N_p/S_p. S_p/z_p is a normal subgroup of $\mathfrak{H}/z_p$ of index n,where z_p is the center of S_p and is different from e. By the induction hypothesis there is a subgroup $\mathfrak{U}/z_p$ of order n in $\mathfrak{H}/z_p$. Let $C_{\sigma,\tau}$ be a factor system of $\mathfrak{U}$ over z_p. Since the order z of z_p is relatively prime to n, we can solve the congruence $nn_1 \equiv 1\,(z)$ and for the factor system $C_{\sigma,\tau}$ of $\mathfrak{U}$ over z_p we find that it is the n_1-th power of the factor system $C^m_{\sigma,\tau}$ which is retracting by Theorem 24. Therefore $C_{\sigma,\tau}$ itself splits over z_p, i.e. $\mathfrak{U}$ contains a subgroup of order n, Q.E.D.

In what follows, let $\mathfrak{F}$ be a finite group of order n.

THEOREM 26: *If $a_\sigma a^\sigma_\tau = a_{\sigma\tau}$ and the a_σ commute with one another, then the equation $a^n_\sigma = \delta^{1-\sigma}$ is solvable, i.e. the mapping $S_\sigma \to a^n_\sigma S_\sigma$ can be accomplished by transformation with an element δ in $\mathfrak{N}$.*

Proof: Form the product over all equations with fixed σ:

$$a^n_\sigma \prod_\tau a^\sigma_\tau = \prod_\tau a_{\sigma\tau} = \prod_\tau a_\tau .$$

We set $\delta = \prod_\tau a_\tau$ and have $a^n_\sigma \delta^\sigma = \delta$, Q.E.D.

It has been conjectured that the following theorem is true in general.

THEOREM 27: *If the order n of the finite factor group $\mathfrak{F}$ is relatively prime to the order m of the finite normal subgroup $\mathfrak{N}$, then two representative groups of $\mathfrak{G}$ over $\mathfrak{N}$ are conjugate in $\mathfrak{G}$. We shall prove the theorem when one of the following additional conditions holds:*

1. *$\mathfrak{N}$ is abelian.*
2. *$\mathfrak{N}$ is solvable.*
3. *$\mathfrak{F}$ is solvable.*

One of the groups $\mathfrak{N}$, $\mathfrak{F}$ is of odd order, and since it is conjectured that groups of odd order are solvable, it is also expected that the above theorem is true. E. Witt reduced the theorem to the case when $\mathfrak{N}$ is simple and the centralizer of $\mathfrak{N}$ in $\mathfrak{G}$ is e. It is believed that the group of outer automorphisms of a finite simple group is solvable, so that we can conjecture the truth of Theorem 27 on this basis also.

Proof of 1: If $\mathfrak{U} = \{S_\sigma\}$ is a representative group and $\mathfrak{V} = \{a_\sigma S_\sigma\}$, a second one, then we have the equations

$$a_\sigma a_\tau^\sigma = a_{\sigma\tau}.$$

By Theorem 26, $a_\sigma^n = \delta^{1-\sigma}(\sigma \in \mathfrak{F})$ is solvable. Since by hypothesis the congruence $n \cdot n_1 \equiv 1 (\mathfrak{N} : 1)$ is solvable, $a_\sigma = a_\sigma^{n\ n_1} = (\delta^{n_1})^{1-\sigma}$, Q.E.D.

Proof of 2: If $\mathfrak{N}$ is abelian, then the theorem is true by 1; let $\mathfrak{N}$ be k-step metabelian and assume the theorem has already been proven for $D^{k-1}(\mathfrak{N}) = e.$. Further, let $\mathfrak{U}$ and $\mathfrak{V}$ be two representative groups of $\mathfrak{G}$ over $\mathfrak{N}$. We apply 1 to $\mathfrak{G}/\mathfrak{N}'$ and find that $\mathfrak{V}\mathfrak{N}' = (\mathfrak{U}\mathfrak{N}')^x$ with $x \in \mathfrak{N}$ is solvable i.e., $\mathfrak{V}^{x^{-1}}\mathfrak{N}' = \mathfrak{U}\mathfrak{N}'$. Since $D^{k-1}(\mathfrak{N}') = e$, then by applying the induction hypothesis to $\mathfrak{U}\mathfrak{N}'$, it follows that $\mathfrak{V}^{x^{-1}} = \mathfrak{U}^y$ is also solvable for $y \in \mathfrak{N}$ and therefore $\mathfrak{V} = \mathfrak{U}^{xy}$, with $xy \in \mathfrak{N}$, Q.E.D.

Proof of 3: Let a principal series of $\mathfrak{G}/\mathfrak{N}$ be of length l, and let $\mathfrak{U}$, $\mathfrak{V}$ be two representative groups of $\mathfrak{G}$ over $\mathfrak{N}$. Let $\mathfrak{u}$ be a minimal normal subgroup of $\mathfrak{U}$; since $\mathfrak{U}$ is solvable, $\mathfrak{u}$ is a p-group. $\mathfrak{u}$ is isomorphic to $\mathfrak{v} = \mathfrak{V} \cap \mathfrak{u}\mathfrak{N}$ where $\mathfrak{v}$ is a normal subgroup of $\mathfrak{V}$.

If $l = 1$, then $\mathfrak{u} = \mathfrak{U}$, $\mathfrak{v} = \mathfrak{V}$. Then $\mathfrak{U}$ and $\mathfrak{V}$ are Sylow p-groups of $\mathfrak{G}$, therefore conjugate in $\mathfrak{G}$. Let $l > 1$, and assume that the theorem has been proven for smaller l. By the induction hypothesis there is an x in $\mathfrak{u}\mathfrak{N} = \mathfrak{v}\mathfrak{N}$, such that $\mathfrak{v} = \mathfrak{u}^x$. We set $\mathfrak{V}_1 = \mathfrak{V}^{x^{-1}}$ and find that $\mathfrak{U}, \mathfrak{V}_1 \leq N_\mathfrak{u}$. Since the principal series of $\mathfrak{U}/\mathfrak{u}$ is of length l-1, it follows by the induction hypothesis applied to $N_\mathfrak{u}/\mathfrak{u}$ that there is a $y \in N_\mathfrak{u}$, such that $\mathfrak{V}_1 = \mathfrak{U}^y$, and therefore $\mathfrak{V} = \mathfrak{U}^{xy}$, Q.E.D.

Exercises

In a finite solvable group, certain generalized Sylow theorems are valid (Hall):

1. For every decomposition $N = n.m$ of the group order into a product of relatively prime factors, there is a subgroup of order m and index n.

2. Let n and m be chosen as in Exercise 1. All subgroups whose order is a divisor of m lie in a subgroup of order m.

3. Let m be as in Exercise 1. All subgroups of order m are conjugate. The normalizer of a subgroup of order m is its own normalizer.

(Proofs of 1 – 3 by induction on the length of the principal series and by use of Theorems 25, 27.)

V. TRANSFERS INTO A SUBGROUP

§ 1. Monomial Representations and Transfers into a Subgroup

We wish to represent the elements of a group $\mathfrak{G}$ as permutations on a set whose objects admit multiplication by the elements of a second group $\mathfrak{H}$.

DEFINITION: A set $\mathfrak{M}$ of elements $u, v, \ldots.$ is called an $(\mathfrak{H}, \mathfrak{G})$-system if for every pair u, G (or H, u) the product uG (or Hu) is defined uniquely as an element of $\mathfrak{M}$, and if moreover for all u in $\mathfrak{M}$:

$$u(GG') = (uG)G', \tag{1}$$

$$(HH')u = H(H'u),$$

$$ue_{\mathfrak{G}} = e_{\mathfrak{H}}u = u, \tag{2}$$

$$H(uG) = (Hu)G. \tag{3}$$

By (1), the correspondence $G \rightarrow \pi_G = \begin{pmatrix} u \\ uG^{-1} \end{pmatrix}$ gives a representation of $\mathfrak{G}$ in single-valued mappings of $\mathfrak{M}$ onto itself. Since $\pi_e = \begin{pmatrix} u \\ ue \end{pmatrix} = \underline{1}$, the π_G form a group $\Delta_{\mathfrak{G}}$ of permutations of $\mathfrak{M}$. We shall assume in addition that $\Delta_{\mathfrak{G}}$ is transitive.

Example: Let $\mathfrak{U}$ be a subgroup of $\mathfrak{G}$, $\mathfrak{u}$ a normal subgroup of $\mathfrak{U}$. Then the right cosets of $\mathfrak{G}$ over $\mathfrak{u}$ form a $(\mathfrak{U}, \mathfrak{G})$-system for which $\Delta_{\mathfrak{G}}$ is transitive.

We shall show that all $(\mathfrak{H}, \mathfrak{G})$-systems with transitive $\Delta_{\mathfrak{G}}$ are of the type described in the preceding example.

First of all, it follows from (1) and (2) that the correspondence

$$H \rightarrow \bar{H} = \begin{pmatrix} u \\ Hu \end{pmatrix}$$

is a representation of $\mathfrak{H}$ in permutations of the elements of $\mathfrak{M}$. All these permutations $\bar{H}$ form a group $\bar{\mathfrak{H}}$, and

$$\overline{HH'} = \bar{H}\bar{H'}. \tag{4}$$

We now define $\bar{H}$ as an operator on $\mathfrak{M}$ by the equation:

$$\bar{H}u = Hu \tag{5}$$

This definition is unambiguous and now $\mathfrak{M}$ is also an $(\bar{\mathfrak{H}}, \mathfrak{G})$-system. Because of (4) and (5) the questions about $(\bar{\mathfrak{H}}, \mathfrak{G})$-systems are equivalent to the questions about $(\mathfrak{H}, \mathfrak{G})$-systems; and so we shall assume that $\mathfrak{H}$ is equal to $\bar{\mathfrak{H}}$, i.e.,

(6) $Hu = u$ for all u implies $H = e_{\mathfrak{H}}$. If $Hu_0 = u_0$, then

$$H(u_0G) = (Hu_0)G = u_0G$$

for all G; therefore because of the transitivity of $\Delta_{\mathfrak{G}}$:

$$Hu = u$$

for all u, and so $H = e_{\mathfrak{H}}$.

(7) Every H is indeed determined completely by the way it operates on only one element of $\mathfrak{M}$.

Let u_0 be a fixed element of $\mathfrak{M}$. All the elements of $\mathfrak{G}$ which leave u_0 fixed form a subgroup $\mathfrak{u}$. We now investigate the complex $\mathfrak{U}$ consisting of all the elements U of $\mathfrak{G}$ for which the equation $u_0U = U^*u_0$ is solvable with some $U^* \in \mathfrak{H}$.

Because of the transitivity of $\Delta_{\mathfrak{G}}$ and because of (7), it follows that the mapping

$$U \to U^*$$

is a single-valued mapping of $\mathfrak{U}$ onto all of $\mathfrak{H}$. If U and V are contained in $\mathfrak{U}$, then

$$u_0(UV) = (u_0U)V = U^*u_0V = U^*V^*u_0,$$

and hence
$$UV \in \mathfrak{U}, \ (UV)^* = U^*V^*.$$

The mapping of U onto U^* maps $\mathfrak{U}$ homomorphically onto $\mathfrak{H}$. Precisely the elements of $\mathfrak{u}$ are mapped onto $e_{\mathfrak{H}}$. Since $\mathfrak{u}$ and $\mathfrak{H}$ are groups, $\mathfrak{U}$ is a group also. $\mathfrak{u}$ is a normal subgroup of $\mathfrak{U}$, and we may, and in fact shall, consider $\mathfrak{H}$ simply as the group of cosets of $\mathfrak{U}$ over $\mathfrak{u}$.

The mapping

(8)
$$\overline{u_0G} = \mathfrak{u}G$$

is single-valued, for from

$$u_0G = u_0G'$$

it follows that
$$u_0GG'^{-1} = u_0, \ GG'^{-1} \subseteq \mathfrak{u},$$

$$\mathfrak{u}G = \mathfrak{u}G',$$

and conversely.

Moreover

$$\overline{(u_0 G) G'} = \overline{u_0 (G G')} = \mathfrak{u} G G' = (\mathfrak{u} G) G' = \overline{u_0 G} \cdot G',$$

$$\overline{U^* u_0 G} = \overline{u_0 U G} = \mathfrak{u}\, U G = U \mathfrak{u} G = U \overline{u_0 G} = U^* \overline{u_0 G}.$$

Therefore, according to (8), the given $(\mathfrak{H}, \mathfrak{G})$-system $\mathfrak{M}$ with transitive $\Delta_{\mathfrak{G}}$ can be identified with the set of right cosets of $\mathfrak{G}$ over a subgroup $\mathfrak{u}$, where $\mathfrak{u}$ is a normal subgroup of a subgroup $\mathfrak{U}$ and the factor group $\mathfrak{U}/\mathfrak{u}$ is isomorphic to $\mathfrak{H}$.

Let $G \rightarrow \overline{G}$ be a representative function belonging to the decomposition $\mathfrak{G} = \sum_1^{\omega} \mathfrak{U} G_i$. If we put $u_i = \mathfrak{u} G_i$, then every coset from $\mathfrak{M}$ has the unique form

$$u = U^* u_i.$$

Accordingly

(9) $$u_i G = U^*_{i,G} u_{iG}.$$

Here the permutation $\binom{i}{iG}$ is determined by the equation

(10) $$\mathfrak{U} G_i G = \mathfrak{U} G_{iG}.$$

Moreover

(11) $$U^*_{i,G} = G_i G \overline{G_i G}^{-1} = G_i G G_{iG}^{-1},$$

for $$U^*_{i,G} u_{iG} = U^*_{i,G} \mathfrak{u} \overline{G_i G} = \mathfrak{u}\, U^*_{i,G} \overline{G_i G} = \mathfrak{u}\, G_i G = u_i G.$$

It is obvious that through (9) a matrix M_G having ω rows and ω columns can be associated with each element G of $\mathfrak{G}$:

(12) $$M_G = (\delta_{iG, k} U^*_{i,G}).$$

That is, M_G is a matrix with the element $U^*_{i,G}$ in the i-th row and iG-th column and with zeros elsewhere M_G is a permutation of the u_i with factors from $\mathfrak{U}^*$ $(= \mathfrak{H})$. From (1) and (3) it now follows that

(13) $$M_G \cdot M_{G'} = M_{GG'},$$

where the product of two matrices is computed in the usual manner. We call the representation (12) of $\mathfrak{G}$ in square matrices of degree ω with coefficients zero, or from the group $\mathfrak{U}^*$, the *monomial representation with* ω *members.*

In going over to another system of representatives of right cosets of

$\mathfrak{G}$ over $\mathfrak{U}$ we change to a new "basis" $v_1, v_2, \ldots, v_\omega$ of $\mathfrak{M}$ which is connected to the old one by equations of the form

(14) $$v_i = U_i^* u_{\tau i} \; ,$$

where $\binom{i}{\tau i}$ is a permutation of the numbers 1, 2 . . . ω. If we put down as transformation matrix

$$T = (\delta_{\tau i, k} U_i^*),$$

where U_i^* stands at the intersection of the i-th row with the τi-th column and there are zeros everywhere else, then the representation with ω members belonging to the v_i and with matrices M_G^* is given by

(15) $$M_G^* = T M_G T^{-1}.$$

If we put $\mathfrak{u} = e$, then we obtain the most general monomial representation of $\mathfrak{G}$ over $\mathfrak{U}$:

(16) $$G \to M_G^{\mathfrak{U}} = (\delta_{iG, k} U_{i, G}),$$

from which the representations with arbitrary normal subgroup of $\mathfrak{U}$ can be obtained by replacing the elements by their cosets.

If $\mathfrak{B}$ is a subgroup of $\mathfrak{U}$, let $\mathfrak{U} = \sum_1^{\nu} \mathfrak{B} U_k$ be a right coset decomposition of $\mathfrak{U}$ over $\mathfrak{B}$. Then $\mathfrak{G} = \sum_{i, k} \mathfrak{B} U_k G_i$ is a right coset decomposition of $\mathfrak{G}$ over $\mathfrak{B}$, and from equations (10) and (12) it follows that:

(17) $M_G^{\mathfrak{B}}$ arises from the matrix $M_G^{\mathfrak{U}}$ upon the replacement of each element U from $\mathfrak{U}$ by the matrix $m_{\mathfrak{U}}^{\mathfrak{B}}$. belonging to the representation of $\mathfrak{U}$; over $\mathfrak{B}$ and by replacing each 0 by a ν-rowed matrix of zero s.

If we replace the normal subgroup $\mathfrak{u}$ by the commutator group $\mathfrak{U}'$ of $\mathfrak{U}$, then there corresponds a representation of $\mathfrak{G}$ in matrices whose coefficients are from an abelian group. Through the construction of determinants we arrive at a new representation. We define:

The *transfer of the element* X from the group $\mathfrak{G}$ into the subgroup $\mathfrak{U}$ is the coset $V_{\mathfrak{G} \to \mathfrak{U}}(X)$ of $\mathfrak{U}$ over its commutator group $\mathfrak{U}$. If $\mathfrak{U}$ is of finite index n and has the system of left representatives $G_1, G_2, \ldots, G_n$ with the representation function $G \to \overline{G}$ then we define

(18) $$V_{\mathfrak{G} \to \mathfrak{U}}(X) = \mathfrak{U}' \cdot \prod_1^n G_i X \overline{G_i X}^{-1}.$$

THEOREM 1: *The transfer is independent of the choice of the system of representatives.*

Proof: $V(X)$ is (to within sign) the determinant of the representation matrix $M_X^{\mathfrak{U}}$, having coefficients from $\mathfrak{U}/\mathfrak{U}'$ or else 0. By transforming to a new system of left representatives we simply transform M_X by a fixed matrix T. This does not alter the value of the determinant.

THEOREM 2: *The transfer of* $\mathfrak{G}$ *to* $\mathfrak{U}$ *is a homomorphy of* $\mathfrak{G}$ *into* $\mathfrak{U}/\mathfrak{U}'$.

Proof: This follows from (13) upon construction of determinants.

The transfer $V_{\mathfrak{G}\to\mathfrak{U}}$ induces an isomorphy between an abelian factor group of $\mathfrak{G}$ and $\mathfrak{U}/\mathfrak{U}'$. Hence $V_{\mathfrak{G}\to\mathfrak{U}}(\mathfrak{G}') = \mathfrak{U}'$. The subgroup of $\mathfrak{U}/\mathfrak{U}'$ consisting of all the cosets $V_{\mathfrak{G}\to\mathfrak{U}}(\mathfrak{G})$ is called the *transferred group of* $\mathfrak{G}$ *to* $\mathfrak{U}$.

THEOREM 3: *For a subgroup* $\mathfrak{V}$ *of* $\mathfrak{U}$ *with finite index it follows that*:

$$V_{\mathfrak{G}\to\mathfrak{V}}(X) = V_{\mathfrak{U}\to\mathfrak{V}}(V_{\mathfrak{G}\to\mathfrak{U}}(X)).$$

Proof: This follows from (17) upon construction of determinants two times.

Remark: If $\mathfrak{G}$ is a group with given automorphism domain, then the transferred group is an admissible group, for when $\overline{\mathfrak{G}}$ is a system of left representatives of $\mathfrak{G}$ over $\mathfrak{U}$, then so is $\overline{\mathfrak{G}}^{\sigma}$ (σ an automorphism of $\mathfrak{G}$) provided $\mathfrak{U}$ is admissible. In particular the transferred group of a transfer into a normal subgroup is itself normal.

In order to compute the transfer of a given element X, it is useful to choose a particular system of representatives. The permutation $\begin{pmatrix}\mathfrak{U}G\\ \mathfrak{U}GX\end{pmatrix}$ of the right cosets of $\mathfrak{G}$ over $\mathfrak{U}$ decomposes into r cycles. From the i-th cycle we choose a representative $\mathfrak{U}T_i$, and the cycle may be written $(\mathfrak{U}T_i, \mathfrak{U}T_iX, \ldots, \mathfrak{U}T_iX^{f_i-1})$. Then for the system of representatives $T_i, T_iX, \ldots, T_iX^{f_i-1}$ $(i=1, 2, \ldots, r)$:

$$V_{\mathfrak{G}\to\mathfrak{U}}(X) = \mathfrak{U}'\prod_1^r T_iX^{f_i}T_i^{-1}, \tag{19}$$

where f_i is the length of the i-th cycle, and hence is a divisor of the order of X. Moreover

$$\sum_1^r f_i = \mathfrak{G}:\mathfrak{U}. \tag{20}$$

Exercise: Prove the three theorems on transfers by direct calculation, on the basis of (18) and the rules about representative functions.

§ 2. The Theorems of Burnside and Grün

LEMMA: If two complexes $\mathfrak{K}$, $\mathfrak{L}$ in a Sylow p-group $\mathfrak{P}$ of the finite group $\mathfrak{G}$ are normal[1] in $\mathfrak{P}$ and conjugate under $\mathfrak{G}$, then they are also conjugate under the normalizer $N_{\mathfrak{P}}$ of $\mathfrak{P}$.

Proof: The hypothesis says $\mathfrak{P}$ is in the normalizer $N_{\mathfrak{K}}$ of $\mathfrak{K}$ and in the normalizer $N_{\mathfrak{L}}$ of $\mathfrak{L}$, and that

$$\mathfrak{L} = T\mathfrak{K}T^{-1} = \mathfrak{K}^T$$

is solvable with T in $\mathfrak{G}$.

From $\mathfrak{P} \leq \mathfrak{K}$ and from $N_{\mathfrak{L}} = N_{\mathfrak{K}}^T$ it follows that: $\mathfrak{P}^T \leq N_{\mathfrak{L}}$. Since $\mathfrak{P}$ and $\mathfrak{P}^T$ are Sylow p-groups of $\mathfrak{G}$ in $N_{\mathfrak{L}}$, they are also Sylow p-groups of $N_{\mathfrak{L}}$, consequently $\mathfrak{P} = \mathfrak{P}^{ST}$ with S in $N_{\mathfrak{L}}$ is solvable by the third Sylow theorem.

Consequently ST is in $N_{\mathfrak{P}}$ and $\mathfrak{K}^{ST} = \mathfrak{L}^S = \mathfrak{L}$, Q.E.D.

THEOREM 4 (Burnside): *If the Sylow p-group $\mathfrak{P}$ of a finite group $\mathfrak{G}$ is in the center of its normalizer, then $\mathfrak{G}$ contains a normal subgroup with $\mathfrak{P}$ as system of representatives.*

Proof: The hypothesis implies that $\mathfrak{P}$ is abelian, so that its commutator subgroup is e. We transfer $\mathfrak{G}$ into $\mathfrak{P}$ and obtain a normal subgroup $\mathfrak{N}$ of all elements which are transferred to e, and a transfer group $V(\mathfrak{G}) \leq \mathfrak{P}$. If we show that $V(\mathfrak{P}) = \mathfrak{P}$, then $V(\mathfrak{G}) = \mathfrak{P}$, hence $\mathfrak{N}\mathfrak{P} = \mathfrak{G}$, and $\mathfrak{P} \cap \mathfrak{N} = e$, and the theorem is proven.

For an element X in $\mathfrak{P}$, by § 1, (19)

$$V(X) = \prod_1^r T_i X^{f_i} T_i^{-1}$$

for certain T_i where $\sum_1^r f_i = \mathfrak{G} : \mathfrak{P}$, and every factor of the product is contained in $\mathfrak{P}$. But the elements X^{f_i}, $T_i X^{f_i} T_i^{-1}$, conjugate under $\mathfrak{G}$, are normal in the abelian Sylow p-group $\mathfrak{P}$; therefore by the lemma they are conjugate under $N_{\mathfrak{P}}$. Therefore by hypothesis they are equal to one another, so that

$$V(X) = \prod_1^r X^{f_i} = X^{\mathfrak{G}:\mathfrak{P}} .$$

Since $\mathfrak{G}:\mathfrak{P}$ is relatively prime to the order of the Sylow group $\mathfrak{P}$, we have $V(\mathfrak{P}) = \mathfrak{P}$, which proves the theorem.

It follows immediately from the Burnside theorem that the order of a finite simple group of composite order is divisible by the cube of its smallest prime factor, or by 12. (See IV, § 3, Exercise 5).

[1] $\mathfrak{K}$ is called normal in $\mathfrak{P}$, if $x\mathfrak{K}x^{-1} = \mathfrak{K}$ for all x in $\mathfrak{P}$.

With the transfer of a finite group $\mathfrak{G}$ into a Sylow p-group $\mathfrak{P}$ we associate the normal subgroup $\mathfrak{G}_1$ which consists of all the elements of $\mathfrak{G}$ which are transferred to the commutator subgroup $\mathfrak{P}'$ of $\mathfrak{P}$. $\mathfrak{G}/\mathfrak{G}_1$ is isomorphic to the transfer of $\mathfrak{G}$ in $\mathfrak{P}$, and therefore is an abelian p-group. By Chapter IV it follows from this that $\mathfrak{G} = \mathfrak{P}\mathfrak{G}_1$ and therefore by the second isomorphy theorem, $\mathfrak{G}/\mathfrak{G}_1 \simeq \mathfrak{P}/\mathfrak{P} \cap \mathfrak{G}_1$.

Can the p-group $\mathfrak{P}_1 = \mathfrak{P} \cap \mathfrak{G}_1$ also be characterized from within? $\mathfrak{P}_1$ is defined from above as the group of those elements of $\mathfrak{P}$ whose transfers in $\mathfrak{P}$ are in $\mathfrak{P}'$. The elements of $\mathfrak{P} \cap \mathfrak{G}'$ are among these elements. In particular, in $\mathfrak{P}_1$ we have the intersection of $\mathfrak{P}$ with the commutator group $N_{\mathfrak{P}}$ of the normalizer $N_{\mathfrak{P}}$ of $\mathfrak{P}$, and the groups $\mathfrak{P} \cap \mathfrak{P}'^{T}$, where $T \in \mathfrak{G}$. Our question is now answered by the FIRST THEOREM OF GRÜN (THEOREM 5): *On transferring a finite group $\mathfrak{G}$ into a Sylow p-group $\mathfrak{P}$, the transferred group is isomorphic to the factor group of $\mathfrak{P}$ over the normal subgroup*

$$(\mathfrak{P} \cap N'_{\mathfrak{P}}) \cdot \prod_{T \in \mathfrak{G}} \mathfrak{P} \cap \mathfrak{P}'^{T}.$$

Proof: We set $V_{\mathfrak{G}\to\mathfrak{P}}(X) = VX$, and $\mathfrak{P}_2 = (\mathfrak{P} \cap N'_{\mathfrak{P}}) \cdot \prod_{T \in \mathfrak{G}} \mathfrak{P} \cap \mathfrak{P}'^{T}$, and then since $\mathfrak{P}_2 \subseteq \mathfrak{P}_1$ and $V\mathfrak{G}/\mathfrak{P}' \simeq \mathfrak{P}/\mathfrak{P}_1$ we must prove that $\mathfrak{P}_2 = \mathfrak{P}_1$.

Assume that $\mathfrak{P}_2 \neq \mathfrak{P}_1$ and then let X be an element of minimal order which is in $\mathfrak{P}_1$ but not in $\mathfrak{P}_2$. We shall be led to a contradiction by showing that $VX \notin \mathfrak{P}'$, and in fact that $VX \notin \mathfrak{P}_2$.

We anticipate the essential argument by remarking that $X^{p^t T} \in \mathfrak{P}$, $t > 0$ implies $X^{p^t T} \in \mathfrak{P}_2$, since $VX^{p^t T} = VT \cdot VX^{p^t} \cdot VT^{-1} = VX^{p^t} = \mathfrak{P}'$; and therefore $X^{p^t T}$ is in $\mathfrak{P}_1$; and since it is of lower order than X, it is in $\mathfrak{P}_2$.

Under the representation $Y \to \begin{pmatrix} \mathfrak{P}G \\ \mathfrak{P}GY^{-1} \end{pmatrix}$ of $\mathfrak{G}$ in permutations of the right cosets of $\mathfrak{G}$ over $\mathfrak{P}$, we also obtain a representation of $\mathfrak{P}$, and the right cosets of $\mathfrak{G}$ over $\mathfrak{P}$ decompose under $\mathfrak{P}$ into systems $\mathfrak{T}_i$ of transitivity having p^{t_i} right cosets. Under multiplication on the right by X, the cosets from $\mathfrak{T}_i = \mathfrak{T}$ are permuted in certain p^m-member cycles. We look for a coset $\mathfrak{P}T$ from $\mathfrak{T}$ which belongs to a cycle of minimal length p^{μ}; then all p^t cosets from $\mathfrak{T}$ are of the form $\mathfrak{P}TP$ with P in $\mathfrak{P}$.

1. $T \notin N_{\mathfrak{P}}$, i.e., $p^t > 1$. Then the cosets of a p^m-member cycle are: $\mathfrak{P}TP, \mathfrak{P}TPX, \ldots, \mathfrak{P}TPX^{p^m-1}$, where m, naturally, depends on P, and $\mathfrak{P}TPX^{p^m} = \mathfrak{P}TP$. If one chooses TPX^i as coset representative, then

the product of associated transfer factors satisfies

$$TPX(TPX)^{-1}\cdot TPX^2(TPX^2)^{-1}\dots TPX^{p^m}(TP)^{-1}=X^{p^m TP}\in \mathfrak{P},$$

and it follows from our determination of T that $\mu \leqq m$, $X^{p^m T}\in\mathfrak{P}$; therefore $X^{p^m TP}\cdot X^{-p^m T}=((X^{p^m})^{P-1})^T=(P,X^{p^m})^T\in\mathfrak{P}'^T$ and by the construction of $\mathfrak{P}_2$:

$$X^{p^m TP}\equiv X^{p^m T}(\mathfrak{P}_2).$$

The product of all transfer factors which belong to the right cosets in $\mathfrak{P}T\mathfrak{P}$, is congruent to $X^{p^t T}(\mathfrak{P}_2)$, and therefore, by the remark at the beginning of the proof, is contained in $\mathfrak{P}_2$.

2. $T\in N_{\mathfrak{P}}$. Then $\mathfrak{P}TX=\mathfrak{P}X^T\cdot T=\mathfrak{P}T=\mathfrak{P}T\mathfrak{P}$. The corresponding transfer factor is X^T. Since $X^T\equiv X(N'_{\mathfrak{P}})$, it follows that:

$$X^T\equiv X(\mathfrak{P}_2)$$

by the construction of $\mathfrak{P}_2$.

1. and 2. together imply the congruence

$$VX\equiv X^{N_{\mathfrak{P}}:\mathfrak{P}}(\mathfrak{P}_2),$$

and since $N_{\mathfrak{P}}:\mathfrak{P}$ is relatively prime to p, we obtain the contradiction $VX\notin\mathfrak{P}_2$, Q.E.D.

Corollary to the First Theorem of Grün:

The normal subgroup $\mathfrak{G}_1$ consisting of all the elements of $\mathfrak{G}$ transferred to $\mathfrak{P}'$ is the p-commutator group $\mathfrak{G}'(p)$ of $\mathfrak{G}$, i.e., the group transferred into a Sylow p-group is isomorphic to the p-factor commutator group.[1]

Proof: Since $\mathfrak{G}/\mathfrak{G}_1$ is an abelian p-group, $\mathfrak{G}'(p)$ is in $\mathfrak{G}_1$. Moreover, by what was shown in the previous proof, $\mathfrak{P}_1=\mathfrak{P}\cap\mathfrak{G}_1\subseteq\mathfrak{P}\cap\mathfrak{G}'$. On the other hand, $\mathfrak{G}'$ is in $\mathfrak{G}_1$; therefore $\mathfrak{P}\cap\mathfrak{G}'$ is in $\mathfrak{P}_1$; therefore $\mathfrak{P}_1=\mathfrak{P}\cap\mathfrak{G}'$. By Chapter IV, $\mathfrak{G}/\mathfrak{G}'(p)\simeq\mathfrak{P}/\mathfrak{P}\cap\mathfrak{G}'$. From this we conclude

$$\mathfrak{G}/\mathfrak{G}'(p)\simeq\mathfrak{P}/\mathfrak{P}_1\simeq\mathfrak{G}/\mathfrak{G}_1,\quad \mathfrak{G}:\mathfrak{G}'(p)=\mathfrak{G}:\mathfrak{G}_1,\quad \mathfrak{G}_1=\mathfrak{G}'(p).$$

Definition: A finite group is said to be *p-normal* if the center of one of its Sylow p-groups is the center of every Sylow p-group in which it is contained.

For example, a finite group with abelian Sylow p-groups is p-normal.

Second Theorem of Grün (Theorem 6): *If the finite group* $\mathfrak{G}$ *is*

[1] See the definitions in IV, § 5, Exercise 1.

p-normal, then the factor commutator group of $\mathfrak{G}$ is isomorphic to the p-factor commutator group of the normalizer of a p-center.

Proof: Let $\mathfrak{z}$ be the center of the Sylow p-group $\mathfrak{P}$; let $\mathfrak{P}_1$ be the intersection of $\mathfrak{G}'$ with $\mathfrak{P}$; let $\mathfrak{P}_2$ be the intersection of $\mathfrak{P}$ with the commutator group $N_{\mathfrak{z}}'$ of the normalizer $N_{\mathfrak{z}}$ of $\mathfrak{z}$. By Chapter IV we know that $\mathfrak{G}/\mathfrak{G}'(p) \simeq \mathfrak{P}/\mathfrak{P}_1$, $N_{\mathfrak{z}}/N_{\mathfrak{z}}'(p) \simeq \mathfrak{P}/\mathfrak{P}_2$. Since $\mathfrak{P}_2$ is contained in $\mathfrak{P}_1$, we have to prove that $\mathfrak{P}_2$ is equal to $\mathfrak{P}_1$. By the first Grün theorem

$\mathfrak{P}_1 = (\mathfrak{P} \cap N'_{\mathfrak{P}}) \cdot \prod_{T \in \mathfrak{G}} \mathfrak{P} \cap \mathfrak{P}'^T$, and therefore we must show:

(a) $\mathfrak{P} \cap N'_{\mathfrak{P}} \leq \mathfrak{P}_2$, (b) $\mathfrak{P} \cap \mathfrak{P}'^T \leq \mathfrak{P}_2$ for all T in $\mathfrak{G}$.

(a) follows from $N_{\mathfrak{P}} \leq N_{\mathfrak{z}}$, $N'_{\mathfrak{P}} \leq N_{\mathfrak{z}}'$.

For the proof of (b) we put $\mathfrak{D} = \mathfrak{P} \cap \mathfrak{P}'^T$ and find that $\mathfrak{z} \leq N_{\mathfrak{D}}$, $\mathfrak{z}^T \leq N_{\mathfrak{D}}$, since $\mathfrak{z}^T$ is the center of $\mathfrak{P}^T$. $\mathfrak{z}$ is in a Sylow p-group $\mathfrak{q}$ of $N_{\mathfrak{D}}$, $\mathfrak{z}^T$ is in a Sylow p-group $\mathfrak{p}$ of $N_{\mathfrak{D}}$ and by the second Sylow theorem there is an S in $N_{\mathfrak{D}}$ such that $\mathfrak{p}^S = \mathfrak{q}$; therefore $\mathfrak{z}^{ST}$ is contained in $\mathfrak{q}$, $\mathfrak{q}$ is contained in a Sylow p-group $\mathfrak{Q}$ of $\mathfrak{G}$, and since by hypothesis $\mathfrak{G}$ is p-normal, both $\mathfrak{z}$ and $\mathfrak{z}^{ST}$ are equal to the center of $\mathfrak{Q}$, and therefore equal to each other. ST is contained in $N_{\mathfrak{z}}$, and $\mathfrak{D} = \mathfrak{D}^S = \mathfrak{P}^S \cap \mathfrak{P}'^{ST}$, $\mathfrak{D} \leq \mathfrak{P}'^{ST} \leq N_{\mathfrak{z}}'$, so finally $\mathfrak{D} \leq \mathfrak{P} \cap N_{\mathfrak{z}}' = \mathfrak{P}_2$, Q.E.D.

Corollary to the Second Theorem of Grün: The transfer of a p-normal group into the Sylow p-group $\mathfrak{P}$ is equal to the transfer $V_{N_{\mathfrak{z}} \to \mathfrak{P}}(N_{\mathfrak{z}})$ of the normalizer $N_{\mathfrak{z}}$ of the center $\mathfrak{z}$ of $\mathfrak{P}$ into the Sylow p-group $\mathfrak{P}$ of $N_{\mathfrak{z}}$.

This is true since by Theorem 3 on transfers, $V_{\mathfrak{G} \to \mathfrak{P}}(\mathfrak{G})$ is contained in $V_{N_{\mathfrak{z}} \to \mathfrak{P}}(N_{\mathfrak{z}})$, and by what has just been proven, these are isomorphic.

In order to obtain results about the case where every Sylow p-group is abelian, we prove the

Lemma: If the index of the finite group $\mathfrak{G}$ over the abelian normal subgroup $\mathfrak{A}$ is relatively prime to the order of $\mathfrak{A}$, then

$$\mathfrak{A} = (\mathfrak{A} \cap \mathfrak{G}') \times (\mathfrak{A} \cap \mathfrak{z}(\mathfrak{G})) \quad \text{and} \quad V_{\mathfrak{G} \to \mathfrak{A}}(\mathfrak{G}) = \mathfrak{A} \cap \mathfrak{z}(\mathfrak{G}).$$

Proof: Let $\mathfrak{G}_1$ be the normal subgroup of all elements which are transferred onto e by $V_{\mathfrak{G} \to \mathfrak{A}} = V$. Then $\mathfrak{G} : \mathfrak{G}_1$ is a divisor of $\mathfrak{A} : 1$; therefore, applying the hypothesis, $\mathfrak{G}_1 : 1$ is divisible by $\mathfrak{G} : \mathfrak{A}$; consequently $\mathfrak{G} = \mathfrak{G}_1 \mathfrak{A}$, i.e., $V\mathfrak{G} = V\mathfrak{A}$.

For an element X in $\mathfrak{A}$,

$$VX = \prod_{T(\mathfrak{A})} X^T = X^{\mathfrak{G} : \mathfrak{A}} \cdot \prod_{T(\mathfrak{A})} X^{T-1} \equiv X^{\mathfrak{G} : \mathfrak{A}} (\mathfrak{A} \cap \mathfrak{G}').$$

We show first that $\mathfrak{A} = (\mathfrak{A} \cap \mathfrak{G}') \times V\mathfrak{A}$. In fact, $\mathfrak{A} \cap \mathfrak{G}' \cap V\mathfrak{A} = e$, since from $X \in \mathfrak{A}$, $VX \in \mathfrak{A} \cap \mathfrak{G}'$, it follows that $X^{\mathfrak{G}:\mathfrak{A}} \equiv e(\mathfrak{A} \cap \mathfrak{G}')$, and by hypothesis we conclude that $X \equiv e(\mathfrak{A} \cap \mathfrak{G}')$; but then $VX = e$. Moreover $(\mathfrak{A} \cap \mathfrak{G}') \cdot V\mathfrak{A} = \mathfrak{A}$, since $(\mathfrak{A} \cap \mathfrak{G}') \cdot V\mathfrak{A}$ contains all the $(\mathfrak{G}:\mathfrak{A})$-th powers of elements in $\mathfrak{A}$, and by hypothesis these form all of $\mathfrak{A}$.

Now we shall show that $V\mathfrak{A} = \mathfrak{A} \cap \mathfrak{z}(\mathfrak{G})$, from which, together with what has already been proved, our assertion follows.

Generally speaking, according to the explicit definition of transfer, we have for an element T of the normalizer of the group $\mathfrak{A}$ to which $\mathfrak{G}$ is transferred, $V(TXT^{-1}) = T \cdot VX \cdot T^{-1}$. Also we have $V(TXT^{-1}) = VT \cdot VX \cdot VT^{-1} = VX$, because the transfer is a homomorphism into an abelian group. Hence VX belongs to the center of $N_A/\mathfrak{A}'$. For an element X in $\mathfrak{z}(\mathfrak{G}) \cap \mathfrak{A}$ the transfer is $X^{\mathfrak{G}:\mathfrak{A}}$, and by hypothesis, the transfer into $\mathfrak{z}(\mathfrak{G}) \cap \mathfrak{A}$ induces an automorphism; consequently $\mathfrak{z}(\mathfrak{G}) \cap \mathfrak{A} = V\mathfrak{G} = V\mathfrak{A}$, Q.E.D.

THEOREM 7: *If a Sylow p-group* $\mathfrak{P}$ *of the finite group* $\mathfrak{G}$ *is abelian, then the transfer of* $\mathfrak{G}$ *into* $\mathfrak{P}$ *maps the p-factor commutator group of* $\mathfrak{G}$ *isomorphically onto the intersection of the Sylow p-group with the center of its normalizer.*

Proof: Since $\mathfrak{z}(\mathfrak{P}) = \mathfrak{P}$, $\mathfrak{G}$ is p-normal; therefore by the corollary to the second Grün theorem, $V_{\mathfrak{G}\to\mathfrak{P}}(\mathfrak{G}) = V_{N_\mathfrak{P}\to\mathfrak{P}}(N_\mathfrak{P})$ and by our lemma $V_{N_\mathfrak{P}\to\mathfrak{P}}(N_\mathfrak{P}) = \mathfrak{z}(N_\mathfrak{P}) \cap \mathfrak{P}$.

FROBENIUS' THEOREM (THEOREM 8): *If the order N of a finite group* $\mathfrak{G}$ *is relatively prime to*

$$k_n = (p^n - 1)(p^{n-1} - 1) \ldots\ldots (p - 1),$$

where p^n *is the order of a Sylow p-group, then the maximal p-factor group*[1] *of* $\mathfrak{G}$ *is isomorphic to every Sylow p-group of* $\mathfrak{G}$.

Proof: If $n = 0$, the theorem is clearly true. Let $n > 0$ and assume the theorem proven for groups whose Sylow p-groups are of order less than p^n.

If $\mathfrak{G}$ is not p-normal, then, by Chapter IV, Theorem 8, there are in $\mathfrak{G}$ a p-group $\mathfrak{D} \neq e$ and an element X such that transforming $\mathfrak{D}$ with X induces an automorphism of order $q > 1$ relatively prime to p. By Chapter IV, q is a divisor of $k_{d(\mathfrak{D})}$, even a divisor of k_n since $d(\mathfrak{D}) \leq n$. Since on the other hand q is a divisor of N, q, by hypothesis, must be equal to 1; therefore $\mathfrak{G}$ is p-normal.

If $\mathfrak{P}$ is abelian, then $\mathfrak{P}$ is in the center of its normalizer since transforming $\mathfrak{P}$ with an element in $N_\mathfrak{P}$ induces an automorphism whose order

[1] Definition see IV, § 5, Exercise 3.

is a divisor of both $k_{d(\mathfrak{P})}$ and N, and therefore is equal to one. Now the assertion follows from Burnside's theorem.

If the center $\mathfrak{z}$ of the Sylow p-group $\mathfrak{P}$ is different from $\mathfrak{P}$, then $\mathfrak{z} \neq e$, and therefore the induction hypothesis is applicable to $N_{\mathfrak{z}}/\mathfrak{z}$. This shows that $N_{\mathfrak{z}}$ has a p-factor group different from e; that the same is true for $\mathfrak{G}$ follows from the second Grün theorem. The maximal p-factor group $\mathfrak{G}/\mathfrak{D}_p$ is now different from e. If p still divides $\mathfrak{D}_p : 1$, then the induction hypothesis would lead to the contradiction $\mathfrak{D}_p(\mathfrak{D}_p(\mathfrak{G})) \neq \mathfrak{D}_p(\mathfrak{G})$. Therefore p is relatively prime to $\mathfrak{D}_p : 1$, and this means that every Sylow p-group of $\mathfrak{G}$ forms a system of representatives of $\mathfrak{G}$ over $\mathfrak{D}_p$, Q.E.D.

COROLLARIES: 1. The order of a finite simple group of even composite order is divisible by 12, 16 or 56.

2. From the proof of the theorem it follows that the number k_n of the theorem may be replaced by k_D, which is at most as large as k_n, where p^D is the order of the maximal abelian factor group of exponent p among all those which are factor groups of p-groups in $\mathfrak{G}$.

3. If a Sylow p-group of $\mathfrak{G}$ contains a cyclic subgroup of index p, and N is relatively prime to p^2-1, then the p-factor group of $\mathfrak{G}$ is isomorphic to a Sylow p-group. For, by Chapter IV, § 3, Exercise 1, $D \leq 2$.

Exercise: A simple finite group whose order is odd and smaller than 1000 is of prime order.

§ 3. Groups whose Sylow Groups are all Cyclic

THEOREM 9: *In the series of higher commutator groups $\mathfrak{G}' \geq \mathfrak{G}'' \geq \ldots$ of a given group $\mathfrak{G}$, two successive factor groups are cyclic only if the latter one is equal to e.*

Proof: It can be assumed that $\mathfrak{G}'/\mathfrak{G}''$ is cyclic, $\mathfrak{G}''$ is generated by A, and $\mathfrak{G}''' = e$. It will be shown that $\mathfrak{G}'' = e$.

The normalizer of $\mathfrak{G}''$ is $\mathfrak{G}$. The factor group of $\mathfrak{G}$ over the centralizer N_A of $\mathfrak{G}''$ is isomorphic to a group of automorphisms of (A), and therefore is abelian. $\mathfrak{G}'$ is in N_A, and since the factor group of $\mathfrak{G}'$ over the normal subgroup $\mathfrak{G}''$ in the center of $\mathfrak{G}'$ is cyclic, $\mathfrak{G}'$ is abelian, and therefore $\mathfrak{G}'' = e$, Q.E.D.

We make the following definition: A group is said to be *metacyclic* if its commutator group and its factor commutator group are cyclic.

As a consequence of Theorem 9, it no longer makes sense to talk of 3-step metacyclic groups. A cyclic group is metacyclic.

THEOREM 10: *If every Sylow group of a finite group $\mathfrak{G}$ is cyclic, then $\mathfrak{G}$ is solvable.*

Proof: If $\mathfrak{G}$ is a p-group then the theorem is clearly true. Let the number r of different prime factors of $\mathfrak{G}:1$ be greater than 1, and assume that the theorem has been proven for all groups whose order is divisible by at most $r-1$ different primes. Let p be the smallest prime factor of $\mathfrak{G}:1$. Since a Sylow p-group is cyclic, the index of its normalizer over its centralizer is a divisor of $p-1$; therefore by the construction of p, there is a Sylow p-group in the center of its normalizer. By Burnside's theorem, $\mathfrak{G}$ contains a normal subgroup $\mathfrak{N}$ with the Sylow p-group as a system of representatives; and we can apply the induction hypothesis to $\mathfrak{N}$. This shows that $\mathfrak{N}$ is solvable, and therefore $\mathfrak{G}$ is solvable, Q.E.D.

THEOREM 11: *A finite group of order N containing only cyclic Sylow groups is metacyclic and has two generators A, B with the defining relations*:

a) $A^m = e,\ B^n = e,\ BAB^{-1} = A^r$, *and the conditions*

b) $0 < m,\ mn = N$,

c) $((r-1).n, m) = 1$,

d) $r^n \equiv 1 (m)$, *and conversely.*

Proof: The conditions imposed on $\mathfrak{G}$ also hold for every subgroup and every factor group of a subgroup. If $\mathfrak{G}$ is abelian, then $\mathfrak{G}$ is cyclic. It follows from Theorem 9 that $\mathfrak{G}$ is metacyclic in any case. Let A be a generating element of the commutator group $\mathfrak{G}'$, of order m. Let $B\mathfrak{G}'$ be a generating coset of the factor commutator group, of order n. Then $BAB^{-1} = A^r$, $B^nAB^{-n} = A^{r^n} = A$, and therefore $r^n \equiv 1\ (m)$. Every commutator is a power of $BAB^{-1}A^{-1} = A^{r-1}$, and therefore $(r\text{-}1, m) = 1$. Since B^n is a power of A which commutes with B, we have $B^n = e$. If a prime p were to divide n and m, then $\left\{B^{\frac{n}{p}}, A^{\frac{m}{p}}\right\}$ would be a non-cyclic subgroup of order p^2 and this contradicts the hypothesis; and therefore $(n, m) = 1$.

Conversely, let $\mathfrak{G}$ be a group with generators A and B which satisfy the defining relations a) and conditions b), c), d). By Hölder's theorem in Chapter III, $\mathfrak{G}$ is of order nm. Since $(r\text{-}1, m) = 1$, $\mathfrak{G}' = (A)$. Since $(n, m) = 1$ and the order N of $\mathfrak{G}$ *is* nm, then for every Sylow group, there is one conjugate to it in (A) or in (B). Therefore every Sylow group of $\mathfrak{G}$ is cyclic, Q.E.D.

§ 4. The Principal Ideal Theorem

First we shall present some considerations about operator domains of abelian groups in pursuance of Chapter III. §§ 3 and 5.

Let $\mathfrak{F}$ be a group of automorphisms $U \to U^\sigma$ of the abelian group $\mathfrak{U}$ with a finite number of generators. All operators $\sum_\sigma c_\sigma \sigma$, with rational integral coefficients c_σ, only a finite number of which are different from zero, form an operator ring Ω with a unit element · Let $\mathfrak{F}_0$ be a normal subgroup of $\mathfrak{F}$ and let $\sigma \to \bar{\sigma}$ be a representation function of $\mathfrak{F}$ over $\mathfrak{F}_0$. Now what does calculation mod $\mathfrak{F}_0$ (i.e., the replacement of σ by $\bar{\sigma}$) mean in Ω? Certain elements in $\mathfrak{U}$ are identified; thus

$$U^\sigma \equiv U^{\sigma'}, \text{ if } \sigma \equiv \sigma'(\mathfrak{F}_0).$$

Instead of calculating in $\mathfrak{U}$, we must now calculate in the factor group of $\mathfrak{U}$ over a subgroup $\mathfrak{U}_0$, where $\mathfrak{U}_0$ must contain at least all $U^{\sigma-\sigma'}$ for $\sigma \equiv \sigma'(\mathfrak{F}_0)$. But all $U^{\sigma-\sigma'}$ with $\sigma \equiv \sigma'(\mathfrak{F}_0)$ generate a subgroup $\mathfrak{U}_0$ of $\mathfrak{U}$ which is admissible with respect to Ω. The automorphisms σ induce automorphisms $\bar{\sigma}$ of $\mathfrak{U}/\mathfrak{U}_0$, and the operator ring Ω goes over into an operator ring $\bar{\Omega}$ of $\mathfrak{U}/\mathfrak{U}_0$.

The order ideal of $\mathfrak{U}/\mathfrak{U}_0$ over $\bar{\Omega}$ is obtained from the order ideal of $\mathfrak{U}$ over Ω by replacing σ by $\bar{\sigma}$ everywhere.

In order to construct the group transferred into the normal subgroup $\mathfrak{U}$, it suffices to calculate in the factor group over $\mathfrak{U}'$, since $\mathfrak{U}'$ is a normal subgroup of $\mathfrak{G}$; thus we assume that $\mathfrak{U}' = e$.

Let $\mathfrak{G}/\mathfrak{U}$ be isomorphic to the abstract group $\mathfrak{F} = \{1, \sigma, \tau, \ldots\}$ and let $(S_\sigma, C_{\sigma,\tau})$ be a factor system of $\mathfrak{G}$ over $\mathfrak{U}$:

$$S_\sigma S_\tau = C_{\sigma,\tau} S_{\sigma\tau}.$$

Every element S in $\mathfrak{G}$ is uniquely of the form $S = US_\tau$ with $U \in \mathfrak{U}$; therefore, using the earlier notation, we form

$$V_{\mathfrak{G}\to\mathfrak{U}}(U) = \prod_\sigma S_\sigma U \overline{S_\sigma U}^{-1} = \prod_\sigma S_\sigma U S_\sigma^{-1} = U^{\sum_\sigma \sigma},$$

$$V_{\mathfrak{G}\to\mathfrak{U}}(S_\tau) = \prod_\sigma S_\sigma S_\tau \overline{S_\sigma S_\tau}^{-1} = \prod_\sigma S_\sigma S_\tau S_{\sigma\tau}^{-1} = \prod_\sigma C_{\sigma,\tau},$$

so that
$$V_{\mathfrak{G}\to\mathfrak{U}}(S) = U^{\sum_\sigma \sigma} \cdot \prod_\sigma C_{\sigma,\tau}.$$

Let $\mathfrak{G}$ be the splitting group (constructed as in Chap. III, § 9) of $\mathfrak{G}$ over the abelian normal subgroup $\mathfrak{U}$; the new normal subgroup $\bar{\mathfrak{U}}$ is the direct product of $\mathfrak{U}$ with the infinite cyclic groups (A_σ), $\sigma \neq 1$, and

$$A_\tau^\sigma = A_\sigma^{-1} A_{\sigma\tau} C_{\sigma,\tau}^{-1}. \tag{1}$$

For $\bar{\mathfrak{G}}$ over $\bar{\mathfrak{U}}$ there is a retracting factor system $(T_\sigma, 1)$, where $T_\sigma = A_\sigma S_\sigma$. Therefore

$$V_{\mathfrak{G}\to\mathfrak{U}}(S) = V_{\bar{\mathfrak{G}}\to\bar{\mathfrak{U}}}(S) = \bar{U}_\sigma^{\Sigma\sigma} \cdot \prod 1 = \bar{U}_\sigma^{\Sigma\sigma}$$

$$V_{\mathfrak{G}\to\mathfrak{U}}(\mathfrak{G}) \leq \bar{\mathfrak{U}}_\sigma^{\Sigma\sigma}. \tag{2}$$

THEOREM 12 (PRINCIPAL IDEAL THEOREM)[1]: *The transfer of a group with a finite factor commutator group into its commutator group is equal to the second commutator group provided the second factor commutator group has a finite number of generators.*

As before we can assume in the proof that $\mathfrak{G}'' = e$. Then it remains to show that $V_{\mathfrak{G}\to\mathfrak{G}'}(\mathfrak{G}) = e$.

The following example shows that the assumption about $\mathfrak{G}'$ is necessary:

Let $\mathfrak{U}$ be the group of all numbers $e^{2\pi i r}$ with rational r, and let $\mathfrak{G} = \{\mathfrak{U}, \mathfrak{j}\}$ be the extension over $\mathfrak{U}$ of index 2 defined by

$$\mathfrak{j}^2 = e^{\pi i} = -1$$

$$\mathfrak{j} e^{2\pi i r} \mathfrak{j}^{-1} = e^{-2\pi i r}.$$

Then $\mathfrak{G}' = \mathfrak{U}$ and $\mathfrak{G}'' = e$ but

$$V_{\mathfrak{G}\to\mathfrak{G}'}(\mathfrak{j}) = -1 \neq e.$$

Instead of the principal ideal theorem, we prove the following slight generalization:

Under the same assumptions, the $(\mathfrak{U} : \mathfrak{G}')$-th power of the transfer of every element G in $\mathfrak{G}$ into an abelian group $\mathfrak{U}$ lying between $\mathfrak{G}$ and $\mathfrak{G}'$ is equal to e: i.e., $(V_{\mathfrak{G}\to\mathfrak{U}}(\mathfrak{G}))^{\mathfrak{U}:\mathfrak{G}'} = e$.

We set $\mathfrak{G} : \mathfrak{U} = n$, $\mathfrak{U} : \mathfrak{G}' = d$; n and d are different from zero; $\mathfrak{U}$ is an abelian normal subgroup of $\mathfrak{G}$. Let $\mathfrak{F}$, $\bar{\mathfrak{G}}$ and $\bar{\mathfrak{U}}$ have the same meaning as previously. By (2) it then suffices to prove

$$\bar{\mathfrak{U}}^{d\Sigma_\sigma \sigma} = e$$

The automorphisms corresponding to σ generate an operator ring Ω of $\bar{\mathfrak{U}}$, which consists of all $\Theta = \sum_\sigma c_\sigma \sigma$ with integral rational c_σ. $\bar{\mathfrak{U}}$ has an order ideal over Ω, since $\mathfrak{G}'$ has a finite number of generators,

[1] The Principal Ideal Theorem of class field theory states that every ideal of an algebraic number field is a principal ideal in the absolute class field. The principal ideal theorem can be stated group-theoretically as Theorem 12.

$\mathfrak{U}$ is finite over $\mathfrak{G}'$, and $\overline{\mathfrak{U}}$ has a finite number of generators A_σ over $\mathfrak{U}$, so that $\overline{\mathfrak{U}}$ has a finite number of generators, hence a finite number of generators over Ω.

If we now show that $d \cdot \sum_\sigma \sigma$ generates the order ideal of $\overline{\mathfrak{U}}$ over Ω, then the theorem is proven. Now let $\Theta = \sum_\sigma c_\sigma \sigma$ be in the Ω-order ideal of $\overline{\mathfrak{U}}$. Then taking note of (1) we have:

$$e = A_\tau^{\Theta} = A_\tau^{\sum_\sigma c_\sigma \sigma} = \prod_\sigma (A_\tau^\sigma)^{c_\sigma} \equiv \prod_\sigma A_\sigma^{-c_\sigma} A_{\sigma\tau}^{c_\sigma} (\mathfrak{U})$$
$$\equiv \prod_\sigma A_\sigma^{-c_\sigma + c_{\sigma\tau^{-1}}} (\mathfrak{U}).$$

Since the A_σ form a basis for $\overline{\mathfrak{U}}/\mathfrak{U}$, we must have $c_\sigma = c_{\sigma\tau^{-1}}$ for all $\sigma \neq 1$, hence $c_\sigma = c_1$ for all σ, whence $\Theta = c_1 \cdot \sum_\sigma \sigma$. Consequently, the order ideal of $\overline{\mathfrak{U}}$ over Ω is a principal ideal which is generated by $c \cdot \sum_\sigma \sigma$ with an integral rational $c \geqq 0$.

If we replace σ by 1, then, as was pointed out at the beginning of the paragraph, we are calculating in the group $\overline{\mathfrak{U}}/\overline{\mathfrak{G}}'$; for since $\overline{\mathfrak{G}}$ is generated by the $\overline{U}$ and the T_σ, $\overline{\mathfrak{G}}'$ is generated by all the elements $\overline{U} T_\sigma \overline{U}^{-1} T_\sigma^{-1} = \overline{U}^{1-\sigma}$, i.e. $\overline{\mathfrak{U}}^{1-\sigma} = \overline{\mathfrak{G}}'$. Here Ω goes over into the ring Ω_0 of rational integers. The Ω_0 order ideal of $\overline{\mathfrak{U}}/\overline{\mathfrak{G}}'$ is obtained from the Ω-order ideal of $\overline{\mathfrak{U}}$ by the same substitution; on the other hand, by Chap. III, § 5, it is generated by the group order $\overline{\mathfrak{U}} : \overline{\mathfrak{G}}'$. Therefore

$$\overline{\mathfrak{U}} : \overline{\mathfrak{G}}' = c \cdot (1 + 1 + \cdots + 1) = cn.$$

In order to show that $c = d$, we prove the isomorphism

$$\overline{\mathfrak{U}}/\overline{\mathfrak{G}}' \simeq \mathfrak{G}/\mathfrak{G}',$$

from which it follows that

$$cn = \overline{\mathfrak{U}} : \overline{\mathfrak{G}}' = \mathfrak{G} : \mathfrak{G}' = dn,$$

$$c = d.$$

Since $\mathfrak{G}'$ is a normal subgroup of each of the four groups, we may set $\mathfrak{G}' = e$ for the proof of the isomorphy. Since 1) $\overline{\mathfrak{G}}' = \overline{\mathfrak{U}}^{1-\sigma}$, 2) $\overline{\mathfrak{U}}$ is generated by $\mathfrak{U}$ and the A_τ, and moreover 3) $\mathfrak{U}^{1-\sigma} = e$, it now follows by (1) that

$$\overline{\mathfrak{G}}' = \{A_\tau^{1-\sigma}\} = \{A_\tau C_{\sigma,\tau} A_{\sigma\tau}^{-1} A_\sigma\}.$$

Therefore we can choose the elements $A_\sigma U$ as representatives of $\overline{\mathfrak{U}}/\overline{\mathfrak{G}}'$. By (1)

$$A_\sigma A_\tau = A_{\sigma\tau} C_{\sigma,\tau}^{-1} A_\tau^{1-\sigma} \equiv A_{\sigma\tau} C_{\sigma,\tau}^{-1} \ (\overline{\mathfrak{G}}').$$

The abelian group $\mathfrak{G}$ consists of the elements $S_\sigma^{-1} U$ with the calculational rule

$$S_\sigma^{-1} S_\tau^{-1} = S_{\sigma\tau}^{-1} C_{\sigma,\tau}^{-1}.$$

The correspondence $A_\sigma U \rightarrow S_\sigma^{-1} U$ therefore gives an isomorphism between $\overline{\mathfrak{U}}/\overline{\mathfrak{G}}'$ and $\mathfrak{G}$, Q.E.D.

COROLLARY OF THE PRINCIPAL IDEAL THEOREM: In a 2-step metabelian group with a finite number of generators and cyclic factor commutator group of order n, every element whose coset generates the factor commutator group is of order n.

Proof: If S is the element described in the above statement, then $\mathfrak{G}/\mathfrak{G}' = (S\mathfrak{G}')$ and therefore the powers $1, S, \ldots, S^{n-1}$ are a system of representatives of $\mathfrak{G}$ over $\mathfrak{G}'$. Consequently

$$V_{\mathfrak{G}\rightarrow\mathfrak{G}'}(S) = \prod_{\nu=0}^{n-1} S^\nu S\, \overline{S^{\nu+1}}^{-1} = S^n,$$

while on the other hand $V_{\mathfrak{G}\rightarrow\mathfrak{G}'}(S) = e$.

APPENDIX A

FURTHER EXERCISES FOR CHAP. II

(For 10 and 11 consult Exx. 15 and 16 at the end of Chap. I.)

10. Let $\mathfrak{g}$ be an abstract group with elements $a_1, \ldots, x_1, \ldots$. Let Σ be a system of groups such that for each member $\mathfrak{g}'$ of Σ there is a given an isomorphism $i(\mathfrak{g}')$ between $\mathfrak{g}$ and $\mathfrak{g}'$. Show that the set of all one-to-one correspondences $c(\mathfrak{g}', \mathfrak{g}'', a)$ between any $\mathfrak{g}'$ and any $\mathfrak{g}''$ which map the element $i(\mathfrak{g}')x$ of $\mathfrak{g}'$ onto the element $i(\mathfrak{g}'')ax$ of $\mathfrak{g}''$ is a groupoid $\mathfrak{G}(\Sigma, \mathfrak{g})$.

11. Let $\mathfrak{G}$ be a groupoid. For any two units e, e' linked by x show that $ex = x = xe'$. Furthermore, show that the mapping of a onto $x^{-1}ax$ is an isomorphism between $\mathfrak{g}_e$ and $\mathfrak{g}_{e'}$. Let Σ be the system of the groups $\mathfrak{g}_e, \mathfrak{g}_{e'}, \ldots$ attached to the units $e, e', \ldots$ of $\mathfrak{G}$. Let $\mathfrak{g}$ be an abstract group mapped by isomorphisms $i(e), i(e'), \ldots$ onto $\mathfrak{g}_e, \mathfrak{g}_{e'}, \ldots$ Show that $\mathfrak{G}$ is isomorphic to $\mathfrak{G}(\Sigma, \mathfrak{g})$.

12. Every homomorphism h of a multiplicative domain $\mathfrak{M}$ into another multiplicative domain defines on $\mathfrak{M}$ the normal multiplicative congruence relation: $aR(h)b$ if and only if $ha = hb$. Conversely, if R is a normal multiplicative congruence relation on $\mathfrak{M}$, then the residue classes form a multiplicative domain $\mathfrak{M}/R$ according to the rule of multiplication $\bar{a}\bar{b} = \overline{ab}$, where $\bar{x}$ denotes the residue class modulo R represented by the element x of $\mathfrak{M}$. The mapping h that maps x onto $\bar{x}$ is a homomorphism (*natural homomorphism*) of $\mathfrak{M}$ onto $\mathfrak{M}/R$ which induces on $\mathfrak{M}$ the congruence relation R in the sense defined above. If j is another homomorphism of $\mathfrak{M}$ inducing the normal multiplicative congruence relation $R = R(j)$ on $\mathfrak{M}$, then j induces the isomorphism between $\mathfrak{M}/R(j)$ and $j(\mathfrak{M})$ which maps $\bar{x}$ onto jx.

13. Show that the intersection of two systems of imprimitivity of a transitive permutation group is itself a system of imprimitivity provided the intersection contains more than one letter.

14. (G. E. Wall.) In the ring I_8 consisting of the eight residue classes of the rational integers modulo 8 show that the mappings $(z, az + b)$ that map the element z of I_8 onto the element $az + b$ (a odd; a, b contained in I_8) form a group $\mathfrak{G}$ of order 32. Show that the mapping of $(z, az + b)$ onto $(z, az + (b + (a^2 - 1)/2))$ is an outer automorphism of $\mathfrak{G}$ that maps each element of $\mathfrak{G}$ onto a conjugate element under $\mathfrak{G}$.

15. If a normal subgroup of a group $\mathfrak{G}$ and its factor group both are solvable, then $\mathfrak{G}$ is solvable.

16. The product of a solvable normal subgroup of a group $\mathfrak{G}$ and a solvable subgroup of $\mathfrak{G}$ is a solvable subgroup of $\mathfrak{G}$. (Use Ex. 15.)

17. The radical $R(\mathfrak{G})$ of a group $\mathfrak{G}$ is defined as a solvable normal subgroup of $\mathfrak{G}$ which is not contained in a larger solvable normal subgroup of $\mathfrak{G}$ (maximal solvable normal subgroup). Show that there is at most one radical of $\mathfrak{G}$ and that it is a characteristic subgroup. If the maximal condition is satisfied for the solvable normal subgroups of $\mathfrak{G}$, then $\mathfrak{G}$ has a radical. (Use Ex. 16.)

18. If the group $\mathfrak{G}$ has a radical $R(\mathfrak{G})$, then the radical of each normal subgroup $\mathfrak{N}$ of $\mathfrak{G}$ is equal to the intersection of $\mathfrak{N}$ and the radical of $\mathfrak{G}$. If $\mathfrak{N}$ is solvable, then $R(\mathfrak{G}/\mathfrak{N}) = R(\mathfrak{G})/\mathfrak{N}$; in particular $R(\mathfrak{G}/R(\mathfrak{G})) = R(\mathfrak{G})/R(\mathfrak{G})$.

19. Each k-step metabelian subgroup of a group $\mathfrak{G}$ is contained in a maximal k-step metabelian subgroup of $\mathfrak{G}$, i. e., a k-step metabelian subgroup not contained in a larger k-step metabelian subgroup.

20. If there are maximal solvable subgroups of a group $\mathfrak{G}$, then the radical of $\mathfrak{G}$ is the intersection of the maximal solvable subgroups.

21. If, for a fixed k, every solvable subgroup of a group is at most k-step metabelian, then every solvable subgroup is contained in a maximal solvable subgroup, and $\mathfrak{G}$ has a radical.

22. Show that for any homomorphism h of a group $\mathfrak{G}$ onto a group $\mathfrak{H}$ we have $h((a, b)) = (ha, hb)$ for a, b contained in $\mathfrak{G}$ and that $h(D^r(\mathfrak{G})) = D^r(h(\mathfrak{G}))$ for $r = 0, 1, 2, \ldots$

23. Let $\mathfrak{S}$ be a semi-group with unit element.

a) The element a is called a *left divisor* (*right divisor*) of the element b of $\mathfrak{S}$ if there is an equation $b = ax$ ($b = ya$), where x and y respectively occur in $\mathfrak{S}$. Show that this relation is reflexive and transitive.

b) Two elements are called *left equivalent* (*right equivalent*) if each is a left divisor (right divisor) of the other. Show the normality of this relation. Show that equivalent elements can be substituted in one-sided divisor relations.

c) If a is a left divisor of b, then ca is a left divisor of cb. If a is a right divisor of b, then ad is a right divisor of bd.

d) We say a divides b if there are equations

$$a = a_0 = a_{01}a_{02}, \quad a_1 = a_{01}x_1a_{02} = a_{11}a_{12}, \quad a_2 = a_{11}x_2a_{12} = a_{21}a_{22}, \quad \ldots,$$
$$b = a_{r+1} = a_{r1}x_{r+1}a_{r2},$$

where all factors belong to $\mathfrak{S}$. Show that the relation a divides b is reflexive and transitive. If a is a left divisor or a right divisor of b, then a divides b. If a divides b and a' divides b', then aa' divides bb'.

e) We say a is equivalent to b if a divides b and b divides a. Show that this equivalence relation is normal. Show that equivalent elements can be substituted in divisor relations.

f) Interpret each of the three relations: a is left divisor of b, a is right divisor of b, a divides b, as ordering relations defining a poset $\mathfrak{S}$. Give for a subset s of $\mathfrak{S}$ the definitions of g. c. (greatest common) left divisor, g. c. right divisor, g. c. divisor of s corresponding to the meet in multiplicative terms. Also, give the definitions of l. c. (least common) left multiple, l. c. right multiple, l. c. multiple of s corresponding to the join in multiplicative terms.

g) An element is called a *unit* if it is both a left divisor and a right divisor of 1. The units form a subgroup $\mathfrak{U}(\mathfrak{S})$ of $\mathfrak{S}$.

h) The element n of $\mathfrak{S}$ is called a zero element of $\mathfrak{S}$ if $nx = xn = n$ for every element x of $\mathfrak{S}$. Show that the divisors of n form a sub-semigroup. Show that there is at most one zero element.

i) The element e is called an *idempotent* if $ee = e$ and if e is not a zero element. An idempotent is a left divisor (right divisor) of the element a if and only if it is a left unit (right unit) of a.

j) If there is no idempotent other than 1, then each divisor of 1 is a unit and all the divisors of 1 form a normal divisor $\mathfrak{U}(\mathfrak{S})$ of $\mathfrak{S}$. Moreover, if $\mathfrak{S}$ is commutative, then congruence modulo $\mathfrak{U}$ coincides with equivalence as defined under e) for every pair of non-divisors of zero.

k) If $\mathfrak{S}$ is commutative, then an element is a divisor of 1 if and only if it is a unit.

24. A sub-semigroup of a group is called a *halfgroup*. Show the following:

a) A finite halfgroup is characterized as a finite semi-group for which the cancellation laws of multiplication are satisfied.

b) An abelian halfgroup is characterized as a semi-group satisfying the commutative and the cancellation laws of multiplication (see § 7, Ex. 4 and also Ex. 25).

c) (O. Ore.) If a semi-group $\mathfrak{S}$ satisfies the cancellation laws of multiplication and also the rule $a\,\mathfrak{S} = \mathfrak{S}a$ for each a contained in $\mathfrak{S}$, then it is a halfgroup. (*Hint*: Form the quotient group of $\mathfrak{S}$ consisting of the formal quotients a/b (a, b any two elements in $\mathfrak{S}$) where $a/b = c/d$ if there are elements e, f such that $ea = fb$, $ec = fd$, and where $a/b \cdot c/d = e/f$ means that there is an element g such that $a/b = e/g$, $c/d = g/f$; show that the mapping of a onto the quotient aa/a gives an isomorphism of $\mathfrak{S}$ into the quotient group.)

d) (Lambek-Mal'cev.) Any halfgroup $\mathfrak{H}$ satisfies, in addition to the cancellation laws of multiplication, certain *polyhedral conditions* given by the following construction: A finite system P of v vertices, e edges, and f faces is called an *abstract Euclidean polyhedron* if 1. every edge is incident with precisely two vertices and with precisely two faces, 2. a vertex is incident with a face if and only if there are precisely two edges incident with both of them, 3. the edges $e_1, e_2, \ldots, e_n$ which are incident with a given vertex (face) form a cycle such that with suitable renumbering e_i and e_{i+1} are incident with the same face (vertex), where n is greater than 1 and $e_{n+1} = e_1$, 4. $v + f = e + 2$. The subset of P formed by an edge e and the face F incident with e is called the *F-side* of e. The subset of P formed by the vertex V incident with the face F is called the *angle* at V on the F-side of a, b where a, b are the two edges incident with both P and F. Assign to each angle and to each side an element of $\mathfrak{H}$ such that the relations $xa = yb$ are satisfied, where x, y are assigned to the two sides of an edge e, say to the F-side and to the G-side, and a, b are assigned to the angles formed at a vertex incident with e and F, G respectively. The polyhedral condition corresponding to P states that any one of the finitely many relations $xa = yb$ explained above is a consequence of all the others. (*Hint*: Apply induction on v; amalgamate adjacent faces.)

e) If a semi-group satisfies the cancellation laws of multiplication and the polyhedral conditions given under d), then it is a halfgroup (see A. Mal'cev, *On the embedding of associative systems in groups*. Mat. Sbornik, Vol. 6 (48) (1939), pp. 331—336; J. Lambek, *The immersibility of a semi-group into a group*, Canadian Journal of Math., Vol. 3 (1951), pp. 34—43).

f) There are semi-groups satisfying any given finite subsystem of the polyhedral conditions given under d) and also the cancellation laws of multiplication which are not halfgroups (see A. Mal'cev, *On the embedding of associative systems in groups II*, Mat. Sbornik, Vol. 8 (50) (1940), pp. 251—264).

25. An element d of a multiplicative semi-group $\mathfrak{S}$ is called a *denominator* if a) $dx = xd$ for any x contained in $\mathfrak{S}$, b) $dx = dy$ implies $x = y$.

a) Assuming denominators exist, show that they form an abelian halfgroup $d(\mathfrak{S})$ (see § 7, Ex. 1).

b) Show that the elements of $\mathfrak{S}$, and the formal quotients a/d where a is contained in $\mathfrak{S}$ and d is contained in $d(\mathfrak{S})$, together with the symbol 1, form a multiplicative semi-group $Q(\mathfrak{S})$ (called the *quotient semi-group* of $\mathfrak{S}$) by introducing the rules: $a = b$ in $Q(\mathfrak{S})$ if $a = b$ in $\mathfrak{S}$, $a = b/d$ if $ad = b$, $b/d = a$ if $b = ad$, $a/d = a'/d'$ if $ad' = da'$, $1 = a$ if a is unit element of $\mathfrak{S}$, $a = 1$ if $1 = a$, $d/d = 1$, $1 = d/d$; ab as in $\mathfrak{S}$, $a \cdot b/d = (ab)/d$, $b/d \cdot a = (ba)/d$, $a/d \cdot a'/d' = (aa')/(dd')$, $1a = a1 = a$, $1 \cdot a/d = a/d \cdot 1 = a/d$, $1 \cdot 1 = 1$.

c) Show that $\mathfrak{S}$ is a sub-semigroup of $Q(\mathfrak{S})$. (As regards b) and c), compare § 7, Ex. 2.)

d) Show that the denominators of $Q(\mathfrak{S})$ form an abelian group $dQ(\mathfrak{S})$ with 1 as unit element and that $dQ(\mathfrak{S}) = Qd(\mathfrak{S})$ (see § 7, Ex. 3).

e) Show that $Q(\mathfrak{S}) = \mathfrak{S}$ if and only if $d(\mathfrak{S})$ is a group (see § 7, Ex. 4).

f) Let $\mathfrak{T}$ be a multiplicative semi-group containing a sub-semigroup $\mathfrak{S}$, where the elements of $d(\mathfrak{S})$ generate a subgroup of $\mathfrak{T}$ whose unit element is the unit element of $\mathfrak{T}$. Show that there is one and only one homomorphism of $Q(\mathfrak{S})$ into $\mathfrak{T}$ leaving every element of $\mathfrak{S}$ invariant, namely the isomorphism which maps a onto a, a/d onto ad^{-1}, and 1 onto the unity element of $\mathfrak{T}$.

26. a) Let $\mathfrak{S}$ be an associative semi-ring. Assume that the multiplicative semi-group belonging to $\mathfrak{S}$ contains denominators. Prove that the quotient semi-group of the multiplicative semi-group belonging to $\mathfrak{S}$ forms an associative semi-ring $Q(\mathfrak{S})$ if the addition is defined as follows: $a + b$ as in $\mathfrak{S}$, $a + (b/d) = (ad + b)/d$, $(b/d) + a = (b + da)/d$, $(a/d) + (a'/d') = (ad' + a'd)/(dd')$.

b) The semi-ring $Q(\mathfrak{S})$ is called the *quotient semi-ring*. Show that it contains $\mathfrak{S}$ as sub-semiring.

c) If $\mathfrak{S}$ is a ring, show that $Q(\mathfrak{S})$ is a ring. $Q(\mathfrak{S})$ is called the *quotient ring* of $\mathfrak{S}$.

d) Let $\mathfrak{T}$ be an associative semi-ring containing $\mathfrak{S}$ as sub-semiring, where the elements of $d(\mathfrak{S})$ generate a multiplicative group in $\mathfrak{T}$ whose unit element is also the unit element of $\mathfrak{T}$. Show that there is one and only one homomorphism of the quotient semi-ring $Q(\mathfrak{S})$ into $\mathfrak{T}$ leaving every element of $\mathfrak{S}$ invariant, namely the isomorphism that maps a onto a and a/d onto ad^{-1}.

27. A subring of a field is called a *half field*. Show that a ring is a half field if and only if the multiplication is commutative and if a product vanishes only if at least one of the factors vanishes. The quotient ring of a half field is a field, which is called the *quotient field* of the half field. A half field with unit element is called an *integral domain* or *domain of integrity*. Give examples.

28. A multiplicative domain $\mathfrak{M}$ is called *ordered* if there is a binary relation $a > b$ on $\mathfrak{M}$, called the *ordering relation* on $\mathfrak{M}$, such that 1. a is not greater than a, 2. if $a > b$, $b > c$, then $a > c$, 3. if a is neither equal to b nor greater than b, then $b > a$, 4. if $a > b$, then $ca > cb$ and $ac > bc$. Show the following:

a) The ordered multiplicative domain $\mathfrak{M}$ satisfies the cancellation laws of multiplication.

b) From $a > b$, $c > d$ it follows that $ac > bd$.

c) The binary relation '$a \geq b$ if b is not greater than a' is both reflexive and transitive. Furthermore, between any two elements a, b of $\mathfrak{M}$ one of the two relations $a \geq b$, $b \geq a$ holds. If $a \geq b$, $b \geq a$ then $a = b$. If $a \geq b$, $c \geq d$ then $ac \geq bd$.

d) For an ordered group $\mathfrak{G}$ which is not 1, the elements > 1 form a semi-group $\mathfrak{H}$ with the following properties: 1. If a, b are contained in $\mathfrak{H}$ and a is not a left divisor of b then either $a = b$ or b is a left divisor of a; 2. $a\mathfrak{H} = \mathfrak{H}a$ for every element a of $\mathfrak{H}$; 3. a is not left divisor of a; 4. the cancellation laws of multiplication are satisfied in $\mathfrak{H}$.

e) A semi-group $\mathfrak{H}$ with the properties 1.-4. mentioned under d) can be embedded into an ordered group so that $\mathfrak{H}$ consists of all the elements > 1. The ordering of the embedding group is uniquely determined (use Ex. 24 c)).

29. A quasi-ring $\mathfrak{S}$ having a binary ordering relation is called an *ordered quasi-ring* if its elements under addition form an ordered module and if the elements > 0 (called the *positive elements*) form an ordered semi-group under multiplication. Show that

a) for each element a one, and only one, of the three relations $a > 0$, $a = 0$, $-a > 0$ is true;

b) if the function $\operatorname{sign} a$ is defined to be $1, 0, -1$ according as $a < 0$, $a = 0$, $-a > 0$, respectively, then $\operatorname{sign}(ab) = \operatorname{sign} a \cdot \operatorname{sign} b$;

c) if the absolute value $|a|$ of a is defined as a or $-a$ according as $a \geq 0$ or $-a > 0$, respectively, then the absolute value also is a multiplicative function: $|a \cdot b| = |a| \cdot |b|$; moreover, the triangle inequality $|a + b| \leq |a| + |b|$ holds;

d) the two inequalities $a > b$, $c > d$ imply the inequality $ac + bd > ad + bc$;

e) the ordering of an ordered ring can be extended to an ordering of its quotient ring as follows: $a > b/c$ if $ac^2 > bc$, $a/b > c$ if $ab > cb^2$, $a/b > c/d$ if $abd^2 > cdb^2$;

f) that there is only one possible way to extend the ordering of an ordered ring to an ordering of the quotient field.

30. A semi-ring $\mathfrak{S}$ having a binary ordering relation is called *ordered* if its elements under addition form an ordered semi-module and if, further, from $a > b$, $c > d$ it follows that $ac + bd > bc + ad$. Show that

a) the ordering of $\mathfrak{S}$ can be extended to an ordering of the difference ring $d(\mathfrak{S})$ by introducing the rules: $a > b - c$ if $a + c > b$, $a - b > c$ if $a > c + b$, $a - b > c - d$ if $a + d > c + b$, $a > 0$ if $a + a > a$, $0 > a$ if $a > a + a$, $a - b > 0$ if $a > b$;

b) there is only one possible way of extending the given ordering of $\mathfrak{S}$ to an ordering of the difference ring.

31. Show that the positive elements of an ordered quasi-ring Q form a sub-semiring P having the property that for any element $a \neq 0$ one and only one of the two elements a, $-a$ belongs to P. If, conversely, Q is a quasi-ring with a sub-semiring P that has the algebraic property outlined in the previous statement, then Q is ordered by the ordering relation: $a > b$ if $a - b$ belongs to P.

32. Define recursively the powers of a semi-ring $\mathfrak{S}$:

$$\mathfrak{S}^1 = \mathfrak{S}, \quad \mathfrak{S}^2 = \mathfrak{S}\mathfrak{S}, \quad \mathfrak{S}^n = \mathfrak{S}\mathfrak{S}^{n-1} + \mathfrak{S}^2\mathfrak{S}^{n-2} + \cdots + \mathfrak{S}^{n-1}\mathfrak{S}$$

and show that they form two-sided ideals satisfying the rule $\mathfrak{S}^n\mathfrak{S}^m \leq \mathfrak{S}^{n+m}$. If $\mathfrak{S}$ is associative, then $\mathfrak{S}^n\mathfrak{S}^m = \mathfrak{S}^{n+m}$. If $\mathfrak{S}$ is a Lie-ring, then $\mathfrak{S}^{n+1} = \mathfrak{S} \bigcirc \mathfrak{S}^n = \mathfrak{S}^n \bigcirc \mathfrak{S}$ (here the convention regarding the definition of $\mathfrak{m}_1\mathfrak{m}_2$ as a sub-semimodule if $\mathfrak{m}_1$, $\mathfrak{m}_2$ are sub-semimodules is to be used).

33. Show that the subsets of a given set S form a commutative ring $B(S)$ if addition and multiplication of the two subsets a, b of S are defined as follows: $a + b$ is the complement of the intersection of a and b relative to the union of a and b; $a \cdot b$ is the intersection of a and b. Show that this ring satisfies the laws of a *Boolean ring*, namely, the laws of a commutative ring and the laws: $aa = a$, $a + a = 0$.

34. A mapping a of a multiplicative domain $\mathfrak{M}$ into the multiplicative domain $\mathfrak{N}$ is called an *anti-homomorphism* if $a(xy) = a(y)a(x)$. Show that

a) homomorphisms and anti-homomorphisms combine in a way analogous to that indicated for lattices in § 5;

b) for a group, the mapping that maps each element onto its inverse is an anti-automorphism;

c) the automorphisms and the anti-automorphisms of a multiplicative domain form a group in which the automorphisms form a normal subgroup of index 1 or 2;

d) every multiplicative domain $\mathfrak{M}$ is anti-isomorphic to its *dual domain* $\mathfrak{M}^{(-1)}$, which is defined to be the multiplicative domain that arises from $\mathfrak{M}$ if multiplication is redefined by taking as the new product of the factors a and b, in this order, the old product ba;

e) a multiplicative domain is isomorphic with its dual if and only if it has an anti-automorphism;

f) every group is isomorphic with its dual;

g) the preceding statements remain true for semi-rings if an anti-homomorphism of a semi-ring $\mathfrak{S}$ into a semi-ring $\mathfrak{T}$ is defined to be a mapping a of $\mathfrak{S}$ into $\mathfrak{T}$ satisfying the conditions: $a(xy) = a(y)a(x)$, $a(x + y) = a(x) + a(y)$;

h) every anti-homomorphism a between an associative semi-ring $\mathfrak{S}$ and an associative semi-ring $\mathfrak{T}$ induces the anti-homomorphism between the matrix rings $M_n(\mathfrak{S})$ and $M_n(\mathfrak{T})$ which maps the matrix (β_{ik}) onto the matrix $(a(\beta_{ki}))$;

i) if $\mathfrak{S}$ is a commutative and associative semi-ring, then the correspondence between the matrix (β_{ik}) and its transpose (β_{ki}), which we denote by $(\beta_{ik})^{\mathsf{T}}$, gives an anti-automorphism of $M_n(\mathfrak{S})$.

35. A *derivation* of a quasi-ring Q is defined to be a mapping d of Q into Q satisfying the following rules of formal differentiation: $d(a + b) = d(a) + d(b)$, $d(ab) = d(a)b + ad(b)$. Show that the derivations of Q form a subring $D(Q)$ of the Lie-ring belonging to the operator ring of the additive group of Q. Assign to any element a of L the mapping a of L into L that maps x onto ax, and show that the quasi-ring L is a Lie-ring if and only if the left multiplication a is a derivation mapping a onto 0. Show that the derivations a of a Lie-ring L associated with the elements a of L (*inner derivations*) form an ideal $I(L)$ of the Lie-ring $D(L)$ of all derivations of L.

36. Let $\mathfrak{S}$ be a ring with unit element. The $\mathfrak{S}$-module $\mathfrak{M}$ is said to be *torsicn-free* if for each denominator d of $\mathfrak{S}$ the equation $dx = 0$ implies $x = 0$. Show that the elements of a torsion free $\mathfrak{S}$-module and the formal quotients u/d, where u belongs to $\mathfrak{M}$ and d is a denominator of $\mathfrak{S}$, form a torsion-free module over the quotient ring $Q(\mathfrak{S})$ of $\mathfrak{S}$ which contains $\mathfrak{M}$ as submodule, if we define: $u = v$ as in $\mathfrak{M}$, $a = v/d$ if $du = v$, $v/d = u$ if $du = v$, $u/d = u'/d'$ if $ud' = u'd$; $u + v$ as in $\mathfrak{M}$, $u + (v/d) = (v/d) + u = (ud + v)/d$, $(u/d) + (u'/d') = (ud' + u'd)/dd'$, βu as in $\mathfrak{M}$; $\beta(v/d) = (\beta v)/d$, $(\beta/d)u = (\beta u)/d$, $(\beta/d)(u/d') = (\beta u)/dd'$.

The $Q(\mathfrak{S})$-module just defined is called the *quotient module* of $\mathfrak{M}$ over $\mathfrak{S}$. Show that the quotient module is generated over $Q(\mathfrak{S})$ by its submodule $\mathfrak{M}$ and thus may be denoted by $Q(\mathfrak{S})\mathfrak{M}$. Every homomorphism over $\mathfrak{S}$ of $\mathfrak{M}$ into a $Q(\mathfrak{S})$-module $\mathfrak{N}$ can be extended in one and only one way to a homomorphism over $Q(\mathfrak{S})$ of the $Q(\mathfrak{S})$-module $Q(\mathfrak{S})\mathfrak{M}$ into the $Q(\mathfrak{S})$-module $\mathfrak{N}$, viz., the mapping that maps u/d onto $d^{-1}h(u)$.

37. Let $\mathfrak{S}$ be a ring with unit element. An $\mathfrak{S}$-submodule $\mathfrak{m}$ of an $\mathfrak{S}$-module $\mathfrak{M}$ is called *primitive in* $\mathfrak{M}$ if the $\mathfrak{S}$-factor module $\mathfrak{M}/\mathfrak{m}$ is torsion free. Show that

a) all elements with torsion in $\mathfrak{M}$, i. e., the elements of $\mathfrak{M}$ that are annihilated by some denominator of $\mathfrak{S}$, form a primitive $\mathfrak{S}$-submodule $\mathfrak{T}(\mathfrak{M})$ of $\mathfrak{M}$;

b) the *torsion submodule* $\mathfrak{T}(\mathfrak{M})$ defined under a) is the smallest primitive $\mathfrak{S}$-submodule of $\mathfrak{M}$;

c) the intersection of any system of primitive $\mathfrak{S}$-submodules is a primitive $\mathfrak{S}$-submodule, and hence for every subset $\mathfrak{m}$ of $\mathfrak{S}$ there is precisely one smallest primitive $\mathfrak{S}$-submodule $\mathfrak{m}'$ containing $\mathfrak{m}$;

d) if the subset $\mathfrak{m}$ occurring in c) is an $\mathfrak{S}$-submodule, then the $\mathfrak{S}$-factor module $\mathfrak{m}'/\mathfrak{m}$ is the torsion module of $\mathfrak{M}/\mathfrak{m}$.

38. Let $\mathfrak{S}$ be a ring with unit element and let $\mathfrak{M}$ be a torsion-free $\mathfrak{S}$-quasi-ring in which $du = ud$ for all denominators d of $\mathfrak{S}$ and all elements u of $\mathfrak{M}$. Show that the quotient module $Q(\mathfrak{S})\,\mathfrak{M}$ defined in Ex. 36 becomes a $Q(\mathfrak{S})$-quasi-ring if the following rules of multiplication are introduced:

$$uv \text{ as in } \mathfrak{M},\ u\cdot(v/d) = (uv)/d, \quad (v/d)u = (vu)/d, \quad (u/d)(u'/d') = (uu')/(dd')$$

(note that we set: $u\beta$ as in $\mathfrak{M}$, $u(\beta/d) = (u\beta)/d$, $(u/d)\beta = (u\beta)/d$, $(u/d)(\beta/d') = (u\beta)/(dd')$.

Show that the $Q(\mathfrak{S})$-quasi-ring just defined contains $\mathfrak{M}$ as an $\mathfrak{S}$-subring and that there is no other way to extend the rules of operation from $\mathfrak{M}$ to $Q(\mathfrak{S})\,\mathfrak{M}$ so as to embed the $\mathfrak{S}$-ring $\mathfrak{M}$ into an $Q(\mathfrak{S})$-quasi-ring. Thus $Q(\mathfrak{S})\,\mathfrak{M}$ can rightly be called the $Q(\mathfrak{S})$-*quotient ring* of the $\mathfrak{S}$-quasi-ring $\mathfrak{M}$.

39. Let $\mathfrak{S}$ be a commutative ring with unit element and let $\mathfrak{M}$ be an $\mathfrak{S}$-quasi-ring without torsion. Prove that all linear transformations of the $\mathfrak{S}$-module $\mathfrak{M}$ that are derivations of the quasi-ring $\mathfrak{M}$ form an $\mathfrak{S}$-Lie-ring $L(\mathfrak{M}, \mathfrak{S})$ without torsion such that $Q(\mathfrak{S})L(\mathfrak{M}, \mathfrak{S}) = L(Q(\mathfrak{S})\,\mathfrak{M}, Q(\mathfrak{S}))$.

APPENDIX B

STRUCTURE THEORY AND DIRECT PRODUCTS

In this appendix, the lattice-theoretical discussion of the ideas of group theory that was begun in Chap. II, § 5 will be continued. It was noted earlier that in a poset formed by subsets of a set the meet operation is not always set-theoretical intersection and the join operation is not always set-theoretical union. In order to emphasize in the ensuing discussions the more abstract role of meet and join, we employ the symbol M for meet and J for join. Thus, we write $a \mathsf{M} b$ for the meet of the lattice elements a, b and $a \mathsf{J} b$ for their join.

1. Projectivities

The isomorphisms occurring in the lemma on four elements are explicitly defined as products of a projection with an anti-projection. They depend on the way in which the four elements are embedded into the given lattice.

Two factor lattices a/b and c/d are *projectively related* if there is a chain of factor lattices

(1) $$a/b = a_0/b_0,\ a_1/b_1,\ \ldots,\ a_{2n}/b_{2n} = c/d$$

connecting a/b and c/d such that

a) $b \mathsf{N} a$, $d \mathsf{N} c$, $b_i \mathsf{N} a_i$ $(i = 0, 1, 2, \ldots, 2n)$ and

b) a_{2i+1}/b_{2i+1} is projective with a_{2i}/b_{2i} and also with a_{2i+2}/b_{2i+2} for $i = 1, 2, \ldots, n - 1$.

According to Ex. 18 of Chap. I, 'projectively related' is the normalized relation of 'projective.' Hence the factor lattices of the given lattice L are distributed among *families of projectively related factor lattices* in such a way that each factor lattice belongs to one and only one family.

Each chain (1) linking a/b and c/d carries with it the specified isomorphism mapping a/b onto c/d which is induced by the projection of a_0/b_0 onto a_1/b_1 followed by the anti-projection of a_1/b_1 onto a_2/b_2, etc., ending with the anti-projection of a_{2n-1}/b_{2n-1} onto a_{2n}/b_{2n}. We call this isomorphism *a projectivity* of a/b onto c/d. Since we can reverse a chain and since we can piece together two chains that have the same factor lattices at the ends at which they are amalgamated, it is clear that all the projectivities linking various members of a family F form a groupoid. The unit elements are the identity automorphisms of the members of F. All the projectivities between a factor lattice a/b and itself form a group which may be denoted by $PP(a/b)$. If a/b and c/d are projectively related, then $PP(a/b)$ is isomorphic to $PP(c/d)$.

Referring to the lattice $S(\mathfrak{G})$ of all subgroups of a group $\mathfrak{G}$, we may observe that every automorphism α of $\mathfrak{G}$ induces an automorphism $\bar{\alpha}$ of $S(\mathfrak{G})$. The correspondence that maps α onto $\bar{\alpha}$ is a homomorphism between the full group $A_{\mathfrak{G}}$ of automorphisms of $\mathfrak{G}$ and a subgroup $PA_{\mathfrak{G}}$ of the group of automorphisms of $S(\mathfrak{G})$.

We indicate by the prefixed letter the transition from any group $\mathfrak{X}$ of automorphisms of $\mathfrak{G}$ to the group $P\mathfrak{X}$ of automorphisms of $S(\mathfrak{G})$ induced by the automorphisms in $\mathfrak{X}$. All the automorphisms in $\mathfrak{X}$ inducing the identity automorphism of $S(\mathfrak{G})$ form a normal subgroup $\mathfrak{X}_P$ of $\mathfrak{X}$ such that $P\mathfrak{X}$ is isomorphic with $\mathfrak{X}/\mathfrak{X}_P$.

We have already seen that there is the specified isomorphism $\Phi_{\mathfrak{A},\mathfrak{B},\mathfrak{C}}$ of $\mathfrak{B}/\mathfrak{C}$ onto $\mathfrak{A}/\mathfrak{A} \cap \mathfrak{C}$ in case $\mathfrak{B}/\mathfrak{C}$ is projective with $\mathfrak{A}/\mathfrak{A} \cap \mathfrak{C}$, and furthermore we have seen that $\Phi_{\mathfrak{A},\mathfrak{B},\mathfrak{C}}$ induces the projection between the factor lattices $\mathfrak{B}/\mathfrak{C}$ and $\mathfrak{A}/\mathfrak{A} \cap \mathfrak{C}$. Hence for every chain $\mathfrak{B}/\mathfrak{C} = \mathfrak{B}_0/\mathfrak{C}_0$, $\mathfrak{B}_1/\mathfrak{C}_1$, $\ldots$, $\mathfrak{B}_{2n}/\mathfrak{C}_{2n} = \mathfrak{D}/\mathfrak{E}$ linking the two factor lattices $\mathfrak{B}/\mathfrak{C}$ and $\mathfrak{D}/\mathfrak{E}$ of the same family of projectively related factor groups in $\mathfrak{G}$ there is the specified isomorphism

$$\Phi_{\mathfrak{B}_{2n},\mathfrak{B}_{2n-1},\mathfrak{C}_{2n-1}} \cdots \Phi_{\mathfrak{B}_2,\mathfrak{B}_1,\mathfrak{C}_1} \Phi^{-1}_{\mathfrak{B}_0,\mathfrak{B}_1,\mathfrak{C}_1}$$

between the factor group $\mathfrak{B}/\mathfrak{C}$ and $\mathfrak{D}/\mathfrak{E}$, which may be termed a *projectivity* between the factor group $\mathfrak{B}/\mathfrak{C}$ and $\mathfrak{D}/\mathfrak{E}$. Each projectivity between the factor groups induces the corresponding projectivity between the related factor lattices.

Again, it is clear that all the projectivities between any two members of a family of factor groups in a group form a groupoid. All the projectivities between a factor group $\mathfrak{B}/\mathfrak{C}$ and itself form a group of automorphisms $P(\mathfrak{B}/\mathfrak{C})$ of $\mathfrak{B}/\mathfrak{C}$. We find that there is a homomorphism between $P(\mathfrak{B}/\mathfrak{C})$ and $PP(\mathfrak{B}/\mathfrak{C})$ with its kernel $P(\mathfrak{B}/\mathfrak{C})_P$ consisting of all the projectivities between $\mathfrak{B}/\mathfrak{C}$ and itself that leave invariant every subgroup of $\mathfrak{B}/\mathfrak{C}$.

All these considerations remain true if applied to a sublattice of $S(\mathfrak{G})$, e. g., to the sublattice formed by all subgroups of $\mathfrak{G}$ that are admissible with respect to a certain operator domain.

At the end of this chapter we will find an interesting example of projectivities between certain factor groups in an abelian group which exhibits the relationship to fundamental notions of projective geometry.

2. Modular and submodular lattices

DEFINITION: A lattice in which each element is a Dedekind element is called a *modular lattice*. All normal subgroups of a group, for example, form a modular lattice. The modular property amounts to the two identities

$$((d \mathsf{J} x) \mathsf{M} c) \mathsf{J} d = (d \mathsf{J} x) \mathsf{M} (c \mathsf{J} d) \tag{2}$$

$$((d \mathsf{M} y) \mathsf{J} c) \mathsf{M} d = (d \mathsf{M} y) \mathsf{J} (c \mathsf{M} d) \tag{3}$$

for any four elements c, d, x, y of the lattice under consideration. Hence the identities are called the *modular law*. Since (3) is the dual of (2), it follows that modularity is a self-dual property. The identities (2) and (3) are not independent. Using only (2) we find

$$((d \mathsf{M} y) \mathsf{J} c) \mathsf{M} d = d \mathsf{M} (c \mathsf{J} (d \mathsf{M} y)) = ((d \mathsf{M} y) \mathsf{J} d) \mathsf{M} (c \mathsf{J} (d \mathsf{M} y))$$
$$= (((d \mathsf{M} y) \mathsf{J} d) \mathsf{M} c) \mathsf{J} (d \mathsf{M} y) = (d \mathsf{M} c) \mathsf{J} (d \mathsf{M} y) = (y \mathsf{M} d) \mathsf{J} (c \mathsf{M} d).$$

Every sublattice of a modular lattice is modular. The Dedekind elements of an arbitrary lattice do not necessarily form a sublattice of the given lattice (cf. the diagram). Since the join of two Dedekind elements is always a Dedekind element, it follows that in a finite lattice the poset formed by the Dedekind elements is a lattice. But, as the diagram shows, this lattice need not be a sublattice of the given lattice. (Doubly encircled elements are Dedekind elements.)

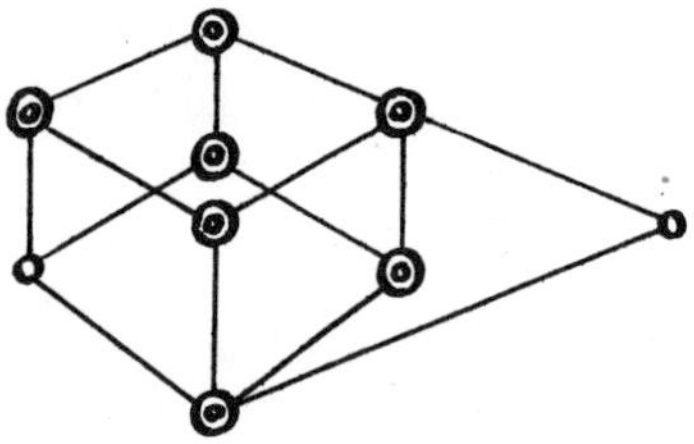

If the Dedekind elements of a lattice form a sublattice, then they form a modular lattice. This is because the modular law holds as an identity for the sublattice of the Dedekind elements.

DEFINITION: In a lattice L with normality relation, the element a is called *subnormal* if it can be connected with L by a normal chain of finite length. We denote this relation by $a \mathsf{N} \mathsf{N} L$ and it means that there is a normal chain $a \mathsf{N} a_1 \mathsf{N} a_2 \mathsf{N} \cdots \mathsf{N} a_{s-1} \mathsf{N} L$ between a and L of finite length s.

More generally, we write $a \mathsf{N} \mathsf{N} b$ if there is a normal chain of finite length s, say $a \mathsf{N} a_1 \mathsf{N} a_2 \mathsf{N} \cdots \mathsf{N} a_s = b$ connecting a with b, where b may either be an element of L or $b = L$. We say 'a is subnormal under b'. In case Kurosh invariance is the normality relation we speak of *subinvariance* and *subinvariant elements*.

As a measure of the degree to which the subnormality $c \mathsf{N} \mathsf{N} b$ deviates from normality we introduce the number $m(b, c)$, which is defined as the minimum of the lengths of all normal chains connecting c and b.

We have $m(b, c) = 0$ if and only if $c = b$; $m(c, b) = 1$ if and only if $c \mathsf{N} b$ and $c \neq b$; $m(c, b) > 1$ if and only if $c \mathsf{NN} b$ and c is not normal in b. For example, the subgroups of a group that occur in normal chains of finite length connecting e and the full group, are the *subnormal subgroups* of the group.

Concerning subnormality we have the following simple facts:

THEOREM 15: *If a is subnormal in the lattice L and if b is subnormal under a, then b is subnormal in L. Also*

$$m(b, L) \leq m(b, a) + m(a, L).$$

Proof: Let $a \mathsf{NN} L$, $b \mathsf{NN} a$; then there are normal chains $b \mathsf{N} b_1 \mathsf{N} b_2 \mathsf{N} \cdots \mathsf{N} b_{m(b,a)} = a$, $a \mathsf{N} a_1 \mathsf{N} a_2 \mathsf{N} \cdots \mathsf{N} a_{m(a,L)} = L$; piecing together the two normal chains, we obtain a normal chain of length $m(b, a) + m(a, L)$ connecting b with L.

THEOREM 16: *If c is subnormal in L, then if a is any element of L, $a \mathsf{M} c$ is subnormal under a and*

$$m(a \mathsf{M} c, a) \leq m(c, L).$$

Proof: There is a normal chain $c \mathsf{N} c_1 \mathsf{N} c_2 \mathsf{N} \cdots \mathsf{N} c_{m(c,L)} = L$, From Rule 4 for normality it follows that $(a \mathsf{M} c) \mathsf{N} (a \mathsf{M} c_1) \mathsf{N} \cdots \mathsf{N} (a \mathsf{M} c_{m(c,L)-1}) \mathsf{N} a$.

By application of Theorems 15 and 16 we obtain

THEOREM 17: *If both a and b are subnormal in L, then $a \mathsf{M} b$ is also subnormal in L and we have the inequality*

$$m(a \mathsf{M} b, L) \leq m(a, L) + m(b, L).$$

Furthermore, there holds

THEOREM 18: *If $b \leq a$, $b \mathsf{NN} L$, then $b \mathsf{NN} a$ and*

$$m(b, a) \leq m(b, L).$$

COROLLARY: *If $b \leq a' \leq a$, $b \mathsf{NN} a$, then $b \mathsf{NN} a'$,*

$$m(b, a') \leq m(b, a).$$

From Theorem 17 it follows that the poset of the subnormal elements of a finite lattice is a lattice; but this lattice need not be a sublattice of the given lattice; the diagram below furnishes a counter-example, if normality is taken as Kurosh invariance. Both a and b are subinvariant, but $a \mathsf{J} b$ is not subinvariant.

Even though we cannot always conclude from $a \mathsf{NN} L$, $b \mathsf{NN} L$ that $(a \mathsf{J} b) \mathsf{NN} L$, at least we have in this direction

THEOREM 19: *If $a \mathsf{N} L$, $b \mathsf{NN} L$, then $(a \mathsf{J} b) \mathsf{NN} L$ and $m(a \mathsf{J} b, L) \leq m(b, L)$.*

Proof: From Rule 5 of normality it follows that there is the normal chain

$$(a \mathsf{J} b) \mathsf{N} (a \mathsf{J} b_1) \mathsf{N} (a \mathsf{J} b_2) \mathsf{N} \cdots \mathsf{N} (a \mathsf{J} b_{m(b,L)-1}) \mathsf{N} L$$

connecting $a \mathsf{J} b$ with L if there is given the normal chain $b \mathsf{N} b_1 \mathsf{N} b_2 \mathsf{N} \cdots \mathsf{N} b_{m(b,L)-1} \mathsf{N} L$ between b and L.

Many normality relations—for example, the normality relation between the subgroups of a group—are *complete with respect to the meet operation*, that is, for any set S of elements of L that are normal in a given element a of L, the meet of the elements in S always exists and is normal in a. Assuming completeness of the normality relation as well as the lattice-theoretical completeness of L, we define the *lower normal series* from L to a given element as follows: Let $S_0(a, L)$ be the all element of L. Assuming that $S_\pi(a, L)$ is already defined as an element of L containing a, define $S_{\pi+1}(a, L)$ as the meet of all the elements of L that are normal in $S_\pi(a, L)$ and contain a. For a limit number ν define $S_\nu(a, L)$ as the meet of all elements $S_\pi(a, L)$ with $\pi < \nu$. Thus, for any ordinal number π the element $S_\pi(a, L)$ of L is so defined that $S_0(a, L)$ is the all element of L, $S_{\pi+1}(a, L) \mathsf{N} S_\pi(a, L)$, $a \leq S_\pi(a, L)$ and $S_\nu(a, L) = \mathsf{M}_{\pi<\nu} S_\pi(a,L)$ for every limit number ν. If T_π is another decreasing well-ordered normal chain with the same properties, then we prove by transfinite induction that $S_\pi(a, L) \leq T_\pi$ for all ordinal numbers π. There is a first ordinal number $m(a, L)$ for which $S_{m(a,L)} = S_{m(a,L)+1}$. It follows that the decreasing well ordered normal chain $S_\pi(a, L)$ $(0 \leq \pi \leq m(a, L))$ is properly decreasing and that $m(a, L)$ is the minimal length of any well-ordered properly decreasing normal chain from L to $S_{m(a,L)}$.

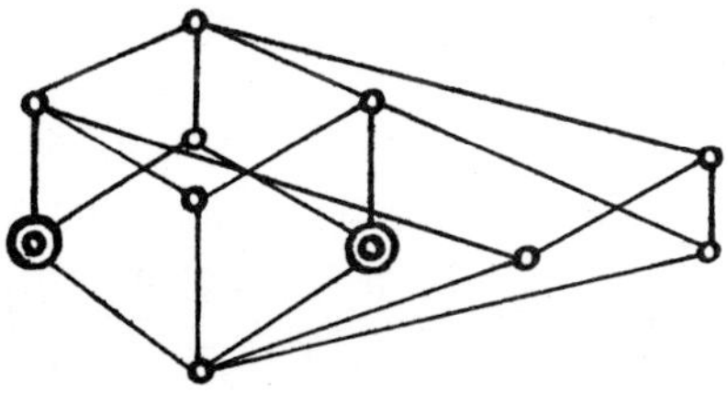

The element a is subnormal if and only if $m(a, L)$ is finite and $S_{m(a,L)}(a, L) = a$, and in this case $m(a, L)$ is indeed equal to the minimal length of a normal chain from a to L as is implied by the notation.

At any rate, the meet of all subnormal elements of L containing a coincides with $S_\omega(a, L)$, where ω denotes the first infinite ordinal number.

If $m(a, L)$ is finite, then the 'subnormal hull' $\tilde{a} = S_{m(a,L)}(a, L)$ of a is subnormal, and it is characterized as the smallest subnormal element of L containing a.

Any lattice automorphism θ that preserves the normality relation maps $S_\pi(a, L)$ onto $S_\pi(\theta a, L)$. Hence $m(a, L) = m(\theta a, L)$, and $\widetilde{\theta(a)} = \theta(\tilde{a})$.

Turning our attention to the subgroup lattice $S(\mathfrak{G})$ of a group $\mathfrak{G}$, we may raise the question of to what extent the subinvariance of a subgroup $\mathfrak{U}$ is determined by the structure of a given normal subgroup $\mathfrak{N}$ and its factor group $\mathfrak{G}/\mathfrak{N}$.

If there is a normal chain $\mathfrak{U} = \mathfrak{U}_0 \mathsf{N} \mathfrak{U}_1 \mathsf{N} \mathfrak{U}_2 \mathsf{N} \cdots \mathsf{N} \mathfrak{U}_r = \mathfrak{G}$, then we can form the normal chains

$$\mathfrak{U}\mathfrak{N}/\mathfrak{N} \,\mathsf{N}\, \mathfrak{U}_1\mathfrak{N}/\mathfrak{N} \,\mathsf{N} \cdots \mathsf{N}\, \mathfrak{U}_r\mathfrak{N}/\mathfrak{N} = \mathfrak{G}/\mathfrak{N}$$

and

$$\mathfrak{U} \,\mathsf{M}\, \mathfrak{N} = (\mathfrak{U}_0 \,\mathsf{M}\, \mathfrak{N}) \,\mathsf{N}\, (\mathfrak{U}_1 \mathsf{M}\, \mathfrak{N}) \,\mathsf{N} \cdots \mathsf{N}\, (\mathfrak{U}_r \mathsf{M}\, \mathfrak{N}) = \mathfrak{N}.$$

Moreover we have

$$(\mathfrak{U}, \mathfrak{U}_i \mathsf{M} \mathfrak{N}) \leq \mathfrak{U}_{i-1} \mathsf{M} \mathfrak{N} \quad \text{for } i = 1, 2, \ldots, r,$$

since plainly $(\mathfrak{U}, \mathfrak{N}) \leq \mathfrak{N}$, $(\mathfrak{U}, \mathfrak{U}_i) \leq \mathfrak{U}_{i-1}$.

Conversely, assume that there are normal chains $\mathfrak{U}\mathfrak{N}/\mathfrak{N} \,\mathsf{N}\, \overline{\mathfrak{V}_1} \,\mathsf{N} \cdots \mathsf{N}\, \overline{\mathfrak{V}_r} = \mathfrak{G}/\mathfrak{N}$ in $\mathfrak{G}/\mathfrak{N}$ and $\mathfrak{U} \mathsf{M} \mathfrak{N} = \mathfrak{W}_0 \mathsf{N} \mathfrak{W}_1 \mathsf{N} \cdots \mathsf{N} \mathfrak{W}_s = \mathfrak{N}$ in $\mathfrak{N}$ such that $(\mathfrak{U}, \mathfrak{W}_i)$ is contained in $\mathfrak{W}_{i-1}$ for $i = 1, 2, \ldots, s$; then we have the normal chain

$$\mathfrak{U} \,\mathsf{N}\, \langle \mathfrak{U}, \mathfrak{W}_1 \rangle \,\mathsf{N}\, \langle \mathfrak{U}, \mathfrak{W}_2 \rangle \,\mathsf{N} \cdots \mathsf{N}\, \langle \mathfrak{U}, \mathfrak{W}_s \rangle = \mathfrak{U}\mathfrak{N} \,\mathsf{N}\, \mathfrak{V}_1 \,\mathsf{N} \cdots \mathsf{N}\, \mathfrak{V}_r = \mathfrak{G},$$

wh ere $\mathfrak{V}_i$ is the subgroup of $\mathfrak{G}$ formed by the cosets $\overline{\mathfrak{V}_i}$.

We apply this remark to the case that $\mathfrak{G}$ is the holomorph of a group $\mathfrak{A}$ of automorphisms of an abelian group $\mathfrak{M}$. It follows that a subgroup $\mathfrak{U}$ of $\mathfrak{A}$ is subinvariant in $\mathfrak{G}$ if and only if

1. $\mathfrak{U}$ is subinvariant in $\mathfrak{A}$;
2. There is a finite number s such that for any s automorphisms $\pi_1, \pi_2, \ldots, \pi_s$ of $\mathfrak{M}$ contained in $\mathfrak{A}$ we have the equation

$$(\pi_1 - \underline{1})(\pi_2 - \underline{1}) \cdots (\pi_s - \underline{1}) = \underline{0}.[1]$$

In Ex. 23 of Appendix D an example will be constructed in which there are two subgroups of $\mathfrak{A}$ both subinvariant in $\mathfrak{G}$ but in which the subgroup generated by them is not subinvariant in $\mathfrak{G}$. Thus in general the subinvariant subgroups of a group do not form a sublattice of $S(\mathfrak{G})$.

Definition: A lattice is called *subnormal* if every element of the lattice is subnormal.

If the poset of all subnormal elements of a lattice with normality relation forms a sublattice, then this sublattice is subnormal. This is because the concept of normality is based on certain identities that remain valid in any sublattice. For groups, the following theorem holds.

[1] This means in ring-theoretic terms that the operator ring of $\mathfrak{M}$ generated by $\mathfrak{A}-\underline{1}$ is nilpotent.

THEOREM 20: *If the subnormal subgroups of a group $\mathfrak{G}$ satisfy the maximal condition, then they form a complete sublattice of the lattice of all subgroups of $\mathfrak{G}$.*

Proof: Since we already know that the intersection of two subnormal subgroups of $\mathfrak{G}$ is subnormal and since the maximal condition is satisfied by the subnormal subgroups of $\mathfrak{G}$, it suffices to prove that the join of finitely many subnormal subgroups of $\mathfrak{G}$ is subnormal. In other words, we have to prove that the subgroup generated by finitely many subnormal subgroups $\mathfrak{U}_1, \mathfrak{U}_2, \ldots, \mathfrak{U}_s$ of $\mathfrak{G}$ satisfying the inequalities

$$m(\mathfrak{U}_i, \mathfrak{G}) \leq n \text{ for } i = 1, 2, \ldots, s$$

is subnormal. This is obvious if $n = 0$, 1 or if $s = 1$. Apply induction on n and on s. Assume $n > 1$, $s > 1$. By the induction hypothesis concerning s, the subgroup $\mathfrak{B}$ generated by the $s - 1$ subnormal subgroups $\mathfrak{U}_2, \mathfrak{U}_3, \ldots, \mathfrak{U}_s$ satisfying $m(\mathfrak{U}_i, \mathfrak{G}) \leq n$ for $i = 2, 3, \ldots, s - 1$, is a subnormal subgroup of $\mathfrak{G}$.

Furthermore, there is a normal chain $\mathfrak{U}_1 \mathsf{N} \mathfrak{B}_1 \mathsf{N} \mathfrak{B}_2 \mathsf{N} \cdots \mathsf{N} \mathfrak{B}_n = \mathfrak{G}$ of length n between $\mathfrak{U}_1$ and $\mathfrak{G}$. Since $\mathfrak{B}_{n-1}$ is a normal subgroup of $\mathfrak{G}$, the subnormal subgroups of $\mathfrak{B}_{n-1}$ are subnormal in $\mathfrak{G}$ and therefore satisfy the maximal condition. For finitely many conjugate subgroups $\mathfrak{U}_1^{x_1}, \mathfrak{U}_1^{x_1}, \ldots, \mathfrak{U}_1^{x_j}$ of $\mathfrak{U}_1$ we find, by application to $\mathfrak{B}_{n-1}$ of the induction hypothesis on n, that each of these subgroups is subnormal under $\mathfrak{B}_{n-1}$, because $n - 1 \geq m(\mathfrak{U}_1, \mathfrak{B}_{n-1}) = m(\mathfrak{U}_1^{x_i}, \mathfrak{B}_{n-1}^{x_i}) = m(\mathfrak{U}_1^{x_i}, \mathfrak{B}_{n-1})$. Since $\mathfrak{B}_{n-1}$ is normal in $\mathfrak{G}$, it follows that the subgroup of $\mathfrak{G}$ that is generated by a finite number of conjugates of $\mathfrak{U}_1$ is subnormal under $\mathfrak{G}$. From the maximal condition it follows that any system of conjugate subgroups of $\mathfrak{U}_1$ generates a subnormal subgroup of $\mathfrak{G}$. For example, the subgroup $\mathfrak{U}_{11}$ of $\mathfrak{G}$ generated by all the subgroups of $\mathfrak{B}_1 = \langle \mathfrak{U}_1, \mathfrak{B} \rangle = \langle \mathfrak{U}_1, \mathfrak{U}_2, \ldots, \mathfrak{U}_s \rangle$ that are conjugate to $\mathfrak{U}_1$ under $\mathfrak{B}_1$, is subnormal under $\mathfrak{G}$. Of course, $\mathfrak{U}_1$ is contained in $\mathfrak{U}_{11}$, $\mathfrak{U}_{11}$ is a normal subgroup of $\mathfrak{B}_1$, and $\mathfrak{B}_1$ is generated by $\mathfrak{U}_{11}$ and $\mathfrak{B}$. Hence each term of the lower normal series $\mathfrak{U}_{11} = S_m \mathsf{N} S_{m-1} \mathsf{N} \cdots \mathsf{N} S_0 = \mathfrak{G}$ (where $S_i = S_i(\mathfrak{U}_{11}, \mathfrak{G})$ and $m = m(\mathfrak{U}_{11}, \mathfrak{G})$) is also invariant under transformation by elements of $\mathfrak{B}$. Hence $S_i \mathsf{N} \langle S_{i-1}, \mathfrak{B} \rangle$. From $\mathfrak{B} \mathsf{N}\mathsf{N} \mathfrak{G}$ it follows that $\mathfrak{B} \mathsf{N}\mathsf{N} \langle S_{i-1}, \mathfrak{B} \rangle$. By Theorem 19 it follows that $\langle S_i, \mathfrak{B} \rangle \mathsf{N}\mathsf{N} \langle S_{i-1}, \mathfrak{B} \rangle$; hence

$$\mathfrak{B}_1 = \langle S_m, \mathfrak{B} \rangle \mathsf{N}\mathsf{N} \langle S_{m-1}, \mathfrak{B} \rangle \mathsf{N}\mathsf{N} \cdots \mathsf{N}\mathsf{N} \langle S_0, \mathfrak{B} \rangle = \mathfrak{G}, \quad \mathfrak{B}_1 \mathsf{N}\mathsf{N} \mathfrak{G},$$

Q. E. D.

COROLLARY: *If $\mathfrak{A}$, $\mathfrak{B}$ are two subgroups of $\mathfrak{G}$ for which there are composition series from $\mathfrak{G}$ to $\mathfrak{A}$ and $\mathfrak{B}$ respectively, then there is also a composition series from $\mathfrak{G}$ to $\langle \mathfrak{A}, \mathfrak{B} \rangle$.*

This theorem, due to Wielandt, can be proved by using the methods of the preceding proof. If S is a system of subnormal subgroups of the group $\mathfrak{G}$ such that the maximal condition holds for the subnormal subgroups of $\mathfrak{G}$ between $\mathfrak{G}$ and one of the members of S, then the union of the members of S is a subnormal subgroup of $\mathfrak{G}$.

If $\mathfrak{A}$, $\mathfrak{B}$ are two subnormal subgroups of a group $\mathfrak{G}$ for which there is a composition series from $\mathfrak{G}$ to $\mathfrak{A}$ and $\mathfrak{B}$ respectively, then by the preceding corollary, $\langle \mathfrak{A}, \mathfrak{B} \rangle$ is subnormal under $\mathfrak{G}$, and hence there is a composition series

$$\mathfrak{A} = \mathfrak{A}_0 \,\mathsf{N}\, \mathfrak{A}_1 \,\mathsf{N} \cdots \mathsf{N}\, \mathfrak{A}_r = \langle \mathfrak{A}, \mathfrak{B} \rangle \tag{4}$$

and a composition series

$$\mathfrak{B} = \mathfrak{B}_0 \,\mathsf{N}\, \mathfrak{B}_1 \,\mathsf{N} \cdots \mathsf{N}\, \mathfrak{B}_s = \langle \mathfrak{A}, \mathfrak{B} \rangle. \tag{5}$$

More generally, let $\mathfrak{A}$, $\mathfrak{B}$ be two subgroups of $\mathfrak{G}$ such that composition series (4) and (5) respectively exists. After elimination of repetitions in the normal chains

$$(\mathfrak{A} \,\mathsf{M}\, \mathfrak{B}) \,\mathsf{N}\, (\mathfrak{A}_1 \,\mathsf{M}\, \mathfrak{B}) \,\mathsf{N}\, (\mathfrak{A}_2 \,\mathsf{M}\, \mathfrak{B}) \,\mathsf{N} \cdots \mathsf{N}\, (\mathfrak{A}_r \,\mathsf{M}\, \mathfrak{B}) = \mathfrak{B}, \tag{6}$$

$$(\mathfrak{A} \,\mathsf{M}\, \mathfrak{B}) \,\mathsf{N}\, (\mathfrak{A} \,\mathsf{M}\, \mathfrak{B}_1) \,\mathsf{N}\, (\mathfrak{A} \,\mathsf{M}\, \mathfrak{B}_2) \,\mathsf{N} \cdots \mathsf{N}\, (\mathfrak{A} \,\mathsf{M}\, \mathfrak{B}_s) = \mathfrak{A}, \tag{7}$$

we obtain composition series (6′), (7′) from $\mathfrak{A} \,\mathsf{M}\, \mathfrak{B}$ to $\mathfrak{B}$ and $\mathfrak{A}$ respectively. We wonder how the composition series (6′), (7′) derived from (4), (5) respectively can be related to the composition series (4), (5).

We call a composition factor group $\mathfrak{A}_i/\mathfrak{A}_{i+1}$ of (4) *abundant* if $\mathfrak{A}_i \,\mathsf{M}\, \mathfrak{B} = \mathfrak{A}_{i-1} \,\mathsf{M}\, \mathfrak{B}$. Similarly, a composition factor group $\mathfrak{B}_j/\mathfrak{B}_{j-1}$ of (5) is called abundant if $\mathfrak{A} \,\mathsf{M}\, \mathfrak{B}_j = \mathfrak{A} \,\mathsf{M}\, \mathfrak{B}_{j-1}$. We ask how the abundant factor groups in (4), (5) respectively are related to the non-abundant factor groups in (4), (5) respectively.

THEOREM 21: *If $\mathfrak{A}$, $\mathfrak{B}$ are two subgroups of a group with the property that there is a composition series* (4) *from $\mathfrak{A}$ to $\langle \mathfrak{A}, \mathfrak{B} \rangle$ and a composition series* (5) *from $\mathfrak{B}$ to $\langle \mathfrak{A}, \mathfrak{B} \rangle$, then*

a) *the non-abundant factor groups of* (4) *are one-to-one projective with a composition series* (6′) *from $\mathfrak{A} \,\mathsf{M}\, \mathfrak{B}$ to $\mathfrak{B}$,*

b) *the abundant factor groups of* (4) *characterized by $\mathfrak{A}_i \,\mathsf{M}\, \mathfrak{B} = \mathfrak{A}_{i-1} \,\mathsf{M}\, \mathfrak{B}$ are abelian,*

c) *every abundant factor group in* (4) *is projectively related with a conjugate of a non-abundant factor group in* (4),

d) *the abundant factor groups in* (4) *and in* (5) *are one-to-one projectively related up to their order.*

Proof: a) The factor group $\mathfrak{A}_i/\mathfrak{A}_{i+1}$ is non-abundant if and only if $\mathfrak{A}_i \mathsf{M} \mathfrak{B}$ is different from $\mathfrak{A}_{i-1} \mathsf{M} \mathfrak{B}$. In this case, because of the simplicity of $\mathfrak{A}_i/\mathfrak{A}_{i-1}$ we find that $\mathfrak{A}_i/\mathfrak{A}_{i-1}$ is isomorphic to $(\mathfrak{A}_i \mathsf{M} \mathfrak{B})/(\mathfrak{A}_{i-1} \mathsf{M} \mathfrak{B})$.

d) The abundant factor groups in (4) and in (5) are one-to-one projectively related up to order, because of the Jordan-Hölder-Theorem.

b) and c) If $r \leq 1$ or $s \leq 1$, no abundant factor groups occur, and there is nothing to prove. Now let $r > 1$, $s > 1$. We apply induction on $n = r + s$. Assume that the statement in question is true when the value of $r + s$ is smaller than n. Since $\mathfrak{A} \leq \mathfrak{A}_{r-1} \leq \langle \mathfrak{A}, \mathfrak{B} \rangle$ we have $\langle \mathfrak{A}, \mathfrak{B} \rangle = \langle \mathfrak{A}_{r-1}, \mathfrak{B} \rangle$ which, by the Second Isomorphism Theorem, implies that $\langle \mathfrak{A}, \mathfrak{B} \rangle/\mathfrak{A}_{r-1}$ is isomorphic with $\mathfrak{B}/(\mathfrak{A}_{r-1} \mathsf{M} \mathfrak{B})$. Hence $\mathfrak{A}/\mathfrak{A}_{r-1}$ is not abundant. By the Corollary to Theorem 20, the subgroup $\langle \mathfrak{A}, \mathfrak{A}_{r-1} \mathsf{M} \mathfrak{B} \rangle$ is subnormal in $\langle \mathfrak{A}, \mathfrak{B} \rangle$, and since $\mathfrak{A} \,\mathsf{N}\,\mathsf{N} \langle \mathfrak{A}, \mathfrak{A}_{r-1} \mathsf{M} \mathfrak{B} \rangle \,\mathsf{N}\,\mathsf{N}\, \mathfrak{A}_{r-1} \,\mathsf{N}\, \langle \mathfrak{A}, \mathfrak{B} \rangle$ there exists a composition series from $\mathfrak{A}$ to $\langle \mathfrak{A}, \mathfrak{B} \rangle$ via $\langle \mathfrak{A}, \mathfrak{A}_{r-1} \mathsf{M} \mathfrak{B} \rangle$ and $\mathfrak{A}_{r-1}$. From the Jordan-Hölder Theorem we deduce that it suffices to prove the theorem only for this composition series instead of (4). There is no loss of generality in taking this new series to be (4); hence for some index j with $0 \leq j \leq r - 1$, $\mathfrak{A}_j = \langle \mathfrak{A}, \mathfrak{A}_{r-1} \mathsf{M} \mathfrak{B} \rangle$ it follows that $\mathfrak{A}_{r-1} \mathsf{M} \mathfrak{B} \leq \mathfrak{A}_j \mathsf{M} \mathfrak{B} \leq \mathfrak{A}_{r-1} \mathsf{M} \mathfrak{B}$ and hence $\mathfrak{A}_j \mathsf{M} \mathfrak{B} = \mathfrak{A}_{r-1} \mathsf{M} \mathfrak{B}$. From the Jordan-Hölder Theorem, the length of a composition series from $\mathfrak{A}_{r-1} \mathsf{M} \mathfrak{B}$ to $\mathfrak{A}_j$ is $1 + s - (r - j) \leq s$. Since the composition series $\mathfrak{A} = \mathfrak{A}_0 \,\mathsf{N}\, \mathfrak{A}_1 \cdots \mathsf{N}\, \mathfrak{A}_j$ is of length $j < r$, we can apply the induction hypothesis to $\mathfrak{A}$, $\mathfrak{A}_{r-1} \mathsf{M} \mathfrak{B}$. It follows that the abundant factor groups between $\mathfrak{A}$ and $\mathfrak{A}_j$ are abelian and are projectively related with conjugates of non-abundant factor groups between $\mathfrak{A}$ and $\mathfrak{A}_j$.

Similarly, we can apply the induction hypothesis to the pair $\mathfrak{A}_j$, $\mathfrak{B}$ provided $\mathfrak{A}_{r-1} \mathsf{M} \mathfrak{B} \neq \mathfrak{A} \mathsf{M} \mathfrak{B}$; then $\mathfrak{A}_j \neq \mathfrak{A}$, and the induction hypothesis yields that the abundant factor groups between $\mathfrak{A}_j$ and $\langle \mathfrak{A}, \mathfrak{B} \rangle$ are abelian and are projectively related with conjugates of non-abundant factor groups between $\mathfrak{A}_j$ and $\langle \mathfrak{A}, \mathfrak{B} \rangle$. Both statements together yield the theorem.

It now remains to consider the case that $\mathfrak{A}_{r-1} \mathsf{M} \mathfrak{B} = \mathfrak{A} \mathsf{M} \mathfrak{B}$. We may assume, similarly, without loss of generality that there is a term $\mathfrak{B}_k$ in (5) with $\mathfrak{B}_k = \langle \mathfrak{A} \mathsf{M} \mathfrak{B}_{s-1}, \mathfrak{B} \rangle$. Then $\mathfrak{B}_k \mathsf{M} \mathfrak{A} = \mathfrak{B}_{s-1} \mathsf{M} \mathfrak{A}$. Hence if $\mathfrak{B}_{s-1} \mathsf{M} \mathfrak{A} \neq \mathfrak{B} \mathsf{M} \mathfrak{A}$, $k > 0$, and the theorem follows when the induction hypothesis is applied to $\mathfrak{A} \mathsf{M} \mathfrak{B}_{s-1}$, $\mathfrak{B}$ and to $\mathfrak{A}$, $\mathfrak{B}_k$.

Finally, we consider the case that $\mathfrak{A}_{r-1} \mathsf{M} \mathfrak{B} = \mathfrak{A} \mathsf{M} \mathfrak{B}_{s-1} = \mathfrak{A} \mathsf{M} \mathfrak{B}$.

From the Second Isomorphism Theorem it follows that $\mathfrak{A}_{r-1} \mathsf{M} \mathfrak{B}$ is a normal subgroup of $\mathfrak{B}$ and $\mathfrak{A} \mathsf{M} \mathfrak{B}_{s-1}$ is a normal subgroup of $\mathfrak{A}$. Hence $\mathfrak{A} \mathsf{M} \mathfrak{B}$ is normal both in $\mathfrak{A}$ and in $\mathfrak{B}$ and therefore is normal in $\langle \mathfrak{A}, \mathfrak{B} \rangle$. We may therefore carry out the remainder of the proof in the factor group

$\langle\mathfrak{A}, \mathfrak{B}\rangle/(\mathfrak{A} \mathsf{M} \mathfrak{B})$ instead of in $\mathfrak{G}$. Let us, then, assume $\mathfrak{A} \mathsf{M} \mathfrak{B} = 1$, $\langle\mathfrak{A}, \mathfrak{B}\rangle = \mathfrak{G}$, $\mathfrak{A}_{r-1} \mathsf{M} \mathfrak{B} = \mathfrak{A} \mathsf{M} \mathfrak{B}_{s-1} = 1$.

Let $\mathfrak{N}$ be the normal subgroup of $\mathfrak{G}$ generated by the conjugates of $\mathfrak{A}$. Then, in view of $\mathfrak{A} \leq \mathfrak{A}_{r-1}$ and $\mathfrak{A}_{r-1} \mathsf{N} \mathfrak{G}$, we conclude that $\mathfrak{N}$ is contained in $\mathfrak{A}_{r-1}$. Therefore $\mathfrak{N} \mathsf{M} \mathfrak{B} \leq \mathfrak{A}_{r-1} \mathsf{M} \mathfrak{B} = 1$, $\mathfrak{A}_{r-1}\mathfrak{B} = \mathfrak{N}\mathfrak{B} = \mathfrak{G}$, $\mathfrak{A}_{r-1}/\mathfrak{N}$ is isomorphic to 1, so that $\mathfrak{A}_{r-1} = \mathfrak{N}$.

$\mathfrak{A}$ is simple, since $\mathfrak{A}/1$ is isomorphic to $\mathfrak{G}/\mathfrak{B}_{s-1}$. Hence the length of any composition series from $\mathfrak{N} = \mathfrak{A}_{r-1}$ to 1 is $1 + r - 1 = r$. The same applies to $\mathfrak{B}$ and to $\mathfrak{B}_{s-1}$, so that $s = r$, $n = 2r$. Thus the induction hypothesis is applicable to any pair of subnormal subgroups of $\mathfrak{N}$.

Set $\overline{\mathfrak{A}_0} = \mathfrak{A}$. We define $\overline{\mathfrak{A}_i}$ recursively. If $\overline{\mathfrak{A}_{i-1}}$ is already defined as a subgroup of $\mathfrak{G}$ generated by certain conjugates of $\mathfrak{A}$ under $\mathfrak{G}$, then by Theorem 20, the subgroup $\overline{\mathfrak{A}_{i-1}}$ is certainly subnormal in $\mathfrak{G}$, and therefore either $\overline{\mathfrak{A}_{i-1}} = \mathfrak{N}$ or $\overline{\mathfrak{A}_{i-1}}$ is not normal in $\mathfrak{G}$. In the first case, set $\overline{\mathfrak{A}_i} = \mathfrak{N}$. In the second case, there is a normal series $\overline{\mathfrak{A}_{i-1}} \mathsf{N} \mathfrak{C}_1 \mathsf{N} \mathfrak{C}_2 \mathsf{N} \cdots \mathsf{N} \mathfrak{C}_j = \mathfrak{G}$ from $\overline{\mathfrak{A}_{i-1}}$ to $\mathfrak{G}$ such that $\overline{\mathfrak{A}_{i-1}}$ is not normal in $\mathfrak{C}_2$. Therefore there is a conjugate of $\overline{\mathfrak{A}_{i-1}}$ under $\mathfrak{C}_2$ which is not contained in $\overline{\mathfrak{A}_{i-1}}$. There must also be a conjugate of $\mathfrak{A}$ under $\mathfrak{C}_2$, say $\mathfrak{X}_i$, such that $\mathfrak{X}_i$ is not contained in $\overline{\mathfrak{A}_{i-1}}$. Since $\mathfrak{A}$ is contained in $\mathfrak{C}_1$ and $\mathfrak{C}_1$ is normal under $\mathfrak{C}_2$, it follows that $\mathfrak{X}_i$ is contained in $\mathfrak{C}_i$. We set $\overline{\mathfrak{A}_i} = \langle\overline{\mathfrak{A}_{i-1}}, \mathfrak{X}_i\rangle$. It follows that $\overline{\mathfrak{A}_{i-1}} \mathsf{N} \overline{\mathfrak{A}_i}$, $\overline{\mathfrak{A}_{i-1}} \neq \overline{\mathfrak{A}_i}$. Since $\mathfrak{X}_i$, a conjugate of $\mathfrak{A}$, is subnormal under $\mathfrak{G}$, and since also $\overline{\mathfrak{A}_{i-1}}$ is subnormal under $\mathfrak{G}$, it follows that $\overline{\mathfrak{A}_{i-1}} \mathsf{M} \mathfrak{X}_i$ is subnormal under $\mathfrak{G}$. But in view of the fact that $\mathfrak{X}_i$, being conjugate to $\mathfrak{A}$, is simple and that furthermore $\mathfrak{X}_i$ is not contained in $\overline{\mathfrak{A}_{i-1}}$, it follows that $\overline{\mathfrak{A}_{i-1}} \mathsf{M} \mathfrak{X}_i = 1$. Hence $\overline{\mathfrak{A}_i}/\overline{\mathfrak{A}_{i-1}}$ is projective with the conjugate $\mathfrak{X}_i$ of $\mathfrak{A}$. It follows that there is the composition series $1 \mathsf{N} \overline{\mathfrak{A}_0} \mathsf{N} \overline{\mathfrak{A}_1} \mathsf{N} \overline{\mathfrak{A}_2} \mathsf{N} \cdots \mathsf{N} \overline{\mathfrak{A}_{r-1}} = \mathfrak{N} = \mathfrak{A}_{r-1}$ from 1 to $\mathfrak{A}_{r-1}$ each composition factor of which is projectively related with some conjugate of $\mathfrak{A}$. Each composition factor of the composition series $1 \mathsf{N} \mathfrak{A} \mathsf{N} \mathfrak{A}_1 \mathsf{N} \cdots \mathsf{N} \mathfrak{A}_{r-1}$ from 1 to $\mathfrak{A}_{r-1}$, also is projectively related, by the Jordan-Hölder Theorem to some conjugate of $\mathfrak{A}$. Similarly, we see that each composition factor of the composition series $\mathfrak{B} \mathsf{N} \mathfrak{B}_1 \mathsf{N} \cdots \mathsf{N} \mathfrak{B}_{r-1}$ is projectively related to a conjugate of $\mathfrak{B}$. Thus c) is proved.

In order to prove b) let us assume that $\mathfrak{A}$ is not abelian. By the induction hypothesis applied to any two simple non-abelian subnormal subgroups $\mathfrak{X}_i$, $\mathfrak{X}_j$ of $\mathfrak{N}$, it follows that there are no abundant composition factor groups between $\mathfrak{X}_i$ and $\langle\mathfrak{X}_i, \mathfrak{X}_j\rangle$. Hence $\mathfrak{X}_i$ is normal in $\langle\mathfrak{X}_i, \mathfrak{X}_j\rangle$, and either $\mathfrak{X}_i = \mathfrak{X}_j$ or $\langle\mathfrak{X}_i, \mathfrak{X}_j\rangle/\mathfrak{X}_i$ is projective with $\mathfrak{X}_j/1$; and since, similarly, $\mathfrak{X}_j$ is normal in $\langle\mathfrak{X}_i, \mathfrak{X}_j\rangle$, it follows that $\langle\mathfrak{X}_i, \mathfrak{X}_j\rangle$ is the direct product of $\mathfrak{X}_i$ and $\mathfrak{X}_j$. This remains true if we set $\mathfrak{X}_0 = \mathfrak{A}$. It follows that $\mathfrak{A}_{r-1}$ is the direct product of $\mathfrak{X}_0, \mathfrak{X}_1, \ldots, \mathfrak{X}_{r-1}$. Since this is a direct product of non-abelian simple groups, it follows[1] that every normal subgroup of $\mathfrak{A}_{r-1}$ is a direct product of some of the $\mathfrak{X}_i$'s. Similarly, $\mathfrak{B}_{s-1} = \mathfrak{B}_{r-1} = \mathfrak{Y}_0 \times \mathfrak{Y}_1 \times \mathfrak{Y}_2 \times \cdots \times \mathfrak{Y}_r$, where the $\mathfrak{Y}_j$ are non-abelian simple groups and $\mathfrak{Y}_0 = \mathfrak{B}$. Since $\mathfrak{B}_{r-1}/(\mathfrak{A}_{r-1} \mathsf{M}\, \mathfrak{B}_{r-1})$ is isomorphic with $\mathfrak{G}/\mathfrak{A}_{r-1}$ and $\mathfrak{G}/\mathfrak{A}_{r-1}$ is isomorphic with $\mathfrak{B}$, it follows, in view of $r > 1$, that $\mathfrak{A}_{r-1} \mathsf{M}\, \mathfrak{B}_{r-1} \neq 1$. Hence $\mathfrak{A}_{r-1} \mathsf{M}\, \mathfrak{B}_{r-1}$ is a direct product of some $\mathfrak{X}$'s as well as of some $\mathfrak{Y}$'s and it is bound to happen at least once that $\mathfrak{X}_i = \mathfrak{Y}_j$. The normalizer of $\mathfrak{X}_i$ thus contains both $\mathfrak{A}_{r-1}$ and $\mathfrak{B}_{r-1}$ and therefore $\mathfrak{G}$ also. But it is impossible for a conjugate of $\mathfrak{A}$ to be normal in $\mathfrak{G}$, because $\mathfrak{A}$ itself is not normal in $\mathfrak{G}$. Hence every composition factor between $\mathfrak{A}$ and $\mathfrak{A}_{r-1}$ is abelian. This completes the proof of the theorem.

3. Direct decompositions of lattices

We assume in this subsection that the lattices that occur are complete and hence have an all element A and a null element Z. Furthermore, in each lattice a normality relation is defined satisfying the conditions given in Chap. II, § 5. The normality relation of a lattice induces a normality relation on each sublattice.

Definition: The lattice L is the *direct join* of the subset B of L if: 1. the all element of L is the join of the subset B of L, 2. for any subset X of B the meet of the join of X and of the join of the complementary subset $B - X$ is the null element of L.

If L is the direct join of a subset B, then as many null elements may be added to B or eliminated from B as one desires. Elimination of all the null elements from a subset of which L is direct join, leads to a *proper direct join*. If L is the proper direct join of B, then the elements of B are all different one from another (this follows from property 2.) and also different from the null element of L.

Trivially, L is the direct join of the subset consisting of the single element A. The lattice L is called *indecomposable* if there is no other proper direct join representation. If the subset B of L is the union of the system S of mutually disjoint subsets X of B, then L is the direct join of B if and only if the join of all the joins over X is direct and is equal to A. We say that the decomposition of L as join of B is obtained *by refinement* from the decomposition of L as join over joins of the subsets X ranging over S.

Definition: A direct join representation of L over the subset B is called *normal* if the join over any subset of B is normal. The terms *normal proper direct join*, *normally indecomposable*, and *normal refinement* are formed

[1] For an independent proof, see Subsection 4 of this chapter.

as above by making use only of normal decompositions. In the sequel we assume all decompositions to be normal, and thus we may omit the adjective normal in connection with decompositions and derived terms.

Let L be the direct join of B. For any subset X of B we define the *decomposition operator*

$$\delta_X = \varphi_{JX}\varphi^{j(B-X)},$$

that is,

$$\delta_X(a) = (aJ(J(B-X))\,\mathsf{M}\,(JX) \text{ for } a \text{ of } L.$$

The element $\delta_X(a)$ is called the *X-component* of a.

It has the properties

(8) $$\delta_X(x\,\mathsf{J}\,y) = \delta_X(x)\,\mathsf{J}\,\delta_X(y),$$

(9) $$\delta_X(A)\,\mathsf{N}\,A;\text{ if } x\,\mathsf{N}\,y \text{ then } \delta_X(x)\,\mathsf{N}\,\delta_X(y),$$

(10) $$\delta_B(a) = a,\quad \delta_{\Lambda(B)}(a) = Z,\quad \delta_X a = a \text{ if and only if } a \leq \mathsf{J}X$$

(where $\Lambda(B)$ is the empty subset of B),

(11) $$\delta_X(a)\,\mathsf{J}\,\delta_Y(a) = \delta_{X\mathsf{J}Y}(a),\quad \delta_X(a)\,\mathsf{M}\,\delta_Y(a) = Z,$$

for any two disjoint subsets X, Y of B.

For the converse, see Ex. 5 at the end of Appendix D.

Proof:

$$\begin{aligned}\delta_X(xJy) &= \varphi_{JX}(x\,\mathsf{J}\,y\,\mathsf{J}(\mathsf{J}(B-X)))\\ &= \varphi_{JX}(x\,\mathsf{J}(\mathsf{J}(B-X))\,\mathsf{J}(y\,\mathsf{J}(\mathsf{J}(B-X))\\ &= \varphi_{JX}(x\,\mathsf{J}(\mathsf{J}(B-X)))\,\mathsf{J}\,\varphi_{JX}(y\,\mathsf{J}(\mathsf{J}(\mathsf{J}(B-X)))\\ &= \delta_X(x)\,\mathsf{J}\,\delta_X(y),\end{aligned}$$

because $A/\mathsf{J}(B-X)$ is projective with $\mathsf{J}X/(\mathsf{J}X)\mathsf{M}(\mathsf{J}(B-X))$, that is, with $\mathsf{J}X/Z$.

Next we consider three elements a, x, y satisfying $x\,\mathsf{N}\,A$, $y\,\mathsf{N}\,A$, $a \leq x\,\mathsf{J}\,y$, and we prove the identity

(12) $$((a\,\mathsf{J}\,y)\mathsf{M}\,x)\,\mathsf{J}((a\,\mathsf{J}\,x)\mathsf{M}\,y) = (a\,\mathsf{J}\,x)\mathsf{M}(a\,\mathsf{J}\,y).$$

Since x is normal in A it follows that x is normal in $x\,\mathsf{J}\,y$ and that $(x\,\mathsf{J}\,y)/x$ is projective with $y/(x\,\mathsf{M}\,y)$, and hence $((a\,\mathsf{J}\,x)\mathsf{M}\,y)\,\mathsf{J}\,x = a\,\mathsf{J}\,x$. Furthermore $(a\,\mathsf{J}\,x)\mathsf{M}\,y \leq (a\,\mathsf{J}\,x)\mathsf{M}(a\,\mathsf{J}\,y) \leq a\,\mathsf{J}\,x$. Since y is normal in A it follows that $y\mathsf{M}(y\,\mathsf{J}\,x)$ is projective with $x/(x\,\mathsf{M}\,y)$ and that

$$\begin{aligned}((a\,\mathsf{J}\,y)\mathsf{M}\,x)\,\mathsf{J}((a\,\mathsf{J}\,x)\mathsf{M}\,y) &= ((a\,\mathsf{J}\,x)\mathsf{M}(a\,\mathsf{J}\,y)\mathsf{M}\,x)\,\mathsf{J}((a\,\mathsf{J}\,x)\mathsf{M}\,y)\\ &= (a\,\mathsf{J}\,x)\mathsf{M}(a\,\mathsf{J}\,y).\end{aligned}$$

From (12) it follows that for any element a' of L we have

$$((a' \mathsf{J}\, y) \mathsf{M}\, x) \mathsf{J}\, ((a \mathsf{J}\, x) \mathsf{M}\, y) \geq (((a' \mathsf{M}\, (x \mathsf{J}\, y)) \mathsf{J}\, y) \mathsf{M}\, x) \mathsf{J}\, (((a' \mathsf{M}\, (x \mathsf{J}\, y)) \mathsf{J}\, x) \mathsf{M}\, y)$$
$$= a' \mathsf{M}\, (x \mathsf{J}\, y),$$

yielding

$$((a' \mathsf{J}\, y) \mathsf{M}\, x) \mathsf{J}\, ((a' \mathsf{J}\, x) \mathsf{M}\, y) \geq a' \mathsf{M}\, (x \mathsf{J}\, y). \tag{13}$$

For three normal elements x, y, z of L satisfying $a \leq x \mathsf{J}\, y \mathsf{J}\, z$, we set $a' = a \mathsf{J}\, z$ and find that

$$((a \mathsf{J}\, y \mathsf{J}\, z) \mathsf{M}\, x) \mathsf{J}\, (a \mathsf{J}\, x \mathsf{J}\, z) \geq (a \mathsf{J}\, z) \mathsf{M}\, (x \mathsf{J}\, y). \tag{14}$$

Now let x be the join of the subset X of B, let y be the join of the subset Y of B, let z be the join of the complement of the union of X and Y with respect to B, and furthermore, let X and Y be disjoint. Then $x \mathsf{J}\, y \mathsf{J}\, z = A \geq a$; hence $\delta_X(a) \mathsf{J}\, \delta_Y(a) \geq \delta_{X \mathsf{J} Y}(a)$. Under the same assumption we have

$$\delta_X \delta_Y(a) = (\delta_Y(a) \mathsf{J}\, y \mathsf{J}\, z) \mathsf{M}\, x = (y \mathsf{J}\, z) \mathsf{M}\, x = (\mathsf{J}(B - X)) \mathsf{M}\, (\mathsf{J}\, X) = Z.$$

If $a \leq \mathsf{J}\, X$, then $\delta_X(a) = (a \mathsf{J}\, (\mathsf{J}\, (B - X)) \mathsf{M}\, (\mathsf{J}\, X) = a$, because $A/\mathsf{J}\, (B-X)$ is projective with $\mathsf{J}\, X/Z$.

DEFINITION: A unique mapping φ of a lattice L into itself is called a *normal operator* if

1. $\varphi(x \mathsf{J}\, y) = \varphi(x) \mathsf{J}\, \varphi(y)$,
2. $x \mathsf{N}\, y$ implies $\varphi(x) \mathsf{N}\, \varphi(y)$,
3. $x \mathsf{N}\, A$ implies $\varphi(x) \mathsf{N}\, A$,
4. φ induces an isomorphic mapping of the factor lattice of A over a certain normal element L_φ of L onto the factor lattice $\varphi(A)/\varphi(Z)$. We note that $\varphi(Z) = \varphi(L_\varphi)$.

This usage of the term normal operator is justified by the fact that the normal operators of a group induce normal operators of the corresponding subgroup lattice.

Every anti-projection φ^z defined by a normal element z of L is a normal operator.

The latter statement is proved as follows. We have

$$\varphi^z(x \mathsf{J}\, y) = z \mathsf{J}\, (x \mathsf{J}\, y) = (z \mathsf{J}\, x) \mathsf{J}\, (z \mathsf{J}\, y) = \varphi^z(x) \mathsf{J}\, \varphi^z(y).$$

If $x \mathsf{N}\, y$, then $(z \mathsf{J}\, x) \mathsf{N}\, (z \mathsf{J}\, y)$. If $x \mathsf{N}\, A$, then $(z \mathsf{J}\, x) \mathsf{N}\, A$. The anti-projection φ^z induces an isomorphic mapping of A/z onto A/z, where $z = \varphi^z(Z)$. Thus $L_{\varphi^z} = z$.

Ordinarily, the decomposition operators of a direct decomposition of L are not normal operators, except for the trivial operators $\underline{1}$ and $\underline{0}$, which map an element a of L onto a and Z respectively. This is because an element normal in the join of some subset of B need not be normal in L.

However, if the normal elements of L form a sublattice of L, then each decomposition operator is normal.

Proof: Let L be the direct join over the subset B, let X be a subset of B, and let a be normal in L. Then $J(B - X)\,\mathsf{N}A$, $(a\,\mathsf{J}(\mathsf{J}(B - X))\,\mathsf{N}A$, $\mathsf{J}X\,\mathsf{N}A$, and hence $((x\,\mathsf{J}(\mathsf{J}(B - X)))\,\mathsf{M}(\mathsf{J}X)\,\mathsf{N}A$, $\delta_X(a)\,\mathsf{N}A$.

Referring to normal operators φ in general the equation $\varphi(x) = \varphi(y)$ implies

$\varphi(x\,\mathsf{J}L_\varphi) = \varphi(x)\,\mathsf{J}\varphi(L_\varphi) = \varphi(x)\,\mathsf{J}\varphi(Z) = \varphi(x\,\mathsf{J}Z) = \varphi(x) = \varphi(y) = \varphi(y\,\mathsf{J}L_\varphi)$

because of 1., $x\,\mathsf{J}L_\varphi = y\,\mathsf{J}L_\varphi$ because of 4. Hence $x\,\mathsf{J}L_\varphi$ is characterized as the maximal element having the image $\varphi(x)$ under φ. In particular, L_φ is characterized as the maximal element having the image $\varphi(Z)$ under φ. In view of 4., the element L_φ of L may be called the *kernel* of the normal operator φ.

The product of two normal operators is a normal operator.

Proof: Let φ, ψ be two normal operators of L. Then we have

$$\varphi\psi(x\,\mathsf{J}y) = \varphi(\varphi(x\,\mathsf{J}y)) = \varphi(\psi(x)\,\mathsf{J}\psi(y)) = (\varphi(\psi(x))\,\mathsf{J}(\varphi(\psi(y)) = \\ = \varphi\psi(x)\,\mathsf{J}\varphi\psi(y).$$

If $x\,\mathsf{N}y$, then $\psi(x)\,\mathsf{N}\psi(y)$, $\varphi\psi(x)\,\mathsf{N}\varphi\psi(y)$. If $x\,\mathsf{N}A$, then $\psi(x)\,\mathsf{N}A$, $\varphi\psi(x)\,\mathsf{N}A$. There is precisely one solution z of the equation $\psi(z) = (\psi(A)\,\mathsf{J}L_\varphi)\,\mathsf{J}\psi(Z)$ satisfying $L_\varphi \leq z \leq A$. Since $\psi(A)\,\mathsf{N}A$, $Z\,\mathsf{N}A$, $L_\varphi\mathsf{N}A$ it follows that $(\psi(A)\,\mathsf{M}L_\varphi)\,\mathsf{N}\psi(A)$, $\psi(Z)\,\mathsf{N}A$, $\psi(z)\,\mathsf{N}\psi(A)$. Since the isomorphism induced by ψ between A/L_φ and $\psi(A)/\psi(Z)$ is supposed to preserve the normality relation in both directions, it follows that $z\,\mathsf{N}A$. Furthermore, it follows that ψ induces an isomorphism of A/z onto

$$\psi(A)/\psi(z) = \psi(A)/(\psi(A)\,\mathsf{M}L_\varphi)\,\mathsf{J}\psi(Z) = \psi(A)/\psi(A)\,\mathsf{M}(L_\varphi\,\mathsf{J}\psi(Z)).$$

Since $L_\varphi\mathsf{N}A$, $\psi(Z)\,\mathsf{N}A$, it follows that $L_\varphi\,\mathsf{J}\psi(Z)\,\mathsf{N}A$, and hence the anti-projection by $L_\varphi\mathsf{J}\psi(Z)$ induces an isomorphism of $\psi(A)/\psi(z)$ onto $(L_\varphi\,\mathsf{J}\psi(A))/(L_\varphi\,\mathsf{J}\psi(Z))$. Since $\psi(Z)$ is contained in $\psi(z)$, it follows that the anti-projection by L_φ induces the same mapping of $\psi(A)/\psi(z)$ as the anti-projection by $L_\varphi\,\mathsf{J}\psi(Z)$. Since φ induces an isomorphism of A/L_φ onto $\varphi(A)/\varphi(Z)$, it follows that φ induces an isomorphism of $(L_\varphi\,\mathsf{J}\psi(A))/(L_\varphi\mathsf{J}\psi(Z)$ onto $\varphi(L_\varphi\,\mathsf{J}\psi(A)/\varphi(L_\varphi\,\mathsf{J}\psi(Z)) = \varphi\psi(A)/\varphi\psi(Z)$. From the equation $\psi(z) = (\psi(A)\,\mathsf{J}L_\varphi\mathsf{J})\,\psi(Z)$ we conclude that $\varphi\psi$ induces an isomorphism of A/z onto $\varphi\psi(A)/\varphi\psi(Z)$. Hence $\varphi\psi$ is a normal operator with z as its kernel.

The following theorems on direct joins presuppose that the normal elements of the given lattice form a sublattice; consequently any decomposition operator is normal. Since we shall be concerned only with normal elements, we may as well speak only of the modular sublattice formed by the normal elements of any given lattice. In addition, the double chain condition will be required. For convenience, a modular lattice satisfying the double chain theorem, will be designated as an MD-lattice.

FITTING'S LEMMA FOR LATTICES: *Let L be an MD-lattice. With a normal operator ω mapping Z onto Z there is associated a direct decomposition*

$$A = L_{\omega^n} \mathsf{J}\, \omega^n(A),$$

where n is a natural number satisfying the condition $L_{\omega^n} = L_{\omega^{n+1}}$.

COROLLARY: *If in addition L is directly indecomposable, then either ω is an automorphism of L or ω^n is the $\underline{0}$-operator.*

Proof: The equation $\omega^j(L_{\omega^j}) = Z$ implies $\omega^{j+1}(L_{\omega^j}) = \omega(\omega^j(L_{\omega^j}) = \omega(Z) = Z$, and hence $L_{\omega^j} \leq L_{\omega^{j+1}}$. Applying the maximal condition to the increasing sequence $L_\omega, L_{\omega^2}, L_{\omega^3}, \ldots$ of normal elements of L, we find an exponent n such that $L_{\omega^n} = L_{\omega^{n+1}}$. For $i > 1$, the equation $\omega^{n+i}(L_{\omega^{n+i}}) = Z$ implies

$$\omega^{n+1}(\omega^{i-1}(L_{\omega^{n+i}})) = Z,\ \omega^{i-1}(L_{\omega^{n+i}}) \leq L_{\omega^{n+1}},\ \omega^{i-1}(L_{\omega^{n+i}}) \leq L_{\omega^n},$$

$$Z = \omega^n(\omega^{i-1}(L_{\omega^{n+i}})) = \omega^{n+i-1}(L_{\omega^{n+i}}), L_{\omega^{n+i}} \leq L_{\omega^{n+i-1}}, L_{\omega^n} = L_{\omega^{n+1}} = L_{\omega^{n+2}} = \cdots.$$

Furthermore, since ω^n is normal, there is a solution z of the equation $\omega^n(z) = \omega^n(A) M L_{\omega^n}$ satisfying the condition $L_{\omega^n} \leq z$. We have

$$\omega^n(z) \leq L_{\omega^n},\quad \omega^{2n}(z) = \omega^n(\omega^n(z)) = Z,\quad z \leq L_{\omega^{2n}},\quad z \leq L_{\omega^n},$$

$$z = L_{\omega^n},\quad \omega^n(z) = Z,\quad \omega^n(A) M L_{\omega^n} = Z.$$

Finally, since L_{ω^n} is normal in A, it follows that the anti-projection by L_{ω^n} induces an isomorphism of $\omega^n(A)/Z$ onto $(L_{\omega^n} J\, \omega^n(A))L/_{\omega^n}$. Since ω^n induces an isomorphism of A/L_{ω^n} onto $\omega^n(A)/Z$, it follows that ω^n induces an isomorphism of $(L_{\omega^n} \mathsf{J}\, \omega^n(A))/L_{\omega^n}$ onto $\omega^n(L_{\omega^n} \mathsf{J} \omega^n(A))/Z = \omega^{2n}(A)/Z$. Hence ω^n induces an isomorphism of $\omega^n(A)/Z$ onto $\omega^{2n}(A)/Z$. But since L satisfies the double chain condition for normal elements, the same is true for the factor lattice L/L_{ω^n} and for the isomorphic images $\omega^n(A)/Z = \omega^{2n}(A)/Z$. Also, an element normal in $\omega^{2n}(A)$ is normal in L and hence normal in $\omega^n A$. Since $\omega^{2n}(A)$ is normal in $\omega^n(A)$ and since the principal series of $\omega^{2n}(A)$ have the same length as the principal series of $\omega^n(A)$, it follows that $\omega^{2n}(A) = \omega^n(A)$. Consequently, $\omega^n(L_{\omega^n} \mathsf{J}\, \omega^n(A)) = \omega^{2n}(A) = \omega^n(A)$, $L_{\omega^n} \mathsf{J}\, \omega^n(A) = A$, Q. E. D.

In lattice theory, Theorem 6 of § 2 becomes the following

LEMMA ON NORMAL OPERATORS: *If $\sigma_1, \sigma_2, \ldots, \sigma_r, \omega_1, \omega_2, \ldots, \omega_r$ are finitely many normal operators of a directly indecomposable MD-lattice L mapping Z onto Z such that*

$$\sigma_1 = \underline{1},[1] \quad \sigma_r = \omega_r, \quad \omega_1 a J \omega_2 a \geq \sigma_1 a = a,$$
$$\omega_2 a J \sigma_3 a \geq \sigma_2 a, \ \ldots, \quad \omega_{r-1} a J \sigma_r a \geq \sigma_{r-1} a,$$

then at least one of the operators $\omega_1, \omega_2, \ldots, \omega_r$ is an automorphism of L.

Proof: This is clear if $r = 1$. Let $r > 1$. If $i < r$, and if the operator ω_i is not an automorphism, then, by Fitting's Lemma and because of the indecomposability of L, we have $\omega_i^n A = Z$ for some n. We have the equation

$$\omega_i(\omega_i^{n-1}(A)) = Z,$$

which implies

$$\sigma_{i+1} \omega_i^{n-1}(A) \geq \omega_i^{n-1}(A), \ \omega_i(\omega_i^{n-2}(A) J \sigma_{i+1} \omega_i^{n-2}(A)) \geq \omega_i^{n-2}(A),$$
$$\omega_i(\omega_i^{n-2}(A)) = \omega_i^{n-1}(A) \leq \sigma_{i+1} \omega_i^{n-1}(A) \leq \sigma_{i+1} \omega_i^{n-2}(A),$$
$$\sigma_{i+1} \omega_i^{n-2}(A) \geq \omega_i^{n-2}(A), \ldots, \ \sigma_{i+1} \omega_i(A) \geq \omega_i(A), \ \sigma_{i+1}(A) \geq A, \ \sigma_{i+1}(A) = A.$$

Hence there is, at any rate, an index j for which $\omega_j(A) = A$. Since ω_j induces an isomorphism between A/L_{ω_j} and $\omega_j(A)/Z = A/Z$, it follows from the double chain condition that $L_{\omega_j} = Z$. Hence ω_j is an automorphism of L, Q. E. D.

DEFINITION: A proper direct decomposition of a lattice into finitely many directly indecomposable components is said to be a *Remak decomposition*. If the lattice is indecomposable, then it is itself the only component of its Remak decomposition. Theorem 7 of § 2 becomes, with minor changes, the following theorem.

THEOREM OF ORE: *Every MD-lattice has a Remak decomposition with components $H_1, H_2, \ldots, H_n$. If there is another Remak decomposition of the same lattice L with components $J_1, J_2, \ldots, J_m$, then the number of components m of one Remak decomposition coincides with the number of components n of the other. Moreover the components J_i can be so numbered that we have the exchange decompositions of L with its Remak components*

$$J_1, \ldots, J_k, \quad H_{k+1}, H_{k+2}, \ldots, H_n \ (k = 1, 2, \ldots, n-1).$$

Finally there is a normal operator of L inducing an isomorphism of H_i/Z onto J_i/Z $(i = 1, 2, \ldots, n)$.

[1] The notation σ_1 for $\underline{1}$ is introduced merely in order to have uniformity of notation in the proof.

Proof: The existence of a Remak decomposition follows as in the proof of Theorem 7, using the minimal condition only. Now consider the decomposition operators $\varphi_i = \varphi_{H_i}\varphi^{H_i'}$, $\omega_j = \varphi_{J_j}\varphi^{J_j'}$, where H_i' is the join of all the H_k with $k \neq i$ and J_j' the join of all the J_k with $k \neq j$. Furthermore let σ_i be the normal operator which is obtained by first applying the anti-projection by the join of $J_1, J_2, \ldots, J_{i-1}$ followed by the projection into the join of $J_i, J_{i+1}, \ldots, J_m$ $(i = 2, 3, \ldots, m-1)$. Since $\omega_i(a) \mathsf{J} \sigma_{i+1}(a) \geq \sigma_i(a)$ for $i = 1, 2, \ldots, m-1$, and for any a of L, we have for $a \leq H_1$ the relation

$$\varphi_1(\omega_i(a) \mathsf{J} \sigma_{i+1}(a)) = \varphi_1 \omega_i(a) \mathsf{J} \varphi_1 \sigma_{i+1}(a) \geq \varphi_1 \sigma_i(a);$$

moreover $\omega_m = \sigma_m$, $\varphi_1 \omega_m = \varphi_1 \sigma_m$, and hence by application of the previous lemma, at least one of the normal operators $\varphi_1 \omega_k$ induces an automorphism of H_1/Z.

The J_k can be so numbered that $\varphi_1 \omega_1$ induces an automorphism of H_1/Z. From $H_1 = \varphi_1\omega_1(H_1) = \varphi_1(A)$ and from the normality of φ_1 we deduce that $A = \omega_1(H_1) \mathsf{J} L_{\varphi_1}$. From the modular law we deduce that $J_1 = A \mathsf{M} J_1 = (\omega_1(H_1) \mathsf{M} L_{\varphi_1}) \mathsf{M} J_1$. Since ω_1 induces a homomorphism of H_1/Z onto $\omega_1(H_1)/Z$, it follows that there is a solution z of the equation $\omega_1(z) = \omega_1(H_1) \mathsf{M} L_{\varphi_1}$ which satisfies the relation $z \leq H_1$. Since $\varphi_1\omega_1(z)$ $\varphi_1(\omega_1(H) \mathsf{M} L_{\varphi_1}) = Z$ and $\varphi_1 \omega_1$ induces an automorphism of H_1, it follows that

$$z = Z, \quad Z = \omega_1(z) = \omega_1(H_1) \mathsf{M} L_{\varphi_1}, \quad J_1 = \omega_1(H_1) J (L_{\varphi_1} \mathsf{M} J_1).$$

But since J_1 is indecomposable, we conclude that $J_1 = \omega_1(H_1)$. Furthermore, if $x \leq y \leq H_1$ and $\omega_1(x) = \omega_1(y)$, then $\varphi_1 \omega_1(x) = \varphi_1 \omega_1(y)$, $x = y$, and hence ω_1 induces an automorphism of H_1/Z onto J_1/Z. Since $\omega_1(H_1) = J_1 = \omega_1(A)$, it follows that $J_1 \mathsf{J} L_{\varphi_1} = A$. Lastly,

$$\varphi_1 \omega_1(H_1 \mathsf{M} L_{\varphi_1}) = \varphi_1(J_1 \mathsf{M} L_{\varphi_1}) = Z, \quad J_1 M L_{\varphi_1} = Z.$$

Thus L is the direct join of $J_1, H_2, H_3, \ldots, H_n$.

Applying the same construction to H_2 in this exchange decomposition, the J_k with $k > 1$ can be re-indexed so that there is a normal operator of L inducing an isomorphism between H_k and J_k such that all of the other exchange decompositions obtain. A natural consequence will be the equation $n = m$. Thus the Theorem of Ore is fully proved.

4. Complemented lattices

In a vector module over a field F each F-submodule and each factor module over an F-submodule has a dimension given by the number of basis elements over F. Generalizing this concept, we make the following definition. A real-valued function $d(a/b)$ defined on all factor lattices a/b of a given lattice L is called a *dimension function on* L if

(13) $$d(a/b) \text{ is a non-negative real number,}$$

(14) $$d(a/b) + d(b/c) = d(a/c) \text{ (whence } d(a/a) = 0),$$

(15) $$d(a/b) = d(c/d) \text{ if } a/b \text{ is projective with } c/d.$$

There is always the trivial dimension function which vanishes on all the factor lattices of L.

In a lattice with normality relation for any three elements a, b, c of L, the statement $a \mathsf{N} b$ implies $d(b/a) \geq d((a \mathsf{J} (b \mathsf{M} c))/a) = d((c \mathsf{M} b)/(c \mathsf{M} a))$. Furthermore, if a is subnormal in $a \mathsf{J} b$, then for any two elements

(16) $$d((a \mathsf{J} b)/a) \geq d(b/(a \mathsf{M} b)).$$

This is because there is a normal chain $a = a_0 \mathsf{N} a_1 \mathsf{N} \cdots \mathsf{N} a_s = a \mathsf{J} b$ projecting into the normal chain $a \mathsf{M} b = a_0 \mathsf{M} b \mathsf{N} (a_1 \mathsf{M} b) \mathsf{N} \cdots \mathsf{N} (a_s \mathsf{M} b) = b$, so that $d(a_{i+1}/a_i) \geq d(a_{i+1} \mathsf{M} b/a_i \mathsf{M} b)$. By adding up these inequalities we obtain (14). In a modular lattice we have, instead of (16), the equality

(17) $$d((a \mathsf{J} b)/a) = d(b/(a \mathsf{M} b))$$

for any two elements a, b; this is a consequence of the fact that $(a \mathsf{J} b)/a$ is projective with $b/(a \mathsf{M} b)$. Conversely, if in a lattice (17) holds for any two elements a, b and if $d(x/y)$ implies $x = y$, then the lattice is modular. In fact, if (13), (14), (15), and (17) is applied to three elements a, b, c which are in the pentagon relation $a \mathsf{M} c \leq b \leq c \leq a \mathsf{J} b$, then we obtain

$$a \mathsf{J} b = a \mathsf{J} c, \quad a \mathsf{M} b = a \mathsf{M} c, \quad d((a \mathsf{J} b)/c) + d(c/b) = d((a \mathsf{J} b)/b) = d(a/(a \mathsf{M} b))$$
$$= d(a/(a \mathsf{M} c)) = d((a \mathsf{J} c)/c) = d((a \mathsf{J} b)/c, \quad d(c/b) = 0, \quad b = c.$$

Condition (15) says that a dimension function is constant on all the members of a family of projectively related factor lattices, so that a dimension function may be interpreted as a non-negative real-valued function on the set $F(L)$ of all families of projectively related factor lattices, subject to the additional condition

(14a) $$d(f_1 + f_2) = d(f_1) + d(f_2),$$

where we define the sum of two families f_1, f_2 as the family represented by a/c whenever it is possible to represent f_1 by a/b, f_2 by b/c for suitable a, b, c of L. Of course, this definition may not be unique. If L satisfies the double chain condition, then any dimension function is uniquely determined by its values on the simple families represented by factor lattices a/b for which b is maximal under a and different from a. Usually there are defining relations between the values taken at the simple families because of the fact that an arbitrary family may be represented in various ways as a sum of simple families. Geometrically speaking, we may represent the dimension functions as the points of a convex cone from the origin in the affine space over the simple families of L. However, if L is a submodular lattice satisfying the double chain condition, the Jordan-Hölder Theorem holds, which says that any composition series of L has the same length n and that the n simple factor lattices corresponding to any composition series of L represent the same system of simple families. Hence in this case any assignment of non-negative real numbers to the simple families can be extended to a dimension function.

The normal dimension function on a submodular lattice with double chain condition is obtained by assigning the value 1 to each simple family. The normal dimension of A/Z is equal to n, being the length of a composition series of L. We also say that L is an *n-dimensional* lattice.

Definition: A lattice is called *complemented* if, for every three elements a, b, c of the lattice satisfying $a \leq b \leq c$, there is a complement b' of b in the lattice relative to c/a such that $b \mathsf{J} b' = c$, $b \mathsf{M} b' = a$.

A modular lattice L having a maximal element A and a minimal element Z is complemented if and only if for each element b of L there is a *complement* b' such that $b \mathsf{J} b' = A$, $b \mathsf{M} b' = Z$. In fact, if a, b, c are three elements of L satisfying $a \leq b \leq c$, then

$$b \mathsf{J}(a \mathsf{J}(b' \mathsf{M} c)) = (b \mathsf{J} a) \mathsf{J}(b' \mathsf{M} c) = b \mathsf{J}(b' \mathsf{M} c) = (b \mathsf{J} b') \mathsf{M} c = A \mathsf{M} c = c,$$
$$b \mathsf{M}(a \mathsf{J}(b' \mathsf{M} c)) = a \mathsf{J}(b' \mathsf{M} c)) = a \mathsf{J}((b \mathsf{M} b') \mathsf{M} c) = a \mathsf{J}(Z \mathsf{M} c) = a \mathsf{J} Z = a,$$

so that $a \mathsf{J}(b' \mathsf{M} c)$ is a complement of b relative to c/a. For example, *the normal subgroups of a group $\mathfrak{G}$ which is generated by its smallest normal subgroups $\neq 1$ form a complemented modular lattice. Moreover, every normal subgroup $\mathfrak{A} \neq 1$ of $\mathfrak{G}$ is the direct product of some of the smallest normal subgroups $\neq 1$ of $\mathfrak{G}$.*

Proof: First of all, we extend the notion of direct product, which previously was defined only for direct products of finitely many factors, to direct products of infinitely many factors, as follows: The group $\mathfrak{G}$ is called the *direct product of its subgroups $\mathfrak{N}$ running over a finite or infinite set B of normal subgroups of* $\mathfrak{G}$ if in the subgroup lattice of $\mathfrak{G}$ the group $\mathfrak{G}$ is the direct join over B.

This definition coincides with the definition in § 1 in the case of a finite set B.

For any direct decomposition of a group $\mathfrak{G}$ into the subgroups belonging to a certain set B and for any subset X of B we have the direct decomposition

$\mathfrak{G} = \mathfrak{X} \times \mathfrak{Y}$, where $\mathfrak{X}$ is generated by all subgroups belonging to X and $\mathfrak{Y}$ is generated by all the subgroups belonging to B, but not to X. The decomposition operator $\delta_{\mathfrak{X}}$ of the given direct decomposition of $\mathfrak{G}$ with respect to $\mathfrak{X}$ is defined as the mapping that maps any element g of $\mathfrak{G}$ onto the element $(g\mathfrak{Y})\mathsf{M}\,\mathfrak{X}$ of $\mathfrak{X}$. It is a normal operator of $\mathfrak{G}$ mapping $\mathfrak{G}$ onto $\mathfrak{X}$ and inducing the decomposition operator δ_X in the subgroup lattice of $\mathfrak{G}$.

A set of necessary and sufficient conditions that a given subset B of subgroups of $\mathfrak{G}$ lead to a decomposition as a direct product over B is the following:

1. The group $\mathfrak{G}$ is generated by the subgroups belonging to B.

2. Each subgroup in B is normal; or: any conjugate of any one member $\mathfrak{N}$ of B under another member of B is contained in $\mathfrak{N}$.

3. The intersection of any one member of B with the subgroups generated by the remainder of B always is 1.

If B is an ordered set, then condition 3. may be replaced by the weaker condition that the intersection of any one member $\mathfrak{N}$ of B with the subgroup generated by the members of B preceding $\mathfrak{N}$ is 1.

Now let us assume that the group $\mathfrak{G}$ is generated by the set S of all the smallest normal subgroups $\neq 1$ of $\mathfrak{G}$. Well order S so that there is a last element $\mathfrak{Z}$. Let $\mathfrak{A}$ be a given normal subgroup of $\mathfrak{G}$ and, for each element $\mathfrak{N}$ of S, form the subgroup $\mathfrak{N}'$ of $\mathfrak{G}$ generated by $\mathfrak{A}$ and by all the members of S preceding $\mathfrak{N}$. It follows that $\mathfrak{Z}'\mathfrak{Z} = \mathfrak{G}$. Let B be the subset of all the members $\mathfrak{N}$ of S for which $\mathfrak{N}$ is not contained in $\mathfrak{N}'$. We conclude, for these members, that $\mathfrak{N}\mathsf{M}\mathfrak{N}'$ is properly contained in $\mathfrak{N}$. Since $\mathfrak{N}$ is a smallest normal subgroup $\neq 1$ of $\mathfrak{G}$ and since $\mathfrak{N}'$ is normal in $\mathfrak{G}$, it follows that $\mathfrak{N}\mathsf{M}\mathfrak{N}' = 1$; hence $\mathfrak{N}\mathfrak{N}'$ is the direct product of $\mathfrak{N}$ and of $\mathfrak{N}'$.

For any members $\mathfrak{N}$ of S let $\mathfrak{N}''$ be the subgroup of $\mathfrak{G}$ generated by $\mathfrak{A}$ and by all members $\mathfrak{Y}$ of B preceding $\mathfrak{N}$. If it happens that $\mathfrak{N}''$ is sometimes not the same as $\mathfrak{N}'$, then let $\mathfrak{N}_1$ be the first element of S satisfying $\mathfrak{N}_1'' \neq \mathfrak{N}_1'$. Hence for all members $\mathfrak{N}$ of S preceding $\mathfrak{N}_1$ we will have $\mathfrak{N}'' = \mathfrak{N}'$. Obviously we have $\mathfrak{N}_1''$ contained in $\mathfrak{N}_1'$. Moreover, if $\mathfrak{N}_1$ is a limit element in the well ordering of S, then $\mathfrak{N}_1'$ is the union of the $\mathfrak{N}'$ with $\mathfrak{N}$ preceding $\mathfrak{N}_1$; similarly, $\mathfrak{N}_1''$ is the union of the $\mathfrak{N}''$ with $\mathfrak{N}$ preceding $\mathfrak{N}_1$. Thus $\mathfrak{N}_1'$ would be equal to $\mathfrak{N}_1''$, which is a contradiction. Since for the first element $\mathfrak{N}_0$ of $\mathfrak{S}$ we have $\mathfrak{N}_0' = \mathfrak{A} = \mathfrak{N}_0''$, it follows that $\mathfrak{N}_0$ precedes $\mathfrak{N}_1$. If $\mathfrak{N}$ is the immediate predecessor of $\mathfrak{N}_1$ in the well ordering of S then we find that $\mathfrak{N}_1' = \mathfrak{N}'\mathfrak{N}$. Either $\mathfrak{N}$ belongs to B—and then $\mathfrak{N}_1'' = \mathfrak{N}''\mathfrak{N} = \mathfrak{N}'\mathfrak{N} = \mathfrak{N}_1'$—or $\mathfrak{N}$ does not belong to B—and then $\mathfrak{N}$ is contained in $\mathfrak{N}'$, $\mathfrak{N}_1' = \mathfrak{N}'\mathfrak{N} = \mathfrak{N}' = \mathfrak{N}'' = \mathfrak{N}_1''$. At any rate, we end up with the contradiction that $\mathfrak{N}_1'$ and $\mathfrak{N}_1''$ coincide.

Consequently, $\mathfrak{N}' = \mathfrak{N}''$ for all $\mathfrak{N}$ of S; in particular, $\mathfrak{G} = \mathfrak{Z}'\mathfrak{Z} = \mathfrak{Z}''\mathfrak{Z} = \mathfrak{A} \times \mathfrak{B}$, where $\mathfrak{B}$ is the direct product of all the subgroups belonging to B. Thus the normal subgroups of $\mathfrak{G}$ form a complemented modular lattice.

Maintaining the notation, we find in a similar way a direct decomposition of $\mathfrak{G}$ into $\mathfrak{B}$ and the direct product over the members of some other subset A of S. Since the mapping $x\mathfrak{B}/\mathfrak{B}$ onto $x\mathfrak{B} \mathsf{M} \mathfrak{A}$ maps $\mathfrak{G}/\mathfrak{B}$ isomorphically onto $\mathfrak{A}$, it follows that the decomposition of $\mathfrak{G}/\mathfrak{B}$ into the direct product over the smallest normal subgroups $\neq 1$ of the form $\mathfrak{Y}\mathfrak{B}/\mathfrak{B}$, with $\mathfrak{Y}$ running over A, is mapped onto a decomposition of $\mathfrak{A}$ into the direct product of the smallest normal subgroups $\neq 1$ of the form $(\mathfrak{B}\mathfrak{Y}) \mathsf{M} \mathfrak{A}$.

In the proof just completed, we have used the following property of groups: If T is a well-ordered set of normal subgroups of a group $\mathfrak{G}$ generating $\mathfrak{G}$ such that the intersection of any member $\mathfrak{X}$ of T with the subgroup generated by all the members of T which precede $\mathfrak{X}$ is 1, then $\mathfrak{G}$ is the direct product over T. This follows from the more general property: If $\mathfrak{U}$ is an increasing set of subgroups of the group $\mathfrak{G}$ and if $\mathfrak{V}$ is an arbitrary subgroup of $\mathfrak{G}$, then the intersection of $\mathfrak{V}$ with the union of the members of $\mathfrak{U}$ is equal to the union of the intersections of the members of $\mathfrak{U}$ with $\mathfrak{V}$. Generalizing this to lattices, we obtain the

Basis theorem of lattice theory: *If in a complete modular lattice L the all element A is the join of the minimal elements different from the zero element Z and if for any increasing subset U of L the meet of an arbitrary element v of L with the join of U is equal to the join of the elements $u \mathsf{M} v$ with u running over U (continuity condition), then L is complemented, and moreover every element of L is the direct join of some minimal elements different from Z.*

For groups, we also have the converse: *If the modular lattice formed by the normal subgroups of a group $\mathfrak{G}$ is complemented, then $\mathfrak{G}$ is generated by its smallest normal subgroups different from* 1.

Proof: Let $\mathfrak{A}$ be the subgroup generated by the smallest normal subgroups different from 1 of $\mathfrak{G}$. If $\mathfrak{A}$ is not $\mathfrak{G}$ then, according to the Maximal Theorem of group theory, there is a maximal normal subgroup $\mathfrak{N}$ of $\mathfrak{G}$ which contains $\mathfrak{A}$ and is not $\mathfrak{G}$. By assumption, there is a direct decomposition of $\mathfrak{G}$ into $\mathfrak{N}$ and another normal subgroup $\mathfrak{M}$ different from 1. Since $\mathfrak{M}$ is isomorphic to $\mathfrak{G}/\mathfrak{N}$, it follows from the maximal property of $\mathfrak{N}$ that $\mathfrak{M}$ is a smallest normal subgroup different from 1 of $\mathfrak{G}$; thus $\mathfrak{M}$ is contained in $\mathfrak{A}$ and is therefore also contained in $\mathfrak{N}$, contrary to the construction of $\mathfrak{M}$. It follows that $\mathfrak{A} = \mathfrak{G}$, Q. E. D.

Definition: In a lattice L we have *unique complementation* if for any three elements a, b, c of L the equations $a \mathsf{J} b = a \mathsf{J} c$, $a \mathsf{M} b = a \mathsf{M} c$ imply $b = c$.

In other words, there is at most one complement of x relative to y/z whenever $z \leq x \leq y$.

DEFINITION: A lattice is called *distributive* if it satisfies the distributive law for lattice operations

(18) $$a \mathsf{M} (b \mathsf{J} c) = (a \mathsf{M} b) \mathsf{J} (a \mathsf{M} c) \text{ for any } a, b, c \text{ of } L.$$

The modular law is a special case of the distributive law, namely the case for which in (18) $a \geq b$ and therefore the simpler form

(18a) $$a \mathsf{M} (b \mathsf{J} c) = b \mathsf{J} (a \mathsf{M} c)$$

obtains.

The distributive law implies the uniqueness of complementation, since from $a \mathsf{J} b = a \mathsf{J} c$, $a \mathsf{M} b = a \mathsf{M} c$ we deduce

$$a' = (a \mathsf{J} (b \mathsf{M} c)) \mathsf{M} (b \mathsf{J} c) = (a \mathsf{M} (b \mathsf{J} c)) \mathsf{J} (b \mathsf{M} c).$$

From the modular law we deduce, furthermore,

$$\begin{aligned} a' \mathsf{M} b &= (a \mathsf{J} (b \mathsf{M} c)) \mathsf{M} (b \mathsf{J} c) \mathsf{M} b = (a \mathsf{J} (b \mathsf{M} c)) \mathsf{M} b = (a \mathsf{M} b) \mathsf{J} (b \mathsf{M} c) \\ &= (a \mathsf{M} c) \mathsf{J} (b \mathsf{M} c) = a' \mathsf{M} c. \end{aligned}$$

From the distributive law we deduce

$$a' = a' \mathsf{M} (b \mathsf{J} c) = (a' \mathsf{M} b) \mathsf{J} (a' \mathsf{M} c) = a' \mathsf{M} b = a' \mathsf{M} c.$$

Finally,

$$\begin{aligned} a' &= (a \mathsf{J} (b \mathsf{M} c)) \mathsf{M} (b \mathsf{J} c) = (a \mathsf{M} (b \mathsf{J} c)) \mathsf{J} (b \mathsf{M} c), \\ b &= a' \mathsf{J} b = (a \mathsf{M} (b \mathsf{J} c)) \mathsf{J} (b \mathsf{M} c) \mathsf{J} b = (a \mathsf{M} (b \mathsf{J} c)) \mathsf{J} b \\ &= (a \mathsf{J} b) \mathsf{M} (b \mathsf{J} c) = (a \mathsf{J} c) \mathsf{M} (b \mathsf{J} c) = c. \end{aligned}$$

Conversely, the uniqueness of complementation in a lattice L implies the modular law, because L cannot contain a pentagon sublattice. Moreover, we demonstrate the distributive law (18) as follows.

Let a, b, c be any three elements of L; then by the modular law,

$$u = (a \mathsf{M} (b \mathsf{J} c)) \mathsf{J} (b \mathsf{M} c) = (a \mathsf{J} (b \mathsf{M} c)) \mathsf{M} (b \mathsf{J} c)$$

and $b \mathsf{M} c \leq u \leq b \mathsf{J} c$,

$$\begin{aligned} (b \mathsf{M} (u \mathsf{J} c)) \mathsf{J} (u \mathsf{M} c) \mathsf{J} u &= (b \mathsf{M} (u \mathsf{J} c)) \mathsf{J} u = (b \mathsf{J} u) \mathsf{M} (c \mathsf{J} u) \\ &= (c \mathsf{M} (u \mathsf{J} b)) \mathsf{J} (u \mathsf{M} b) \mathsf{J} u. \end{aligned}$$

By duality,

$$(b \mathsf{J} (u \mathsf{M} c)) \mathsf{M} (u \mathsf{J} c) \mathsf{M} u = (c \mathsf{J} (u \mathsf{M} b)) \mathsf{M} (u \mathsf{J} b) \mathsf{M} u.$$

By the modular law, we obtain

$$(19)\qquad \begin{aligned}(b\,\mathsf{M}\,(u\,\mathsf{J}\,c))\,\mathsf{J}\,(u\,\mathsf{M}\,c) &= (b\,\mathsf{J}\,(u\,\mathsf{M}\,c))\,\mathsf{M}\,(u\,\mathsf{J}\,c),\\ (c\,\mathsf{M}\,(u\,\mathsf{J}\,b))\,\mathsf{J}\,(u\,\mathsf{M}\,b) &= (c\,\mathsf{J}\,(u\,\mathsf{M}\,b))\,\mathsf{M}\,(u\,\mathsf{J}\,b).\end{aligned}$$

From the uniqueness of complementation it follows that

$$(b\,\mathsf{M}\,(u\,\mathsf{J}\,c))\,\mathsf{J}\,(u\,\mathsf{M}\,c) = (c\,\mathsf{M}\,(u\,\mathsf{J}\,b))\,\mathsf{J}\,(u\,\mathsf{M}\,b).$$

Furthermore,

$$(20\text{a})\qquad \begin{aligned}(b\,\mathsf{M}\,(u\,\mathsf{J}\,c))\,\mathsf{J}\,(u\,\mathsf{M}\,c)\,\mathsf{J}\,(c\,\mathsf{M}\,(u\,\mathsf{J}\,b))\,\mathsf{J}\,(u\,\mathsf{M}\,b) &= (b\,\mathsf{M}\,(u\,\mathsf{J}\,c))\,\mathsf{J}\,(c\,\mathsf{M}\,(u\,\mathsf{J}\,b))\\ &= ((b\,\mathsf{M}\,(u\,\mathsf{J}\,c))\,\mathsf{J}\,c)\,\mathsf{M}\,(u\,\mathsf{J}\,b) = (b\,\mathsf{J}\,c)\,\mathsf{M}\,(u\,\mathsf{J}\,c)\,\mathsf{M}\,(u\,\mathsf{J}\,b)\\ &= (b\,\mathsf{J}\,u)\,\mathsf{M}\,(c\,\mathsf{J}\,u) = (b\,\mathsf{M}\,(u\,\mathsf{J}\,c))\,\mathsf{J}\,(u\,\mathsf{M}\,c)\,\mathsf{J}\,u.\end{aligned}$$

By duality,

$$(20\text{b})\qquad \begin{aligned}(b\,\mathsf{J}\,(u\,\mathsf{M}\,c))\,\mathsf{M}\,(u\,\mathsf{J}\,c)\,\mathsf{M}\,(c\,\mathsf{J}\,(u\,\mathsf{M}\,b))\,\mathsf{M}\,(u\,\mathsf{J}\,b) &= (u\,\mathsf{M}\,b)\,\mathsf{J}\,(u\,\mathsf{M}\,c)\\ &= (b\,\mathsf{J}\,(u\,\mathsf{M}\,c))\,\mathsf{M}\,(u\,\mathsf{J}\,c)\,\mathsf{M}\,u.\end{aligned}$$

Applying the uniqueness of complementation to (20a, b) and (19), it follows that

$$(b\,\mathsf{J}\,(u\,\mathsf{M}\,c))\,\mathsf{M}\,(u\,\mathsf{J}\,c) = u = (c\,\mathsf{J}\,(u\,\mathsf{M}\,b))\,\mathsf{M}\,(u\,\mathsf{J}\,c),$$

$$\begin{aligned}u = (u\,\mathsf{M}\,b)\,\mathsf{J}\,(u\,\mathsf{M}\,c) &= ((a\,J\,(b\,\mathsf{M}\,c))\,\mathsf{M}\,b)\,\mathsf{J}\,((a\,\mathsf{J}\,(b\,\mathsf{M}\,c))\,\mathsf{M}\,c)\\ &= (a\,\mathsf{M}\,b)\,\mathsf{J}\,(a\,\mathsf{M}\,c)\,\mathsf{J}\,(b\,\mathsf{M}\,c),\end{aligned}$$

$$\begin{aligned}a\,\mathsf{M}\,(b\,\mathsf{J}\,c)\,\mathsf{M}\,(b\,\mathsf{M}\,c) = a\,\mathsf{M}\,b\,\mathsf{M}\,c &= (a\,\mathsf{M}\,b\,\mathsf{M}\,c)\,\mathsf{J}\,(a\,\mathsf{M}\,b\,\mathsf{M}\,c)\\ &= ((a\,\mathsf{M}\,b)\,\mathsf{J}\,(a\,\mathsf{M}\,c))\,\mathsf{M}\,(b\,\mathsf{M}\,c);\end{aligned}$$

(18) now follows from the uniqueness of complementation.

Since the uniqueness of complementation is a self-dual property, the same is true for the distributive law. In other words, (18) is equivalent to

$$(18\text{a})\qquad a\,\mathsf{J}\,(b\,\mathsf{M}\,c) = (a\,\mathsf{J}\,b)\,\mathsf{M}\,(a\,\mathsf{J}\,c) \text{ for } a,\ b,\ c \text{ in } L.$$

It is not difficult to show that all elements x of a lattice L that satisfy the distributive rules

$$\begin{aligned}x\,\mathsf{M}\,(b\,\mathsf{J}\,c) &= (x\,\mathsf{M}\,b)\,\mathsf{J}\,(x\,\mathsf{M}\,c), & x\,\mathsf{J}\,(b\,\mathsf{M}\,c) &= (x\,\mathsf{J}\,b)\,\mathsf{M}\,(x\,\mathsf{J}\,c),\\ b\,\mathsf{M}\,(x\,\mathsf{J}\,c) &= (b\,\mathsf{M}\,x)\,\mathsf{J}\,(b\,\mathsf{M}\,c), & b\,\mathsf{J}\,(x\,\mathsf{M}\,c) &= (b\,\mathsf{J}\,x)\,\mathsf{M}\,(b\,\mathsf{J}\,c)\end{aligned}$$

for all pairs of elements b, c of L form a distributive sublattice $D(L)$ of L which coincides with L if and only if L is distributive. Moreover, if L is complemented, then $D(L)$ is also complemented.

In order to prove the last statement, let b be an element of $D(L)$, and let u, v be two elements of L for which $b\,\mathsf{M}\,u = b\,M\,v$, $b\,\mathsf{J}\,u = b\,\mathsf{J}\,v$; then

$u = u \mathsf{J}(b\mathsf{M}u) = u\mathsf{J}(b\mathsf{M}v) = (u\mathsf{J}b)\mathsf{M}(u\mathsf{J}v) = (v\mathsf{J}b)\mathsf{M}(v\mathsf{J}u) = v$. Hence for every element b of $D(L)$ there is only one complement relative to any factor lattice c/a containing b. Let a, b, c in $D(L)$, and let $a \leq b \leq c$. Since L is complemented, there is an element b' in L for which $b\mathsf{J}b' = c$, $b\mathsf{M}b' = a$. It follows that for any two elements x, y

$$\begin{aligned} b\mathsf{J}((b'\mathsf{M}x)\mathsf{J}(b'\mathsf{M}y)) &= (b'\mathsf{M}x)\mathsf{J}b\mathsf{J}(b'\mathsf{M}y) \\ &= ((b\mathsf{J}b')\mathsf{M}(b\mathsf{J}x))\mathsf{J}((b\mathsf{J}b')\mathsf{M}(b\mathsf{J}y)) = (c\mathsf{M}(b\mathsf{J}x))\mathsf{J}(c\mathsf{M}(b\mathsf{J}y)) \\ &= c\mathsf{M}((b\mathsf{J}x)\mathsf{J}(b\mathsf{J}y)) = (b\mathsf{J}b')\mathsf{M}(b\mathsf{J}(x\mathsf{J}y)) = b\mathsf{J}(b'\mathsf{M}(x\mathsf{J}y)), \end{aligned}$$

$$\begin{aligned} b\mathsf{M}((b'\mathsf{M}x)\mathsf{J}(b'\mathsf{M}y)) &= (b\mathsf{M}b'\mathsf{M}x)\mathsf{J}(b\mathsf{M}b'\mathsf{M}y) = (a\mathsf{M}x)\mathsf{J}(a\mathsf{M}y) \\ &= a\mathsf{M}(x\mathsf{J}y) = (b\mathsf{M}b')\mathsf{M}(x\mathsf{J}y) = b\mathsf{M}(b'\mathsf{M}(x\mathsf{J}y)). \end{aligned}$$

Since there is only one complement of b relative to the factor lattice $(b\mathsf{J}((b'\mathsf{M}x)\mathsf{J}(b'\mathsf{M}y)))/(b\mathsf{M}((b'\mathsf{M}x)\mathsf{J}(b'\mathsf{M}y)))$, we find $(b'\mathsf{M}x)\mathsf{J}(b'\mathsf{M}y) = b'\mathsf{M}(x\mathsf{J}y)$. By duality, $(b'\mathsf{J}x)\mathsf{M}(b'\mathsf{J}y) = b'\mathsf{J}(x\mathsf{M}y)$. Moreover,

$$\begin{aligned} b\mathsf{J}(x\mathsf{M}(b'\mathsf{J}y)) &= (b\mathsf{J}x)\mathsf{M}(b\mathsf{J}b'\mathsf{J}y) = (b\mathsf{J}x)\mathsf{M}(c\mathsf{J}y) \\ &= ((c\mathsf{J}x)\mathsf{M}(c\mathsf{J}y))\mathsf{M}(b\mathsf{J}x) = (c\mathsf{J}(x\mathsf{M}y))\mathsf{M}(b\mathsf{J}x) = (c\mathsf{M}(b\mathsf{J}x))\mathsf{M}(x\mathsf{M}y) \\ &= ((b\mathsf{J}x)\mathsf{M}(b\mathsf{J}b'))\mathsf{J}(x\mathsf{M}y) = (b\mathsf{J}(x\mathsf{M}b'))\mathsf{J}(x\mathsf{M}y) \\ &= b\mathsf{J}((x\mathsf{M}b')\mathsf{J}(x\mathsf{M}y)), \end{aligned}$$

$$\begin{aligned} b\mathsf{M}(x\mathsf{M}(b'\mathsf{J}y)) &= x\mathsf{M}g\mathsf{M}(b'\mathsf{J}y) = x\mathsf{M}((b\mathsf{M}b')\mathsf{J}(b\mathsf{M}y)) \\ &= x\mathsf{M}(a\mathsf{J}(b\mathsf{M}b)) = (a\mathsf{M}x)\mathsf{J}(x\mathsf{M}(b\mathsf{M}y)) = (b\mathsf{M}(x\mathsf{M}b'))\mathsf{J}(b\mathsf{M}(x\mathsf{M}y)) \\ &= b\mathsf{M}((x\mathsf{M}b')\mathsf{J}(x\mathsf{M}y)). \end{aligned}$$

Since there is only one complement of b relative to $(b\mathsf{J}(x\mathsf{M}(b'\mathsf{J}y)))/(b\mathsf{M}(x\mathsf{M}(b'\mathsf{J}y)))$, it follows that $x\mathsf{M}(b'\mathsf{J}y) = (x\mathsf{M}b')\mathsf{J}(x\mathsf{M}y)$, and by duality, $x\mathsf{J}(b'\mathsf{M}y) = (x\mathsf{J}b')\mathsf{J}(x\mathsf{M}y)$. Hence b' belongs to $D(L)$, and this shows that $D(L)$ is complemented.

If L is a complemented modular lattice with all element A and zero element Z, then for every element of $D(L)$ there is precisely one complement in L. Conversely, let a be an element of L with precisely one complement a' in L for which $a\mathsf{J}a' = A$, $a\mathsf{M}a' = Z$. The investigation to follow will show that a belongs to $D(L)$.

For any x of L we have $(a\mathsf{M}x)\mathsf{J}(z'\mathsf{M}x) \leq x$; hence there is a complement y of $(a\mathsf{M}x)\mathsf{J}(a'\mathsf{M}x)$ relative to x/Z such that $(a\mathsf{M}x)\mathsf{J}(a'\mathsf{M}x)\mathsf{J}y = x$, $((a\mathsf{M}x)\mathsf{J}(a'\mathsf{M}x))\mathsf{M}y = Z$. Hence

$$x\mathsf{M}y = y,\ a\mathsf{M}y = a\mathsf{M}x\mathsf{M}y = (a\mathsf{M}x)\mathsf{M}((a\mathsf{M}x)\mathsf{J}(a'\mathsf{M}x))\mathsf{M}y = a\mathsf{M}x\mathsf{M}z = Z.$$

There is a complement z of $(a \mathsf{J} y) \mathsf{M} a'$ relative to a'/Z such that $((a \mathsf{J} y) \mathsf{M} a') \mathsf{J} z = a'$, $(a \mathsf{J} y) \mathsf{M} a' \mathsf{M} z = Z$. Hence

$$(a \mathsf{J} y) \mathsf{M} z = Z,$$

$$a \mathsf{J} y \mathsf{J} z = (a \mathsf{J} y))((a \mathsf{J} y) \mathsf{M} a') \mathsf{J} z = a \mathsf{J} y \mathsf{J} a' = y \mathsf{J} a \mathsf{J} a' = y \mathsf{J} A = A,$$

$$\begin{aligned} a \mathsf{M}(y \mathsf{J} z) &= a \mathsf{M}(a \mathsf{J} y) \mathsf{M}(y \mathsf{J} z) = a \mathsf{M}(y \mathsf{J}((a \mathsf{J} y) \mathsf{M} z)) = a \mathsf{M}(y \mathsf{J} Z) \\ &= a \mathsf{M} y = Z. \end{aligned}$$

Since there is only one complement of a in L, it follows that

$$y \mathsf{J} z = a', \quad y = a' \mathsf{M} x \mathsf{M} y = z' \mathsf{M} x \mathsf{M} y \mathsf{M}((a \mathsf{M} x) \mathsf{J}(z' \mathsf{M} x)) = a' \mathsf{M} x \mathsf{M} Z = Z,$$
$$x = (a \mathsf{M} x) \mathsf{J}(a' \mathsf{M} x).$$

By duality we obtain $x = (a \mathsf{J} x) \mathsf{M}(a' \mathsf{J} x)$.

If $x = u \mathsf{J} u'$, $u \leq a$, $u' \leq a'$, then

$$a \mathsf{M} x = a \mathsf{M}(u \mathsf{J} u') = u \mathsf{J}(a \mathsf{M} u') = u \mathsf{J}(a \mathsf{M} a' \mathsf{M} u') = u \mathsf{J} Z = u,$$

and similarly $s' \mathsf{M} x = u'$. Hence if $x = x_1 \mathsf{J} x_2$, then

$$\begin{aligned} x &= (a \mathsf{M} x_1) \mathsf{J}(a' \mathsf{M} x_1) \mathsf{J}(a \mathsf{M} x_2) \mathsf{J}(a' \mathsf{M} x_2) \\ &= ((a \mathsf{M} x_1) \mathsf{J}(a \mathsf{M} x_2)) \mathsf{J}((a' \mathsf{M} x_1 \mathsf{J}(a' \mathsf{M} x_2)), \\ a \mathsf{M} x &= (a \mathsf{M} x_1) \mathsf{J}(a \mathsf{M} x_2), \quad a' \mathsf{M} x = (a' \mathsf{M} x_1) \mathsf{J}(a' \mathsf{M} x_2). \end{aligned}$$

Moreover, if $x = x_1 \mathsf{M} x_2$, then we have

$$a \mathsf{M} x = a \mathsf{M} x_1 \mathsf{M} x_2 = (a \mathsf{M} x_1) \mathsf{M}(a \mathsf{M} x_2), \quad a' \mathsf{M} x = (a' \mathsf{M} x_1) \mathsf{M}(a' \mathsf{M} x_2).$$

The last statements suggest the following definition.

DEFINITION: The *vector sum* of two lattices L_1, L_2 is the lattice $DU(L_1, L_2)$ consisting of all the ordered pairs (x_1, x_2) with the component x_i in L_i, subject to the rules of operation

$$(x_1, x_2) \mathsf{J}(y_1, y_2) = (x_1 \mathsf{J} y_1, x_2 \mathsf{J} y_2)$$
$$(x_1, x_2) \mathsf{M}(y_1, y_2) = (x_1 \mathsf{M} x_2, x_2 \mathsf{M} y_2).$$

It is clear that $L = DU(L_1, L_2)$ is also a lattice. Moreover, $D(L)$ consists of all the pairs (x_1, x_2) with x_i contained in $D(L_i)$ such that $D(L) = DU(D(L_1,), D(L_2))$.

If L is a complemented modular lattice with all element A and zero element Z, then for every element a of L with unique complement a' in L it was shown above that there is an isomorphism between L and $DU(a/Z, a'/Z)$, namely, the isomorphism that maps the element x of L onto the pair $(z \mathsf{M} x, a' \mathsf{M} x)$. Hence both a and a' belong to $D(L)$.

DEFINITION: A lattice is called *irreducible* if it is not isomorphic to the vector sum of two lattices of more than one element each. An irreducible finite-dimensional complemented modular lattice is called a *projective geometry*.[1]

The minimal elements $\neq Z$ of a projective geometry are called the *points*. In the scale of dimensionality the points are followed by the lines, planes, 3-hyperplanes, etc. An example of a projective geometry is provided by the n-dimensional lattice $\Gamma(n-1, F)$ formed by the F-submodules of an n-dimensional vector module over a division ring F. There is precisely one 1-dimensional projective geometry that consists only of the all element and the zero element. The only projective geometries of dimension 2 are, q being any ordinal greater than 1, the lattices $\Gamma(1, q)$ consisting of $q + 3$ elements $a_0, a_1, \ldots, a_{q+2}$ subject to the composition rules:

$$\text{(i)} \qquad a_0 \mathsf{M} a_i = a_0,$$

$$\text{(ii)} \qquad a_0 \mathsf{J} a_i = a_i,$$

$$\text{(iii)} \qquad a_{q+2} \mathsf{M} a_i = a_i,$$

$$\text{(iv)} \qquad a_{q+2} \mathsf{J} a_i = a_{q+2},$$

$$\text{(v) if } 0 < i < j < q + 2, \text{ then } a_i \mathsf{M} a_j = a_0,\ a_i \mathsf{J} a_j = a_{q+2}.$$

A projective geometry that is isomorphic to a factor lattice of a projective geometry of dimension greater than 3 is called *Desarguean*. $\Gamma(n, F)$, for example, is Desarguean. It is a fundamental theorem of projective geometry that every Desarguean projective geometry is isomorphic to a $\Gamma(n, F)$ where, in case n is greater than 1, the division ring F is uniquely determined up to isomorphism. Also, every projective geometry of more than 3 dimensions is Desarguean. There are non-Desarguean 3-dimensional geometries. But even the finite ones cannot be completely classified yet.

We define the vector sum of an arbitrary set S of lattices as the lattice $DU(S)$ consisting of all functions f defined on S for which $f(L)$ is contained in L for each member L of S and $f \mathsf{J} g(L) = f(L) \mathsf{J} g(L), f \mathsf{M} g(L) = f(L) \mathsf{M} g(L)$. The operation DU is associative and commutative in the widest possible sense. It coincides with the previously defined operation in the case of a set of two lattices.

Each of the following is a property that is satisfied by a vector sum if and only if it is satisfied by each vector summand: That of being comple-

[1] Note that the geometrical dimension is obtained from the lattice-theoretical dimension by subtracting 1. In what follows only the lattice-theoretical dimension is mentioned.

mented, distributivity, modularity, that of having an all element or a zero element, and completeness.

An isomorphism of a lattice L onto a vector sum is called a *vector decomposition of* L. If L has a maximal and a minimal element, a vector decomposition of L leads to a direct decomposition of $D(L)$ into the elements of $D(L)$ which are mapped onto those vectors that have all but one of its components zero, the remaining component being the all element. Conversely, any representation of a lattice L as the direct join of finitely many elements of $D(L)$ is derived from a vector decomposition.

A *Remak vector decomposition* is a vector decomposition into irreducible components each consisting of at least two elements. If L has an all element and a zero element, then there is at most one Remak vector decomposition. It corresponds to the Remak decomposition of $D(L)$. Every finite-dimensional complemented modular lattice has a Remak vector decomposition into projective geometries.

If the modular lattice formed by the normal subgroups of a group $\mathfrak{G}$ is complemented and distributive, then the subgroup lattice of $\mathfrak{G}$ is the Remak vector sum of the lattices $\mathfrak{N}/1$ with $\mathfrak{N}$ running over the smallest normal subgroups different from 1 of $\mathfrak{G}$. The lattice formed by the normal subgroups of such a group is isomorphic to the subset lattice of the set of all smallest normal subgroups different from 1.

A set of necessary and sufficient conditions that the lattice of the normal subgroups of a group $\mathfrak{G}$ be complemented and distributive, is the following:

1. The group $\mathfrak{G}$ is generated by its smallest normal subgroups different from 1.

2. A smallest normal subgroup different from 1 either is non-abelian, or is abelian and non-isomorphic to any other smallest normal subgroup different from 1 of $\mathfrak{G}$.

A particular case is that of the *semi-simple groups*, which are defined as the groups that have a decomposition into the direct product of non-abelian simple groups. They are characterized by the extreme rigidity of the structure formed by the normal subgroups, viz., by the property that in any semi-simple group and in any direct product of semi-simple groups there is for any normal subgroup precisely one complementary normal subgroup. A simple group is semi-simple if and only if it is non-abelian. The direct product of semi-simple groups is semi-simple.

For any complete lattice L we define the *Frattini element* $\Phi(L)$ of L to be the meet of A and of all the maximal elements of L that are different from A. The Frattini element in the subgroup lattice of a group coincides with the Frattini subgroup of the given group. The Frattini element of a complete lattice L is the join of all elements x of L with the property that $x \mathsf{J} y = A$ always implies $y = A$.

The *descending Frattini series* is recursively defined as

$$\Phi_0(L) = A,\ \Phi_1(L) = \Phi(L),\ \ldots,\ \Phi_n(L) = \Phi(\Phi_{n-1}(L)/Z),\ \ldots.$$

It follows that

$$\Phi(\Phi_{n-1}(L)/\Phi_n(L)) = \Phi_n(L).$$

The dual concept is the *ascending Frattini series:* $\Phi^0(L) = Z$, $\Phi^1(L) =$ the join of Z and of all minimal elements $\neq Z, \ldots, \Phi^n(L) = \Phi^1(A/\Phi^{n-1}(L)), \ldots$. If L is a modular lattice, then the Frattini series is called a *Loewy series.* From the modularity it follows that $\Phi_1(a/b) \leq (\Phi_1(L) \mathsf{M} a) \mathsf{J} b$, $\Phi^1(a/b) \geq (\Phi^1(L) \mathsf{J} a) \mathsf{M} b$. If L is of finite dimension, then all three statements '$\Phi_1(L) = Z$', '$\Phi^1(L) = A$', and 'L is complemented' are equivalent, as follows from the basis theorem.

For a complete modular lattice L we define a *Loewy chain of length* r as a chain $A = a_0 \geq a_1 \geq a_2 \geq \cdots \geq a_r = Z$ in which all the factor lattices a_i/a_{i+1} are complemented. It follows that

(21) $$\Phi_i(L) \leq a_i, \quad \Phi^i(L) \geq a_{r-i}.$$

The Loewy series become Loewy chains by elimination of repetitions, provided both A and Z are members of the given Loewy series. From (21) there follows the

THEOREM OF LOEWY: *For a complete modular lattice, the two statements*

$$A = \Phi_0(L) > \Phi_1(L) > \cdots > \Phi_\lambda(L) = Z,$$
$$A = \Phi^{\lambda'}(L) > \Phi^{\lambda'-1}(L) > \cdots > \Phi^0(L) = Z$$

are equivalent; that is, if the descending Loewy series without repetitions is a Loewy chain then the ascending Loewy series without repetitions is also a Loewy chain, and vice versa. Moreover

(22) $$\Phi_i(L) \leq \Phi^{\lambda'-1}(L),$$

and therefore $\lambda = \lambda'$.

Example: The *descending Loewy series without repetitions* of a group $\mathfrak{G}$ with a composition series is defined as the characteristic chain

(23) $$\mathfrak{G} = \Lambda_0(\mathfrak{G}) > \Lambda_1(\mathfrak{G}) > \cdots > \Lambda_\lambda(\mathfrak{G}) = 1,$$

where $\Lambda_1(\mathfrak{G})$ is the intersection of all the maximal normal subgroups $\neq \mathfrak{G}$ of $\mathfrak{G}$, and $\Lambda_{i+1}(\mathfrak{G}) = \Lambda_1(\Lambda_i(\mathfrak{G}))$. Since the intersection of all the subgroups of a normal subgroup $\mathfrak{N}$ of $\mathfrak{G}$ that are conjugate to a maximal normal subgroup of $\mathfrak{N}$ under $\mathfrak{G}$ is a normal subgroup of $\mathfrak{G}$ maximal under $\mathfrak{N}$, it follows that $\Lambda_{i+1}(\mathfrak{G})$ can also be defined as the intersection of all the normal subgroups of $\mathfrak{G}$ that are maximal under $\Lambda_i(\mathfrak{G})$.

Similarly, the ascending Loewy series without repetitions of $\mathfrak{G}$ is defined as the characteristic chain

$$\mathfrak{G} = \Lambda^{\lambda'}(\mathfrak{G}) > \Lambda^{\lambda'-1}(\mathfrak{G}) > \cdots > \Lambda^0(\mathfrak{G}) = 1, \tag{23}$$

where $\Lambda^1(\mathfrak{G})$ is the subgroup generated by all the smallest normal subgroups $\neq 1$ of $\mathfrak{G}$, and $\Lambda^{i+1}(\mathfrak{G})/\Lambda^i(\mathfrak{G}) = \Lambda^1(\mathfrak{G}/\Lambda^i(\mathfrak{G}))$. As a consequence of Loewy's Theorem, we note the equality $\lambda = \lambda'$ of the length of the two characteristic chains (22) and (23) and the relation

$$\Lambda_i(\mathfrak{G}) \leq \Lambda^{\lambda-i}(\mathfrak{G}) \tag{24}$$

between its members.

APPENDIX C

FREE PRODUCTS AND GROUPS GIVEN BY A SET OF GENERATORS AND A SYSTEM OF DEFINING RELATIONS

An Introduction to §§ 3—9 of Chap. III

Let us consider the subgroups of a given group $\mathfrak{G}$. It may happen that some of them, say $\mathfrak{h}_1, \mathfrak{h}_2, \ldots, \mathfrak{h}_r$, generate the full group. We are interested in knowing to what extent the structure of $\mathfrak{G}$ is determined by the structure of the generating subgroups $\mathfrak{h}_1, \mathfrak{h}_2, \ldots, \mathfrak{h}_r$. With this in mind, we make the following definition:

A semi-group $\mathfrak{S}$ with a unit element $1_{\mathfrak{S}}$ is called a *product over the system* H of semi-groups with unit element if for each semigroup $\mathfrak{h}$ contained in H there is given a homomorphism $\sigma_{\mathfrak{h}}$ of $\mathfrak{h}$ into $\mathfrak{S}$ such that $\sigma_{\mathfrak{h}}(1_{\mathfrak{h}}) = 1_{\mathfrak{S}}$ and $\mathfrak{S}$ is generated by all the images $\sigma_{\mathfrak{h}}(h)$ with $\mathfrak{h}$ running over H and h running over the elements of the semi-group $\mathfrak{h}$ of H.

A homomorphic mapping Θ of one product over H, say $\mathfrak{S}$, onto another such product, say $\overline{\mathfrak{S}}$, is called a *homomorphism over* H if $\Theta\sigma_{\mathfrak{h}}(h) = \bar{\sigma}_{\mathfrak{h}}(h)$ for any element h of the semi-group $\mathfrak{h}$ running over H. Here, of course, $\bar{\sigma}_{\mathfrak{h}}$ denotes the homomorphism corresponding to $\sigma_{\mathfrak{h}}$ in the definition of $\overline{\mathfrak{S}}$ as a product over H. It is clear that there can be at most one homomorphism over H between any two given products over H, since every element x of $\mathfrak{S}$ can be written as

$$x = \sigma_{\mathfrak{h}_1}(h_1) \cdot \sigma_{\mathfrak{h}_2}(h_2) \cdots \sigma_{\mathfrak{h}_r}(h_r),$$

where h_i belongs to the semi-group $\mathfrak{h}_i$ of H for $i = 1, 2, \ldots, r$, and thus

$$\begin{aligned}\Theta(x) &= \Theta(\sigma_{\mathfrak{h}_1}(h_1) \cdots \sigma_{\mathfrak{h}_r}(h_r)) = \Theta\sigma_{\mathfrak{h}_1}(h_1) \cdot \Theta\sigma_{\mathfrak{h}_2}(h_2) \cdots \Theta\sigma_{\mathfrak{h}_r}(h_r) \\ &= \bar{\sigma}_{\mathfrak{h}_1}(h_1) \cdot \bar{\sigma}_{\mathfrak{h}_2}(h_2) \cdots \bar{\sigma}_{\mathfrak{h}_r}(h_r).\end{aligned}$$

The relation 'homomorph over H' is reflexive and transitive, but not symmetric. In fact, the identity mapping provides an isomorphism over H for any product over H. If Θ_1 is a homomorphism over H of the product $\mathfrak{S}_1$ onto the product $\mathfrak{S}_2$ over H and if Θ_2 is a homomorphism over H of the product $\mathfrak{S}_2$ onto the product $\mathfrak{S}_3$ over H, then $\Theta_2\Theta_1$ is a homomorphism over H of $\mathfrak{S}_1$ onto $\mathfrak{S}_3$. The group of one element together with the set of mappings of each element h of each member $\mathfrak{h}$ of H onto the unit element provides the *trivial product over* H. For every product over H we obtain a homomorphism over H onto the trivial product over H by mapping each element onto the unity element. But there is no converse homomorphism over H unless both products over H are trivial. More generally, two products over H are mutually homomorphic over H if and only if they are isomorphic over H in one direction.

DEFINITION: A product over H is called a *free product* over H if it can be mapped homomorphically over H onto any product over H.

It is immediate that two free products over H are isomorphic over H. *There always is a free product over H.* As a first step in constructing a free product over H, we denote by $\mathfrak{W}(H)$ the system of all 'words' over H, i. e. all expressions $h_1 h_2 \cdots h_r$—denoted for brevity by W—where the length r ranges over all the natural numbers and the letters h_i of the word W denote any element of any semi-group $\mathfrak{h}_i$ belonging to H. Furthermore, we denote by Z the *empty word*, which has no letters and which, by definition, is the only word of length 0. Two words are called *equal* if they are of equal length and if corresponding letters denote equal elements of equal semi-groups. This notion of equality has the usual three properties.

Two words W_1, W_2 are multiplied by juxtaposition, e. g., for $W_1 = h_1 h_2 \cdots h_r$, $W_2 = h_{r+1} h_{r+2} \cdots h_{r+s}$ we define the product by $W_1 W_2 = h_1 h_2 \cdots h_{r+s}$; in particular, $ZW = WZ = W$ for any word W.

It is important to note that the product of a word of length r and a word of length s is uniquely defined as a word of length $r + s$.

From the definition it is clear that the associative law of multiplication holds. Thus the words over H form a semi-group $\mathfrak{W}(H)$, with the empty word as unit element.

In a word $W = h_1 h_2 \cdots h_r$, it may happen

1. in case $\mathfrak{h}_i = \mathfrak{h}_{i+1}$ that the product h' of the two elements h_i, h_{i+1} is defined within the semi-group $\mathfrak{h}_i$, or

2. that $h_i = 1_{\mathfrak{h}_i}$.

In the first case, we replace the two letters concerned by h'; in the second case, we simply omit $1_{\mathfrak{h}_i}$. Either process will be called a *reduction*, and we will write

Case 1: $W = \cdots h_i h_{i+1} \cdots \to W' = \cdots h' \cdots,$

Case 2: $W = \cdots \quad 1_{\mathfrak{h}_i} \cdots \to W' = \cdots \cdots,$

where the dots always refer to unaltered letters.

The reverse process we will call an *anti-reduction*. Both processes are referred to as *elementary transformations*. Writing $W \to W'$ if the word W' is obtained by a reduction from the word W, we establish a binary relation in the set $\mathfrak{W}(H)$ of words over H. The normalized relation is the *congruence relation between words*, defined as follows:

$$W \equiv W'$$

if there is a chain of words $W = W_0, W_1, \ldots, W_s = W'$ such that for each index $i = 0, 1, 2, \ldots, s - 1$ either $W_i = W_{i+1}$ or $W_i \to W_{i+1}$ or $W_{i+1} \to W_i$. This congruence relation is normal and multiplicative. In fact, the normality follows from the definition as normalized relation (see Chap. I, Ex. 17). The substitution law of multiplication follows by repeated application of the following statement:

$$\text{If } W \to W', \text{ then } W_1 W \to W_1 W' \text{ and } W W_1 \to W' W_1,$$

which can be verified directly without any difficulty.

We denote the class of words congruent to the word W by $|W|$. The factor semi-group $\mathfrak{F}$ of $\mathfrak{W}(H)$ over the normal multiplicative relation defined above (for its definition, see Chap. II, Ex. 12) formed by the classes of congruent words that are multiplied by multiplication of the representatives, is a product over H. The correspondence of the element h of the member $\mathfrak{h}$ of H with the class $|h|$ represented by the one-letter word h, in fact, defines a homomorphism of the semi-group $\mathfrak{h}$ into $\mathfrak{F}$, since for $hh' = h''$ in $\mathfrak{h}$ it follows that $hh' \to h''$ in $\mathfrak{W}(H)$ and hence $|hh'| = |h''|$, $|h| \cdot |h'| = |h''|$. Furthermore, for each word $W = h_1 h_2 \cdots h_r$ of positive length we have $|W| = |h_1| \cdot |h_2| \cdot |h_3| \cdots |h_r|$. Finally, the empty word Z is congruent to each of the words $1_\mathfrak{h}$, $\mathfrak{h}$ being any member of H, so that $|Z| = |1_\mathfrak{h}|$. Hence the totality of the homomorphic images of $\mathfrak{h}$ in H generates $\mathfrak{F}$.

The product $\mathfrak{F}$ over H is free. To see this, let the semi-group $\mathfrak{S}$ with unit element be an arbitrary product over H for which the given homomorphism of each semi-group $\mathfrak{h}$ contained in H into $\mathfrak{S}$ is $\sigma_\mathfrak{h}$. Note that $\sigma_\mathfrak{h}(1_\mathfrak{h}) = 1_\mathfrak{S}$. Furthermore, the semi-group $\mathfrak{S}$ is generated by the sub-semigroups $\sigma_\mathfrak{h}(\mathfrak{h})$, with $\mathfrak{h}$ a member of H. Let us construct the homomorphism θ of $\mathfrak{F}$ onto $\mathfrak{S}$ over H which maps the class $|h_1 h_2 \cdots h_r|$ onto $\sigma_{\mathfrak{h}_1}(h_1)\sigma_{\mathfrak{h}_2}(h_2) \cdots \sigma_{\mathfrak{h}_r}(h_r)$ and $|Z|$ onto $1_\mathfrak{S}$. This mapping is unique, since if

$$\cdots h_i h_{i+1} \cdots \to \cdots h' \cdots,$$

then

$$\cdots \sigma_{\mathfrak{h}_i}(h_i)\sigma_{\mathfrak{h}_{i+1}}(h_{i+1}) \cdots = \cdots \sigma_{\mathfrak{h}_i}(h') \cdots$$

and if

$$\cdots 1_{\mathfrak{h}_i} \cdots \to \cdots \; \cdots$$

then

$$\cdots \sigma_{\mathfrak{h}_i}(1_{\mathfrak{h}_i}) \cdots = \cdots \; \cdots.$$

The preservation of multiplication follows from considering two words $W_1 = h_1 h_2 \cdots h_r$ and $W_2 = h_{r+1} \cdots h_{r+s}$. We have

$$\Theta(W_1) = \sigma_{\mathfrak{h}_1}(h_1)\sigma_{\mathfrak{h}_2}(h_2) \cdots \sigma_{\mathfrak{h}_r}(h_r),$$
$$\Theta(W_2) = \sigma_{\mathfrak{h}_{r+1}}(h_{r+1})\sigma_{\mathfrak{h}_{r+2}}(h_{r+2}) \cdots \sigma_{\mathfrak{h}_{r+s}}(h_{r+2}),$$
$$\Theta(|W_1||W_2|) = \Theta(|W_1 W_2|) = \sigma_{\mathfrak{h}_1}(h_1) \cdots \sigma_{\mathfrak{h}_{r+s}}(h_{r+s})$$
$$= \Theta(|W_1|)\Theta(|W_2|), \quad \text{Q. E. D.}$$

We denote the free product over H, as constructed in the preceding, by

$$\prod_{\mathfrak{h} \in \mathfrak{h}}^{*} \mathfrak{h}$$

or, more briefly, by

$$\prod^{*} H.$$

If H consists of finitely many semi-groups $\mathfrak{h}_1, \mathfrak{h}_2, \ldots, \mathfrak{h}_r$, then we also write $\mathfrak{h}_1 * \mathfrak{h}_2 * \cdots * \mathfrak{h}_r$ for the free product. It is independent of the order of the factors, and it is associative.

For many applications it is necessary to solve the *word problem* for a given product, that is, to determine a general procedure by which it can be decided in a finite number N of computational steps whether two given words W_1, W_2 interpreted as elements of the product are equal. This is understood to imply that the number N be not greater than a certain recursive function that depends only on the length of W_1 and W_2; we speak in that case of an *effective solution of the word problem.*

It has been shown that an effective solution of the word problem does not always exist. However, in free products we can solve it quite easily. For the most elegant solution, see Ex. 1, Appendix D. A method of solution that lends itself to other applications will now be given.

We call a word *irreducible* if no reduction can be made. Since any reduction diminishes the length by 1, it follows that every word of length r can be reduced to an irreducible word by at most r reductions.

LEMMA: *If* $W \to W_1$, $W \to W_2$, $W_1 \neq W_2$, *then there is a word* W_3 *such that* $W_1 \to W_3$, $W_2 \to W_3$.

Proof: a) If no unit elements are eliminated, then we distinguish two cases.

1.
$$W = \cdots h_i h_{i+1} \cdots h_j h_{j+1} \cdots;$$
$$W \to W_1 = \cdots h' \cdots h_j h_{j+1} \cdots;$$
$$W \to W_2 = \cdots h_i h_{i+1} \cdots h'' \cdots.$$

Set $W_3 = \cdots h' \cdots h'' \cdots$.

2.
$$W = \cdots h_i h_{i+1} h_{i+2} \cdots;$$
$$W \to W_1 = \cdots h' h_{i+2} \cdots;$$
$$W \to W_2 = \cdots h_i h'' \cdots.$$

Thus the semi-groups $\mathfrak{h}_i$, $\mathfrak{h}_{i+1}$, $\mathfrak{h}_{i+2}$ coincide, and within $\mathfrak{h}_i$ the equations $h_i h_{i+1} = h'$, $h_{i+1} h_{i+2} = h''$, $h' h'' = h$ hold. Set $W_3 = \cdots h \cdots$.

b) In the event that unit elements are to be eliminated, we have

1.
$$W = \cdots h_i h_{i+1} \cdots 1_{\mathfrak{h}_j} \cdots;$$
$$W \to W_1 = \cdots h' \cdots 1_{\mathfrak{h}_j} \cdots;$$
$$W \to W_2 = \cdots h_i h_{i+1} \cdots.$$

Set $W_3 = \cdots h' \cdots$.

2.
$$W = \cdots 1_{\mathfrak{h}_j} \cdots 1_{\mathfrak{h}_j} \cdots;$$
$$W_1 = \cdots\cdots 1_{\mathfrak{h}_j} \cdots;$$
$$W_2 = \cdots 1_{\mathfrak{h}_j} \cdots\cdots.$$

Set $W_3 = \cdots\cdots\cdots$.

Note that there is no loss of generality in considering the particular orders we have selected and that in each case $W_1 \to W_3$, $W_2 \to W_3$.

On the basis of the preceding observations, let us investigate the problem of reduction in general.

To every binary relation $a \to b$ in a set S there belongs a poset, where $a \geq b$ signifies the fact that there is a chain $a = a_0, a_1, \ldots, a_r = b$ of length $r \geq 0$ linking a and b such that either (i) $r = 0$ and $a = b$ or (ii) $r > 0$ and $a_0 \to a_1$, $a_1 \to a_2, \ldots, a_{r-1} \to a_r$.

That the $\geq$ relation is reflexive and transitive is immediate. Also, if $a \to b$, then $a \geq b$. The two relations coincide if and only if the relation $\to$ is itself reflexive and transitive.

An element x of a $\geq$ poset is called *minimal* if $x \geq y$ implies $y \geq x$. Any element equivalent to a minimal element is itself minimal.

An element of a set S is called *irreducible* with respect to a given binary relation $a \to b$ if it is minimal in the corresponding poset.

A chain of elements $a = a_0, a_1, \ldots, a_r = b$ of S leading from a to b such that b is irreducible and either $r = 0$ and $a = b$ or $r > 0$, $a_0 \to a_1, \ldots, a_{r-1} \to a_r$ is called a *complete reduction of a to its result b*. The element a is called *completely reducible* if there is a complete reduction of a. Every irreducible element, for example, is completely reducible.

As another example, in our special case of the set of all words over a system of semi-groups with unit element, every word is completely reducible. A word is irreducible if and only if it has no unit letter and adjacent letters belong to different semi-groups.

A *set with reduction* is defined as a set S with a binary relation $a \to b$, where:

1. The poset belonging to the relation $\to$ satisfies the minimal condition: In any monotonic decreasing sequence $a_1 \geq a_2 \geq a_3 \geq \cdots$ there is an index n for which all members $a_n, a_{n+1}, a_{n+2}, \ldots$ of the sequence from the n-th member on are equivalent;

2. (Birkhoff condition.) If $a \to b$, $a \to c$, then there is an element d such that $b \geq d$, $c \geq d$.

For example, the set of all words over a system of semi-groups with unit element in which the $\to$ relation is defined as above is a set with reduction.

For such sets we have the

Principle of Reduction: *Every element is completely reducible, and the result of a complete reduction is uniquely determined up to equivalence.*

Proof of the Principle of Reduction: Owing to the minimal condition for each element a, there is an irreducible element b satisfying $a \geq b$ which can be obtained by a complete reduction from a.

It is convenient to write $a > b$ in place of '$a \geq b$ but not $b \geq a$.' This relation is transitive, but not reflexive.

We form the subset S' of S which consists of all the elements p of S having the property that $p \geq x$, $p \geq y$ in S implies the existence of an element z of S satisfying $x \geq z$, $y \geq z$. All irreducible elements of S, for example, belong to S'. If x, y are the results of two complete reductions of the element p of S', then it follows that there is an element z in S satisfying $x \geq z$, $y \geq z$, and since both x and y are irreducible, we find that x is equivalent to z, z equivalent to y, and thus x equivalent to z. Hence for each element of S' the result of a complete reduction is uniquely determined up to equivalence. We now wish to show that the difference set $S - S'$ is empty.

If a is in S but not in S', then there are two elements x, y of S such that $a \geq x$, $a \geq y$, and for any element z of S satisfying $x \geq z$ we never have $y \geq z$. Since $y \geq y$, it follows that $x \geq y$ cannot hold. Hence $x \geq a$ cannot hold, and thus $a > x$. Similarly, $a > y$. There are chains $a = a_0 \to a_1 \to \to \cdots \to a_r = x$, $a = b_0 \to b_1 \to \cdots \to b_s = y$ and indices i, j satisfying

$0 \leq i < r, 0 \leq j < s$ such that (i) $a_0, a_1, \ldots, a_i$ are equivalent, but $a_i > a_{i+1}$ (ii) $b_0, b_1, \ldots, b_j$ are equivalent, but $b_j > b_{j+1}$. Hence $a > a_{i+1}$, $a > b_{j+1}$. Using the Birkhoff condition, we deduce from $a_i \to a_{i+1}$, a_i equivalent to b_j, and $b_j \to b_{j+1}$ the existence of an element z of S satisfying $a_{i+1} \geq z$, $b_{j+1} \geq z$. Let $\bar{z}$ be the result of a complete reduction of z, and similarly let $\bar{x}$ and $\bar{y}$ be the result of a complete reduction of x and y respectively. Since $a_{i+1} \geq x$, it follows that $\bar{x}$ also is the result of a complete reduction of a_{i+1}. Similarly, we deduce from $a_{i+1} \geq z$ that $\bar{z}$ is the result of a complete reduction of a_{i+1}.

Now, either a_{i+1} does not belong to S', in which case we set $a' = a_{i+1}$, or a_{i+1} belongs to S', in which case the elements $\bar{x}$, $\bar{z}$, being the results of complete reductions of a_{i+1}, must be equivalent. From $x \geq \bar{x}$, $\bar{x}$ equivalent to $\bar{z}$ we have $x \geq \bar{z}$, and hence $y \geq \bar{z}$ does not hold. Using the same argument applied to y, b_{j+1} instead of x, a_{i+1}, we come to the conclusion that b_{j+1} does not belong to S'. In this case we set $a' = b_{j+1}$.

At any rate, for every element a of S not belonging to S' there is a successor a' satisfying $a > a'$ and not belonging to S'. Since repetition of this construction leads to a strictly monotonically decreasing sequence, we find a contradiction with the minimal condition; and hence every element of S belongs to S', Q. E. D.

For a partial converse of the principle of reduction see Ex. 2 of Appendix D.

Corollary to the principle of reduction: *The normalized relation of the relation $\to$ on a set with reduction is the relation:*

'$a \equiv b$ if the complete reduction of a, b leads to equivalent results.'

Proof: If $a \equiv b$, then there are full reductions $a = a_0 \to a_1 \to \cdots \to a_r$, $b = b_0 \to b_1 \to \cdots \to b_s$ such that a_r, b_s are equivalent irreducible elements. Hence there is a chain $a_r \to a_{r+1} \to \cdots \to a_t = b_s$. But from $a \to a_1 \to \cdots \to a_t$, $b \to b_1 \to \cdots \to b_{s-1} \to a_t$ it follows that a, b satisfy the normalized relation of the relation $\to$. Conversely, if a, b satisfy the normalized relation of the relation $\to$, then there is a chain $a = a_0, a_1, \ldots, a_r = b$ such that either $a_i = a_{i+1}$ or $a_i \to a_{i+1}$ or $a_{i+1} \to a_i$ for $i = 0, 1, 2, \ldots, r - 1$. At any rate, complete reduction of a_i and a_{i+1} leads to equivalent results. Hence complete reduction of a and b also leads to equivalent results.

Applying these concepts and conclusions to the relation $\to$ previously studied, we find that poset equivalence of two words is the same as their equality and that, furthermore, the words form a set with reduction. The reduction principle yields: *Each word over a system H of semi-groups with unit element is congruent to the uniquely determined result of any complete* *reduction of the given word. The free product over H may be formed by taking the set of all irreducible words in which the combination of two irreducible words to a third irreducible word is obtained by juxtaposition followed by complete reduction.*

It quite often happens that a product over H is required to satisfy a system $\mathfrak{R}$ of relations between the generating elements of the form

$$h_1 h_2 \cdots h_r = h_{r+1} h_{r+2} \cdots h_s, \tag{1}$$

where h_i is contained in a member $\mathfrak{h}_i$ of H and i runs from 1 to s. We call a semi-group $\mathfrak{S}$ a *product over H defined by the system $\mathfrak{R}$ of defining relations* if:

1. $\mathfrak{S}$ is a product over H, i. e., to each member $\mathfrak{h}$ of H there is assigned a homomorphism $\sigma_{\mathfrak{h}}$ of $\mathfrak{h}$ into $\mathfrak{S}$, mapping $1_{\mathfrak{h}}$ onto $1_{\mathfrak{S}}$, such that $\mathfrak{S}$ is generated by the sub-semigroups $\sigma_{\mathfrak{h}}(\mathfrak{h})$, with $\mathfrak{h}$ running over H,

2. each relation (1) holds in $\mathfrak{S}$, i. e., in $\mathfrak{S}$ there hold all of the equations

$$\sigma_{\mathfrak{h}_1}(h_1)\sigma_{\mathfrak{h}_2}(h_2) \cdots \sigma_{\mathfrak{h}_r}(h_r) = \sigma_{\mathfrak{h}_{r+1}}(h_{r+1}) \cdots \sigma_{\mathfrak{h}_s}(h_s); \tag{2}$$

3. $\mathfrak{S}$ is homomorphic over H with every product over H having the properties 1 and 2.

A product over H defined by $\mathfrak{R}$ can be constructed as follows. Two words are called *congruent modulo* $\mathfrak{R}$ if one word can be obtained from the other word by a combination of

1. a finite number of elementary transformations,

2. a finite number of replacements of $W_1 h_1 h_2 \cdots h_r W_2$ by $W_1 h_{r+1} \cdots h_s W_2$ or of $W_1 h_{r+1} \cdots h_s W_2$ by $W_1 h_1 \cdots h_r W_2$, in accordance with the relations (1) comprising $\mathfrak{R}$.

This congruence relation is normal and multiplicative, and hence the residue classes form a semi-group $\mathfrak{W}(H)/\mathfrak{R}$. Denoting by $W(\mathfrak{R})$ the residue class represented by the word W, we find that the correspondence between h and $\sigma_{\mathfrak{h}}(h) = h(\mathfrak{R})$ defines a homomorphism of the element h of the member $\mathfrak{h}$ of H onto a sub-semigroup $\mathfrak{h}(\mathfrak{R})$ of $\mathfrak{W}(H)/\mathfrak{R}$ such that $\sigma_{\mathfrak{h}}(1_{\mathfrak{h}}) = Z(\mathfrak{R})$ is the unit element of $\mathfrak{W}(H)/\mathfrak{R}$ and such that for each relation (1) we have

$$\begin{aligned}\sigma_{\mathfrak{h}_1}(h_1)\sigma_{\mathfrak{h}_2}(h_2) \cdots \sigma_{\mathfrak{h}_r}(h_r) &= h_1 h_2 \cdots h_r(R) \\ &= h_{r+1} \cdots h_s(R) = \sigma_{\mathfrak{h}_{r+1}}(h_{r+1}) \cdots \sigma_{\mathfrak{h}_s}(h_s).\end{aligned}$$

Hence $\mathfrak{W}(H)/\mathfrak{R}$ is a product over H satisfying the relations comprising $\mathfrak{R}$.

Now let $\mathfrak{S}$ be another product over H satisfying the relations comprising $\mathfrak{R}$; that is, the mapping of any word W onto $\overline{W}$ defines a homomorphism of $\mathfrak{W}(H)$ onto $\mathfrak{S}$ for which $\overline{W}_1 = \overline{W}_2$ whenever W_1 is congruent to W_2 modulo $\mathfrak{R}$. From this definition, which is equivalent to the statement that $\mathfrak{S}$

is a product over H satisfying the relations $\mathfrak{R}$, it follows that the mapping of $W(\mathfrak{R})$ onto $\overline{W}$ defines a homomorphism over H of $\mathfrak{W}(H)/\mathfrak{R}$ onto $\mathfrak{S}$. This shows that $\mathfrak{W}(H)/\mathfrak{R}$ is a product over H defined by $\mathfrak{R}$. It is clear that any two products over H defined by $\mathfrak{R}$ are isomorphic over H.

Let us study some examples.

1. $\mathfrak{R}$ is empty. Then $\mathfrak{W}(H)/\mathfrak{R}$ is the free product over H.

2. $\mathfrak{R}$ consists of relations of the type $h = h'$, with h, h' contained in the same member $\mathfrak{h}$ of H.

The subset $\mathfrak{R}_{\mathfrak{h}}$ of all relations in $\mathfrak{R}$ pertaining to one member $\mathfrak{h}$ of H defines a factor semi-group $\mathfrak{h}/\mathfrak{R}_{\mathfrak{h}}$ of $\mathfrak{h}$. We now show that the free product $\mathfrak{G}$ of all factor semi-groups $\mathfrak{h}/\mathfrak{R}_{\mathfrak{h}}$ is a product over H defined by $\mathfrak{R}$. For a_i contained in the member $\mathfrak{h}_i$ of H $(i = 1, 2, \ldots, t)$ the correspondence between $a_1(\mathfrak{R}_{\mathfrak{h}_1})a_2(\mathfrak{R}_{\mathfrak{h}_2}) \cdots a_t(\mathfrak{R}_{\mathfrak{h}_t})$ and $\bar{a}_1\bar{a}_2 \cdots \bar{a}_t$ defines a homomorphism of $\mathfrak{G}$ over H onto each product $\mathfrak{S}$ over H defined by $\mathfrak{R}$, and certainly all the relations in $\mathfrak{R}$ are satisfied in $\mathfrak{G}$.

3. $\mathfrak{R}$ consists of all relations

$$hh' = h'h$$

with h, h' belonging to different members of H. Often the permutability of any two elements x, y of a multiplicative domain expressed by the equation $xy = yx$ is also denoted by $x \leftrightarrow y$. Thus in our case we have the defining relations

$$h \leftrightarrow h' \tag{4}$$

for any pair of elements belonging to different members of H. The product over H defined by elementwise permutability of different factors is called the *direct product over* H and is denoted by $\overset{\times}{\prod\limits_{\mathfrak{h} \in H}} \mathfrak{h}$ or, more concisely, by $\overset{\times}{\prod} H$. The direct product can be constructed by taking the semi-group $\mathfrak{S}$ of all functions f defined on H with the properties

a) $f(\mathfrak{h})$ is an element of $\mathfrak{h}$;

b) $f(\mathfrak{h}) = 1_{\mathfrak{h}}$ for all but a finite number of members of H;

c) $fg(\mathfrak{h}) = f(\mathfrak{h})g(\mathfrak{h})$.

It is easy to verify that $\mathfrak{S}$ is a semi-group and that the correspondence between the element h of the member $\mathfrak{h}$ of H and the function $\underline{h}$ on H which assumes the value h on $\mathfrak{h}$, but the value $1_{\mathfrak{h}'}$ on all members $\mathfrak{h}'$ of H other than $\mathfrak{h}$, defines a homomorphism of $\mathfrak{h}$ onto a sub-semigroup $\underline{\mathfrak{h}}$ of $\mathfrak{S}$ such that $\mathfrak{S}$ appears as a product over H satisfying all relations in $\mathfrak{R}$. On the other hand, given any word W, we can form its $\mathfrak{h}$-*component* by taking the product in $\mathfrak{h}$ of all the letters of W belonging to $\mathfrak{h}$. If no letter belonging to $\mathfrak{h}$ occurs, then the $\mathfrak{h}$-component is defined to be $1_{\mathfrak{h}}$. The $\mathfrak{h}$-component remains unchanged under all elementary transformations and all substitutions derived from one of the relations in $\mathfrak{R}$. Also, after imposing an order on H, each word over H is congruent modulo $\mathfrak{R}$ to the word made up from the $\mathfrak{h}$-components different from the unit element, with the $\mathfrak{h}$'s concerned following in the same order as they occur in the ordering imposed on H. We may call the word thus constructed the *direct normal form* of the given word. It is uniquely determined by the given word. Two words are congruent modulo $\mathfrak{R}$ if and only if they have the same direct normal form, that is, if they coincide in each component. Of course, all but a finite number of the $\mathfrak{h}$-components are equal to $1_{\mathfrak{h}}$. There are no other restrictions. The $\mathfrak{h}$-component of a product of two words is equal to the product of the $\mathfrak{h}$-components of the factors. Hence there is the homomorphic mapping of $\mathfrak{S}$ onto $\mathfrak{W}(H)/\mathfrak{R}$ over H which maps the function f onto the residue class characterized by having its $\mathfrak{h}$-component equal to $f(\mathfrak{h})$, with $\mathfrak{h}$ running over H. This shows that $\mathfrak{S}$ is a direct product over $\mathfrak{S}$. We verify easily that the new definition of the direct product coincides with the one given in § 1 in the case of a finite number of direct factors.

4. If every member of H is a group, then every product over H is a group. The inverse element of $h_1 h_2 \cdots h_r$ is the element $h_r^{-1} \cdots h_2^{-1} h_1^{-1}$. The word $W^{-1} = h_r^{-1} h_{r-1}^{-1} \cdots h_1^{-1}$ is called the *inverse word* of the word $W = h_1 h_2 \cdots h_r$. The inverse word of the empty word is the empty word. It holds true that $(W^{-1})^{-1} = W$, $(W_1 W_2)^{-1} = W_2^{-1} W_1^{-1}$. By elementary transformations the word $W W^{-1}$ can be carried over into the empty word, for any given word W.

The product over H, a system of groups, defined by a system of relations $\mathfrak{R}$ can be obtained as the factor group of the free product $\mathfrak{F} = \overset{*}{\prod} H$ over the normal subgroup $\mathfrak{N}(\mathfrak{R})$ generated by all quotients (5)

$$h_1 h_2 \cdots h_r h_{s+r}^{-1} h_{s+r-1}^{-1} \cdots h_{r+1}^{-1} \tag{5}$$

derived from (1). The elements of this normal subgroup are often called the *consequence relations* of $\mathfrak{R}$. This name is chosen because, as a consequence of (1), in any product over H in which all of the relations (1) hold, all of the consequence relations become 1. A consequence relation may be characterized as an element of the free product $\mathfrak{F}$ over H which is of the form $W_1 R_1^{a_1} W_1^{-1} W_2 R_2^{a_2} W_2^{-1} \cdots W_r R_r^{a_r} W_r^{-1}$ with a_i either 1 or -1, R_i one of the quotients (5) derived from (1), and W_i an arbitrary element of $\mathfrak{F}$. But since elementary transformations do change the appearance of the elements of F, even though they do not change the elements themselves, it is often extremely difficult to recognize a given element of F as a consequence relation of a given system of defining relations.

5. If each member of H is an infinite cyclic group, say the group $\mathfrak{h}$ generated by the element $x_{\mathfrak{h}}$, then the free product over H is called the *free group with generators* $x_{\mathfrak{h}}$. Its elements can be uniquely represented by expressions $x_{\mathfrak{h}_1}^{a_1} x_{\mathfrak{h}_2}^{a_2} \cdots x_{\mathfrak{h}_r}^{a_r}$ ($a_i \neq 0$, $\mathfrak{h}_i$ a member of H where $\mathfrak{h}_i$ is different from $\mathfrak{h}_{i+1}$ if $i < r$; $i = 1, 2, \ldots, r$; $r = 0, 1, 2, \ldots$), and the product of two such expressions is formed by juxtaposition and the subsequent cancelling of adjacent factors as often as possible, e. g.

$$x_{\mathfrak{h}_1}^3 x_{\mathfrak{h}_2}^5 \cdot x_{\mathfrak{h}_2}^{-3} x_{\mathfrak{h}_3}^{-1} = x_{\mathfrak{h}_1}^3 x_{\mathfrak{h}_2}^2 x_{\mathfrak{h}_3}^{-1}, \quad x_{\mathfrak{h}_1}^{-3} x_{\mathfrak{h}_2}^5 \cdot x_{\mathfrak{h}_2}^{-5} x_{\mathfrak{h}_1}^3 = x_{\mathfrak{h}_1}^{-3} x_{\mathfrak{h}_1}^3 = Z.$$

6. The group $\mathfrak{G}$ given by generators $S_1, S_2, \ldots, S_s$ and defining relations $R_i(S_1, S_2, \ldots, S_s) = 1$ ($i = 1, 2, \ldots, r$) is obtained as the factor group of the free group with the generators $S_1, S_2, \ldots, S_s$ over the normal subgroup generated by the r elements $R_i(S_1, S_2, \ldots, S_s)$. A correspondence between the generators of $\mathfrak{G}$ and some elements $S'_1, S'_2, \ldots, S'_r$ of another group $\mathfrak{J}$ which maps S_i onto S'_i, can be extended to a homomorphism of $\mathfrak{G}$ onto $\mathfrak{J}$ if and only if $R_i(S'_1, S'_2, \ldots, S'_s) = 1$ for $i = 1, 2, \ldots, r$.

The problem of determining a method whereby it can be effectively recognized whether a group $\mathfrak{G}$ generated by finitely many generators $S_1, S_2, \ldots, S_s$ satisfying the finitely many defining relations

$$R_i(S_1, S_2, \ldots, S_s) = 1 \quad (i = 1, 2, \ldots, r) \tag{6}$$

is isomorphic to another group $\mathfrak{h}$ generated by the finitely many elements $U_1, U_2, \ldots, U_u$ and satisfying the finitely many defining relations

$$T_j(U_1, U_2, \ldots, U_u) = 1 \tag{7}$$

is called the *isomorphism problem*.

A necessary and sufficient condition that $\mathfrak{G}$ be isomorphic to $\mathfrak{H}$ is the existence of some words $W_1, W_2, \ldots, W_u$ in $S_1, S_2, \ldots, S_s$ and $X_1, X_2, \ldots, X_s$ in $U_1, U_2, \ldots, U_u$ such that the words

$$T_j(W_1(S_1, \ldots, S_s), \ldots, W_u(S_1, \ldots, S_s)) \quad (j = 1, 2, \ldots, t)$$

and the words

$$S_i^{-1} X_i(W_1(S_1, \ldots, S_s), \ldots, W_u(S_1, \ldots, S_s)) \quad (i = 1, 2, \ldots, s)$$

are consequence relations of the relations (6). In the special case $s = u$, $S_i = U_i$ ($i = 1, 2, \ldots, s$) and $W_i = S_i$, $X_j = U_j$, we call (6) and (7) *equivalent systems of defining relations* if each relation of one system is a consequence relation of the other system, i. e., if they define isomorphic groups with the same set of generators.

For free groups the isomorphism problem is easily solved. Two free groups with the same number of generators are isomorphic. Furthermore, for a free group $\mathfrak{G}$ of s generators $S_1, S_2, \ldots, S_s$ the index of the subgroup by all squares is 2^s, the representatives of the cosets being the elements $S_1^{a_1} S_2^{a_2} \cdots S_s^{a_s}$ with a_i either 0 or 1 for $i = 1, 2, \ldots, s$. For isomorphic groups the index over the subgroup generated by the squares must be the same; hence if $\mathfrak{G}$ is isomorphic with a free group of u generators, then $2^s = 2^u$ if s is finite, and $s = u$ if s is infinite. At any rate, $s = u$.

7. The factor commutator group of a group $\mathfrak{G}$ generated by the elements $S_1, S_2, \ldots, S_s$ with the defining relations (6) is obtained by including among the defining relations the additional relations

$$S_i \leftrightarrow S_k, \text{ i. e., } S_i S_k S_i^{-1} S_k^{-1} = 1 \quad (1 \leq i < k \leq s). \tag{8}$$

Proof: Denote by $\mathfrak{F}_s$ the free group generated by $S_1, S_2, \ldots, S_s$, and let $\mathfrak{N}$ be the normal subgroup formed by the consequence relations of (6); then $\mathfrak{G}$ is isomorphic to $\mathfrak{F}_s/\mathfrak{N}$; hence $\mathfrak{G}/D\mathfrak{G}$ is isomorphic to $(\mathfrak{F}_s/\mathfrak{N})/D(\mathfrak{F}_s/\mathfrak{N}) = (\mathfrak{F}_s/\mathfrak{N})/(D\mathfrak{F}_s\mathfrak{N}/\mathfrak{N})$; and hence $\mathfrak{G}/D\mathfrak{G}$ is also isomorphic to the factor group of $\mathfrak{F}_s$ over the normal subgroup $\mathfrak{N} \cdot D\mathfrak{F}_s$ consisting of the consequence relations of (6) and (8).

We may find an equivalent system of defining relations for $\mathfrak{G}/D\mathfrak{G}$ by permuting the order of the letters in each relation so that the letters are ordered lexicographically; let the group $\mathfrak{G}_n$, for example, be generated by $A_1, A_2, \ldots, A_{n-1}$ subject to the defining relations

$$\begin{gathered} A_1^2 = 1,\ A_2^2 = 1,\ A_3^2 = 1,\ \ldots,\ A_{n-1}^2 = 1, \\ (A_1A_2)^3 = 1,\ (A_2A_3)^3 = 1,\ \ldots,\ (A_{n-2}A_{n-1})^3 = 1, \\ A_i \leftrightarrow A_k \text{ if } 1 \leq i < k - 1 < n - 1. \end{gathered} \tag{9}$$

Then $\mathfrak{G}_n/D\mathfrak{G}_n$ is generated by $A_1, A_2, \ldots, A_{n-1}$, subject to the defining relations

$$\begin{gathered} A_1^2 = A_2^2 = \cdots = A_{n-1}^2 = 1, \\ A_1^3A_2^3 = 1,\ \ldots,\ A_{n-2}^3A_{n-1}^3 = 1, \\ A_i \leftrightarrow A_k \text{ if } 1 \leq i < k < n, \end{gathered} \tag{10}$$

which are equivalent to

$$A_1 = A_2 = \cdots = A_{n-1},\ A_1^2 = 1. \tag{11}$$

Hence the generators $A_2, A_3, \ldots, A_{n-1}$ can be eliminated, and $\mathfrak{G}_n/D\mathfrak{G}_n$ is generated by A_1, subject to the defining relation $A_1^2 = 1$, giving $\mathfrak{G}_n : D\mathfrak{G}_n = 2$.

8. To give another example of the reduction principle, let H be the system of the $n - 1$ infinite cyclic groups (A_1), (A_2), ..., (A_{n-1}). Define a reduction R in $\mathfrak{W}(H)$ as follows:

$$\cdots A_i^n A_i^m \cdots R \cdots A_i^{n+m} \cdots,$$
$$\cdots A_i^{2n+1} \cdots R \cdots A_i \cdots \quad \text{in case } n \neq 0,$$
$$\cdots A_i^{2n} \cdots R \cdots \cdots,$$
$$\cdots A_k A_i \cdots R \cdots A_i A_k \cdots \text{ if } 1 \leq i < k - 1 < n - 1,$$
$$\cdots A_{i+1} A_i A_{i+1} \cdots R \cdots A_i A_{i+1} A_i \cdots \text{ if } 1 \leq i < n - 1.$$

We easily prove that $\mathfrak{W}(H)$ is a set with reduction R. In the corresponding poset, equivalence is equality. For example, if

$$W = \cdots A_{i+1} A_{i+1} A_i A_{i+1} \cdots R W_1 = \cdots A_{i+1}^2 A_i A_{i+1} \cdots$$

and

$$W = \cdots A_{i+1} A_{i+1} A_i A_{i+1} \cdots R W_2 = \cdots A_{i+1} A_i A_{i+1} A_i \cdots,$$

then we have

$$W_1 R \cdots A_i A_{i+1} \cdots, \quad W_2 R \cdots A_i A_{i+1} A_i A_i \cdots R \cdots A_i A_{i+1} \cdots.$$

For every word $a_1 a_2 \cdots a_r$ that is irreducible with respect to R, each section $a_i a_{i+1} \cdots a_j$ is also irreducible with respect to R. The letter A_{n-1} occurs at most once. If it occurs, then the section beginning with A_{n-1} and terminating with a_r is one of the words A_{n-1}, $A_{n-1}A_{n-2}$, $A_{n-1}A_{n-2}A_{n-3}$, ..., $A_{n-1}A_{n-2} \cdots A_1$. Conversely, if A_{n-1} does not occur in the given R-irreducible word, then any of the above $n - 1$ expressions may be affixed to the given word to give another irreducible word. By induction on n, the number of R-irreducible words turns out to be $1 \cdot 2 \cdot \ldots \cdot n = n!$. The normalized relation of the relation R is normal and multiplicative. The corresponding factor group is the group $\mathfrak{G}_n$ occurring in 7. It has the order $n!$.

Since the mapping of A_i onto the transposition $(i, i + 1)$ of the n digits 1, 2, ..., n, which is defined for $i = 1, 2, \ldots, n - 1$, preserves the defining relations of $\mathfrak{G}_n$, it follows that the mapping can be extended to a homomorphism of $\mathfrak{G}_n$ onto $\mathfrak{S}_n$. But since the order of $\mathfrak{G}_n$ is finite and coincides with the order of the symmetric permutation group of n digits $\mathfrak{S}_n$, it follows that $\mathfrak{G}_n$ is isomorphic with $\mathfrak{S}_n$, where we made use of the fact that the transpositions $(1, 2)$, $(2, 3)$, ..., $(n - 1, n)$ generate $\mathfrak{S}_n$. Moreover a set of defining relations for the generators $A_i = (i, i + 1)$ $(i = 1, 2, \ldots, n - 1)$ is given by (9).

APPENDIX D

FURTHER EXERCISES FOR CHAP. III

1. (Van der Waerden, Artin.) Let H be a system of semi-groups with unit element. Let $I(H)$ be the set of all irreducible words over H, i. e. the set of all expressions $a_1 a_2 \cdots a_r$, where a_i is an element of the member $\mathfrak{h}_i$ of H different from the unit element of $\mathfrak{h}_i$ for $i = 1, 2, \ldots, r$, where $\mathfrak{h}_i$ is different from its neighbor $\mathfrak{h}_{i+1}$ for $i = 1, 2, \ldots, r - 1$, and where r ranges over 0 and all the natural numbers. Assign to each element h of a member $\mathfrak{h}$ of H a unique mapping $\underline{h}$ of $I(H)$ into itself defined as follows

$$\underline{h}(a_1 a_2 \cdots a_r) = \begin{cases} h a_1 a_2 \cdots a_r \text{ if } h \neq 1_{\mathfrak{h}},\ \mathfrak{h} \neq \mathfrak{h}_1 \\ a_1' a_2 \cdots a_r \text{ if } \mathfrak{h} = \mathfrak{h}_1,\ h a_1 = a_1' \neq 1_{\mathfrak{h}} \text{ in } \mathfrak{h} \\ a_1 a_2 \cdots a_r \text{ if } h = 1_{\mathfrak{h}} \\ a_2 \cdots a_r \text{ if } \mathfrak{h} = \mathfrak{h}_1,\ h a_1 = 1_{\mathfrak{h}} \text{ in } \mathfrak{h}. \end{cases}$$

Show that

a) $\underline{h}\,\underline{h}' = \underline{hh'}$ if h and h' belong to the same member $\mathfrak{h}$ of H;

b) $\underline{1_{\mathfrak{h}}} = \underline{1}$;

c) if $a_1 a_2 \cdots a_r$ and $b_1 b_2 \cdots b_s$ both belong to $I(H)$, then the equation $\underline{a_1}\,\underline{a_2} \cdots \underline{a_r} = \underline{b_1}\,\underline{b_2} \cdots \underline{b_s}$ implies $r = s$, $a_i = b_i$ for $i = 1, 2, \ldots, r$;

d) by forming all possible product mappings of the mappings $\underline{h}$ defined above, a free product over H is obtained. Each element of this free product is equal to one and only one mapping of the form $\underline{a_1}\,\underline{a_2} \cdots \underline{a_r}$, where $a_1 a_2 \cdots a_r$ is an element of $I(H)$.

2. Let S be a set with a binary relation such that the corresponding poset satisfies the minimal condition. If the result of a complete reduction of any given element a is uniquely determined up to equivalence, then the Birkhoff condition is satisfied.

3. (von Neumann.) Show that for any game with full information between two partners A and B there exists a strategy. To elucidate this statement note that such a game consists essentially of a set S of possible positions, where the term 'position' denotes a set of data concerning the board, pieces, etc., and concerning which of the two players is to move, together with a binary relation $a \to b$ indicating that the player who is to move from position a may move to position b. There are at most three types of irreducible positions: a draw, A wins, B wins. There is a starting position s. The game consists in a finite chain $s = a_0 \to a_1 \to a_2 \to \cdots \to a_r = c$ of moves linking the starting position with an irreducible position c. It is assumed that no infinite sequence $a_0 \to a_1 \to a_2 \to a_3 \to \cdots$ exists. An A-strategy is a list of relations $a \to b$, where a runs over all the positions obtainable from s from which A is to move, such that if A plays in conformity with the list, B never wins. A B-strategy is defined similarly.

4. Let H be a system of groups one of which is $\mathfrak{g}$. To every member $\mathfrak{h}$ of H there may be assigned an isomorphism $\theta_{\mathfrak{h}}$ of $\mathfrak{g}$ into $\mathfrak{h}$ such that $\theta_{\mathfrak{g}} = \underline{1}_{\mathfrak{g}}$, together with a

left representative system $\bar{h}$ of $\mathfrak{h}$ modulo $\theta_{\mathfrak{h}}(\mathfrak{g})$ such that $\bar{\bar{h}} = \bar{h}$ and $h\bar{h}^{-1}$ is contained in $\theta_{\mathfrak{h}}(\mathfrak{g})$ for each element h of $\mathfrak{h}$. Let the binary relation R be defined on $\mathfrak{W}(H)$ as follows:

$$\ldots\ldots hh' \ldots\ldots R \ldots\ldots h'' \ldots\ldots$$

if h, h', h'' belong to the same member $\mathfrak{h}$ of H and if $hh' = h''$ in $\mathfrak{h}$,

$$\ldots\ldots 1_{\mathfrak{h}} \ldots\ldots R \ldots\ldots\ldots\ldots, \ldots\ldots hh' \ldots\ldots R \ldots\ldots h''\bar{h}' \ldots\ldots$$

if h, h' belong to different members $\mathfrak{h}$ and $\mathfrak{h}'$ of H respectively and if

$$h' \neq \bar{h}',\ h'' = h\theta_{\mathfrak{h}}\theta_{\mathfrak{h}'}^{-1}(h'\bar{h}'^{-1}) \text{ in } \mathfrak{h}',$$

$$h \ldots\ldots\ldots\ldots R\theta_{\mathfrak{h}}^{-1}(h\bar{h}^{-1})\bar{h} \ldots\ldots$$

if h belongs to a member $\mathfrak{h}$ of H other than $\mathfrak{g}$. Prove that R defines $\mathfrak{W}(H)$ as a set with reduction and that the corresponding normalized relation is normal and multiplicative and defines a product over H. This product is called the product over H with *identified* (or *amalgamated*) subgroup $\mathfrak{g}$. Solve the word problem for this product.

5. (L. E. Dickson.) Denote by $GL(n, K)$ (which stands for 'general linear group of degree n over K') the group formed by all the units (or: non-singular matrices; or: regular matrices) of the ring of matrices of degree n over the division ring K.

a) Prove that $GL(n, K)$ is generated by the *translation matrices* $T_{ik}^{\lambda} = I_n + \lambda e_{ik}$, where $i \neq k$ and λ is an element of K, and by the *special diagonal matrices* $D_i(\alpha) = I_n + (\alpha - 1)e_{ii}$, where α is an element of K different from 0, where $I_n = (\delta_{ik})$ is the unit matrix of degree n, and where $e_{rs} = (\delta_{ir}\delta_{ks})$, for r, s running independently over $1, 2, \ldots, n$ are the n^2 basic matrix units. (*Hint:* Apply induction on n, using the fact that the subgroup $\mathfrak{U}$ of all matrices (α_{ik}) satisfying $\alpha_{ll} = 1$, $\alpha_{li} = \alpha_{il} = 0$ for $i > 1$ is isomorphic to $GL(n - 1, K)$).

b) Prove that the defining relations of $GL(n, K)$ with respect to the generators given in a) are obtained as follows:

(1) $T_{ik}^{\lambda}T_{ik}^{\mu} = T_{ik}^{\lambda+\mu}$,

(2) $T_{ik}^{\lambda} \leftrightarrow T_{rs}^{\mu}$ if $i = r$, or if $k = s$, or if i, k, r, s are distinct,

(3) $D_i(\alpha)D_i(\beta) = D_i(\alpha\beta)$,

(4) $D_i(\alpha) \leftrightarrow D_k(\beta)$ if $i \neq k$,

(5) $D_i(\alpha)T_{ik}^{\lambda}D_i(\alpha)^{-1} = T_{ik}^{\alpha\lambda}$, $D_k(\alpha)T_{ik}^{\lambda}D_k(\alpha)^{-1} = T_{ik}^{\lambda\alpha^{-1}}$,

(6) if $i < j$, $i < k$, $j \neq k$ then

a) $T_{jk}^{\lambda}T_{ij}^{\mu}T_{jk}^{-\lambda} = T_{ij}^{\mu}T_{ik}^{-\mu\lambda}$

b) $T_{kj}^{\lambda}T_{ji}^{\mu}T_{kj}^{-\lambda} = T_{ji}^{\mu}T_{ki}^{\lambda\mu}$

c) $T_{ij}^{\lambda}T_{ji}^{\mu} = T_{ji}^{\mu(1+\lambda\mu)^{-1}}D_i(1 + \lambda\mu)T_{ij}^{(1+\lambda\mu)^{-1}\lambda}D_j(1-\mu(1+\lambda\mu)^{-1}\lambda)$ if $1 + \lambda\mu \neq 0$,

d) $T_{ji}^{\mu}T_{ij}^{-\lambda^{-1}}T_{ji}^{\lambda} = T_{ij}^{-\lambda^{-1}}T_{ji}^{\lambda}T_{ij}^{-\lambda^{-1}\mu\lambda^{-1}}$ if $\lambda \neq 0$,

e) $T_{ki}^{\mu}T_{ij}^{-\lambda^{-1}}T_{ji}^{\lambda} = T_{ij}^{-\lambda^{-1}}T_{ji}^{\lambda}T_{kj}^{-\mu\lambda}$ if $\lambda \neq 0$.

(*Hint:* Show that the elements $T_{12}^{\alpha_2} T_{13}^{\alpha_3} \cdots T_{1n}^{\alpha_n}$ form a right representative system modulo $\mathfrak{U}$ [for the definition see a)] of the subgroup $\mathfrak{V}$ which is generated by $\mathfrak{U}$ and the elements T_{1i}^{λ}. Show that the elements

$$D_1(\beta_1)\, T_{21}^{\beta_2} T_{31}^{\beta_3} \cdots T_{n1}^{\beta_n}$$

with $\beta_1 \neq 0$ and the elements

$$T_{i+1,i}^{\gamma_{i+1}} \cdots T_{n-1,i}^{\gamma_{n-1}} T_{n,i}^{\gamma_n} T_{li}^{-\gamma_i^{-1}} T_{il}^{\gamma_i}$$

with $\gamma_i \neq 0$ for $i = 2, 3, \ldots, n$ form a right representative system of $GL(n, K)$ over $\mathfrak{V}$; apply induction on n).

c) The commutator group of $GL(n, K)$ is generated by the elements $D_i(\alpha\beta\alpha^{-1}\beta^{-1})$ and T_{ik}^{λ}. It is often denoted by $SL(n, K)$ (which stands for 'special linear group of degree n over K'). The factor commutator group of $GL(n, K)$ is isomorphic with the factor commutator group of the multiplicative group of K.

d) Develop determinant theory. Show that $SL(n, K)$ coincides with the group of all matrices of degree n over K with determinant unity and that it is generated by the elements T_{ik}^{λ} if $n > 1$.

6. Let $\mathfrak{M}$ be a vector module with basis $u_1, u_2, \ldots, u_n$ over the field F. Show that the projectivities of the F-submodule $\mathfrak{m}_j$ of $\mathfrak{M}$ with the basis $u_1, u_2, \ldots, u_j$ $(0 < j < n)$ in $\mathfrak{M}$ coincide with the group of all regular linear transformations of $\mathfrak{m}_j$, where $\mathfrak{M}$ is to be interpreted as an additive group with operators. Furthermore, two factor F-modules formed within $\mathfrak{M}$ are projectively related in $\mathfrak{M}$ if and only if they have the same dimension over F. (*Hint:* Use the preceding exercise.)

7. Let B be a subset of the complete lattice L with normality relation. If there is assigned to each subset X of B a unique mapping δ_X of L into L such that

a) $\delta_X(A)$ is equal to the join of X and is normal in L,

b) the mapping $\delta_{A(B)}$ corresponding to the empty subset of B maps each element of L onto the zero element Z of L,

c) for any two elements x, y of L, $\delta_X(x \mathsf{J} y) = \delta_X(x) \mathsf{J} \delta_X(y)$,

d) $\delta_X(a) \mathsf{J} \delta_Y(a) \geq \delta_{X \mathsf{J} Y}(a)$ for any a of L and any pair of subsets X, Y of B,

e) $\delta_X \delta_Y(a) = \delta_X(a) \mathsf{M} \delta_Y(a) = Z$ for any a of L and any pair of disjoint subsets X, Y of B,

then L is the direct join of B and the δ_X are the decomposition operators.

8. Let B be a subset of the lattice $S(\mathfrak{G})$ formed by the subgroups of the group $\mathfrak{G}$. Show that $S(\mathfrak{G})$ is the direct join of B if and only if $\mathfrak{G}$ is the direct product of the subgroups in B.

9. Let $\mathfrak{u}$, $\mathfrak{U}$, $\mathfrak{v}$, $\mathfrak{V}$ be four normal subgroups of a group $\mathfrak{G}$ such that $\mathfrak{u} \leq \mathfrak{U}$, $\mathfrak{v} \leq \mathfrak{V}$ and $\mathfrak{U}/\mathfrak{u}$ is projectively related with $\mathfrak{V}/\mathfrak{v}$ in the modular lattice of the normal subgroups of $\mathfrak{G}$.

a) Show that every projectivity of $\mathfrak{U}/\mathfrak{u}$ onto $\mathfrak{V}/\mathfrak{v}$ maps $(\mathfrak{G}, \mathfrak{U})\,\mathfrak{u}/\mathfrak{u}$ onto $(\mathfrak{G}, \mathfrak{V})\,\mathfrak{v}/\mathfrak{v}$ and $\mathfrak{z}(\mathfrak{G}/\mathfrak{u})\, M\, \mathfrak{G}/\mathfrak{u}$ onto $\mathfrak{z}(\mathfrak{G}/\mathfrak{v})\, M\, \mathfrak{V}/\mathfrak{v}$.

b) A factor group $\mathfrak{U}/\mathfrak{u}$ is called *central* if $(\mathfrak{G}, \mathfrak{U})$ is contained in $\mathfrak{u}$. Show that every factor group formed between subgroups of the center or between subgroups of $\mathfrak{G}$ containing the commutator group is central. Prove that the central factor groups in a principal series of $\mathfrak{G}$ are uniquely determined up to order and projectivity.

c) If $\mathfrak{G} = \mathfrak{B}_1 \times \mathfrak{B}_2 \times \cdots \times \mathfrak{B}_r$ is a Remak decomposition of the finite group $\mathfrak{G}$, if b_i denotes the modular lattice of the normal subgroups of $\mathfrak{B}_i$, and if B denotes the system of the lattices $b_1, b_2, \ldots, b_r$, then the modular lattice of the normal subgroups of $\mathfrak{G}$ is isomorphic to $DU(B)$ (for the definition, see § 3, Subsection 4) if and only if the order of each central principal factor group of $\mathfrak{B}_i$ is prime to the order of any central principal factor group of $\mathfrak{B}_k$ for $i \neq k$.

10. An element b of the modular lattice L is called *join-irreducible* if it is not the join of two elements different from b. A join representation $A = b_1 \mathsf{J} b_2 \mathsf{J} \cdots \mathsf{J} b_r$ of length r of the all element A of L is called *irreducible* if each constituent b_j is join irreducible and if each complement c_i, which is defined as the join of all b_j with $j \neq i$, is different from A for $i = 1, 2, \ldots, r$.

a) Show that for every join representation the mapping $\delta_i = \varphi_{b_i} \varphi^{c_i}$ of L into L is a normal operator and that $\delta_1(a) \mathsf{J} \delta_2(a) \mathsf{J} \cdots \mathsf{J} \delta_r(a) \geq a$.

b) Theorem of Kurosh: If $A = b_1 \mathsf{J} b_2 \mathsf{J} \cdots \mathsf{J} b_r = \mathsf{J} b'_1 \mathsf{J} b'_2 \mathsf{J} \cdots \mathsf{J} b'_s$ are two irreducible join representations of A, then $r = s$; and after suitable numeration of $c_1, c_2, \ldots, c_s$, there are the irreducible join representations $A = b'_1 \mathsf{J} b'_2 \mathsf{J} \cdots \mathsf{J} b'_k \mathsf{J} b_{k+1} \mathsf{J} \cdots \mathsf{J} b_r$ for $k = 1, 2, \ldots, r-1$. (*Hint:* Use ideas from the proof of Ore's Theorem.)

c) State the dual of Kurosh's Theorem.

11. In a distributive lattice L, the factor lattices b/a, d/c are projectively related if and only if $a \mathsf{J} d = b \mathsf{J} c$, $a \mathsf{M} d = b \mathsf{M} c$. In particular, a/Z is projectively related with b/Z if and only if $a = b$, where Z denotes the zero element of L.

12. Let L be a distributive lattice, and let $F(L)$ be the system of all families of projectively related factor lattices of L.

a) Prove that for any two families f_1, f_2 represented by b/a and d/c respectively, the family $f_1 \mathsf{M} f_2$ represented by $(b \mathsf{M} d)/(a \mathsf{M} d) \mathsf{J} (b \mathsf{M} c)$ is uniquely determined and that $f \mathsf{M} f = f$, $f_1 \mathsf{M} f_2 = f_2 \mathsf{M} f_1$, $(f_1 \mathsf{M} f_2) \mathsf{M} f_3 = f_1 \mathsf{M} (f_2 \mathsf{M} f_3)$.

b) With the same notation as in a), define $f = f_1 \mathsf{J} f_2$ in case the families f, f_1, f_2 can be represented by factor lattices b/a, c/a, b/d, respectively, where $a \leq d \leq c \leq b$. Show that $f \mathsf{J} f = f$, and that $f = f_1 \mathsf{J} (f_2 \mathsf{J} f_3)$ is equivalent to $f = (f_1 \mathsf{J} f_2) \mathsf{J} f_3$. Furthermore, if $f = f_1 \mathsf{J} f_2$, then $f \mathsf{M} g = (f_1 \mathsf{M} g) \mathsf{J} (f_2 \mathsf{M} g)$ for any family g.

c) Show that if $f = f_1 \mathsf{J} f_2 \mathsf{J} \cdots \mathsf{J} f_n$ and if $f' = f_{P1} \mathsf{J} f_{P2} \mathsf{J} \cdots \mathsf{J} f_{Pn}$ is defined for some permutation P of $1, 2, \ldots, n$, then $f = f'$. (*Hint:* Use Ex. 11.)

13. With the same notation as in Ex. 12, construct a lattice L' consisting of all formal unions $\mathsf{J}(f_1, f_2, \ldots, f_r)$ of finitely many families, r being any natural number. The formal union $\mathsf{J}(f_1, f_2, \ldots, f_r)$ is called *equal* to the formal union $\mathsf{J}(g_1, g_2, \ldots, g_s)$ if and only if there are equations

$$f_i = f_{i1} \mathsf{J} f_{i2} \mathsf{J} \cdots \mathsf{J} f_{im_i} \qquad (i = 1, 2, \ldots, r)$$

and

$$g_j = g_{j1} \mathsf{J} g_{j2} \mathsf{J} \cdots \mathsf{J} g_{jn_j} \qquad (j = 1, 2, \ldots, s)$$

such that the f_{ij}'s coincide with the g_{ij}'s up to order and multiplicity. Define the lattice operations as follows:

$$\mathsf{J}(f_1, f_2, \ldots, f_r) \mathsf{J} \mathsf{J}(f_{r+1}, \ldots, f_{r+s}) = \mathsf{J}(f_1, \ldots, f_{r+s})$$

$$\mathsf{J}(f_1, f_2, \ldots, f_r) \mathsf{M} \mathsf{J}(g_1, g_2, \ldots, g_s) = \mathsf{J}(f_1 \mathsf{M} g_1, f_1 \mathsf{M} g_2, \ldots, f_r \mathsf{M} g_s).$$

a) Show that L' is a distributive complemented lattice.

b) Assuming that L has a zero element Z show that the mapping of the element a of L onto the family represented by a/Z is an isomorphism of L onto a sublattice of L'.

14. Let L be a distributive complemented lattice with zero element Z. A non-empty subset S of L may be called a *point* if Z does not belong to S and if the following conditions hold:

a) S is closed under the meet operation;

b) If x and y are any two elements of L for which the join $x \mathsf{J} y$ belongs to S but x does not, then y belongs to S.

c) If x belongs to S and y belongs to L, then $x \mathsf{J} y$ belongs to S.

We say that the point S is 'on' the element a of L if a belongs to the subset S of L. Show that the correspondence between the elements a of L and the set of all points belonging to a is an isomorphism of L onto a sublattice of the lattice formed by the subsets of the set of all points in L, taking union and intersection as lattice operations.

Exx. 11—14 give a proof of the

THEOREM OF STONE: *Every distributive lattice is isomorphic with a sublattice of the lattice formed by all subsets of a certain set, taking union and intersection as lattice operations.*

(*Hint for the solution of* 14: The main difficulty is the construction of a point S on an element $a \neq Z$. Well-order the elements of L, taking a as first element. Define S_a as the set of all elements y of L satisfying $y \geq a$. Define S_b for elements other than a by transfinite induction. Assume S_x already defined for all elements x preceding b in the well ordering of L. Let S'_b be the union of all subsets S_x already defined. If, for some y contained in S'_b, we have $b \mathsf{M} y = Z$, then let $S_b = S'_b$. If, however, $b \mathsf{M} y \neq Z$ for all elements y contained in S'_b, then let S_b be the set of all elements $(b \mathsf{J} z) \mathsf{M} y$ with y contained in S'_b, z contained in L. Define S to be the union of all subsets S_b of L with b running over L.)

15. Show that a Boolean ring (for the definition, see Ex. 33 at the end of Chap. II) becomes a complemented distributive lattice if we define $a \mathsf{M} b = ab$, $a \mathsf{J} b = a + b + ab$. Show that, conversely, a complemented distributive lattice L becomes a Boolean ring if we define $a \mathsf{M} b$ to be ab and $a + b$ to be the complement of $a \mathsf{M} b$ in $a \mathsf{J} b/Z$.

16. (Wielandt.) Show that in a finite group the subgroup generated by two subnormal subgroup $\mathfrak{U}_1$, $\mathfrak{U}_2$ with mutually prime orders is a direct product of $\mathfrak{U}_1$, $\mathfrak{U}_2$. (Apply Theorem 21.)

17. Show that the descending Loewy series of the lattice formed by the normal subgroups of a finite group coincides with the descending Loewy series of the lattice formed by the subnormal subgroups.

18. (Wielandt.) A group with precisely one maximal proper normal subgroup is called *one-headed*. Show that, in finite groups,

a) A one-headed group coincides with its commutator group if and only if the factor group over its maximal normal subgroup is not abelian;

b) If $\mathfrak{A}$ is a one-headed subnormal subgroup of a finite group and $\mathfrak{B}$ an arbitrary subnormal subgroup, then at least one of the following is true: $\mathfrak{A}$ is contained in $\mathfrak{B}$, or $D\mathfrak{A}$ is a proper subgroup of $\mathfrak{A}$, or $D\mathfrak{A}$ coincides with $\mathfrak{A}$ and is normal in the subgroup generated by $\mathfrak{A}$ and $\mathfrak{B}$. (Apply Theorem 21.)

19. (Wielandt.) Let $\mathfrak{G}$ be a group with a composition series. The r subnormal subgroups $\mathfrak{U}_1, \mathfrak{U}_2, \ldots, \mathfrak{U}_r$ different from 1 determine a *Wielandt decomposition* of $\mathfrak{G}$ if

1. $\mathfrak{G}$ is generated by $\mathfrak{U}_1, \mathfrak{U}_2, \ldots, \mathfrak{U}_r$;
2. For any system of subnormal subgroups $\mathfrak{V}_i$ of $\mathfrak{U}_i (i = 1, 2, \ldots, r)$, the group generated by $\mathfrak{V}_1, \mathfrak{V}_2, \ldots, \mathfrak{V}_r$ coincides with $\mathfrak{G}$ if and only if $\mathfrak{V}_i = \mathfrak{U}_i$.

Show that

a) Any Wielandt constituent $\mathfrak{U}_i$ is one-headed, and the intersection of $\mathfrak{U}_i$ with the first member of the descending Loewy series is the maximal proper normal subgroup of $\mathfrak{U}_i$;

b) The factor group of $\mathfrak{G}$ over the first member $\Lambda_1(\mathfrak{G})$ has the Remak decomposition

$$\mathfrak{G}/\Lambda_1 = \mathfrak{U}_1\Lambda_1/\Lambda_1 \times \cdots \times \mathfrak{U}_r\Lambda_1/\Lambda_1;$$

c) Conversely, for any Remak decomposition $\mathfrak{G}/\Lambda_1 = \overline{\mathfrak{U}}_1 \times \overline{\mathfrak{U}}_2 \times \cdots \times \overline{\mathfrak{U}}_r$ there is a Wielandt decomposition $\mathfrak{G} = \langle \mathfrak{U}_1, \mathfrak{U}_2, \ldots, \mathfrak{U}_r \rangle$ such that $\overline{\mathfrak{U}}_i = \mathfrak{U}_i\Lambda_1/\Lambda_1$;

d) If in c) the Remak constituent $\overline{\mathfrak{U}}_i$ is non-abelian, then $\mathfrak{U}_i$ is uniquely determined and normal in $\mathfrak{G}$. (Apply Theorem 21.)

20. Let $\mathfrak{U}/\mathfrak{u}$ and $\mathfrak{V}/\mathfrak{v}$ be two non-abelian composition factor groups of a group $\mathfrak{G}$ with a composition series which are projectively related in the lattice formed by the subnormal subgroups of $\mathfrak{G}$. Show that both factor groups are projective with $(\mathfrak{U} \mathsf{M} \mathfrak{V})/(\mathfrak{u} \mathsf{M} \mathfrak{V})(\mathfrak{U} \mathsf{M} \mathfrak{v})$. Hence either they are identical or they cannot belong to the same composition series of $\mathfrak{G}$. (*Hint:* Show that 1. there is a chain $\mathfrak{U}/\mathfrak{u} = \mathfrak{U}_0/\mathfrak{u}_0, \mathfrak{U}_1/\mathfrak{u}_1, \ldots, \mathfrak{U}_m/\mathfrak{u}_m = \mathfrak{V}/\mathfrak{v}$ such that $\mathfrak{U}_i$ is subnormal in $\mathfrak{G}$, $\mathfrak{u}_i$ is normal in $\mathfrak{U}_i$ and either (i) $\mathfrak{U}_{i-1}$ is normal in $\mathfrak{U}_i$, (ii) $\mathfrak{U}_i/\mathfrak{U}_{i-1}$ is simple and $\neq 1$, and (iii) $\mathfrak{U}_i = \mathfrak{U}_{i-1}\mathfrak{u}_i$, or (i′) $\mathfrak{U}_i$ is normal in $\mathfrak{U}_{i-1}$, (ii′) $\mathfrak{U}_{i-1}/\mathfrak{U}_i$ is simple and $\neq 1$, and (iii′) $\mathfrak{U}_{i-1} = \mathfrak{u}_{i-1}\mathfrak{U}_i$ $(i = 1, 2, \ldots, m)$, 2. in case $\mathfrak{U}_{i-1}$ is normal in $\mathfrak{U}_i$, $\mathfrak{U}_{i+1}$ is normal in $\mathfrak{U}_i$, and $\mathfrak{U}_{i+1}$ is not $\mathfrak{U}_{i-1}$, then we can replace the factor group $\mathfrak{U}_i/\mathfrak{u}_i$ by the factor group $(\mathfrak{U}_{i-1} \mathsf{M} \mathfrak{U}_{i+1})/(\mathfrak{u}_{i-1} \mathsf{M} \mathfrak{u}_{i+1})$.

21. Using the preceding exercise, show that in the lattice of all subnormal subgroups of a group with composition series, every projectivity of a semi-simple factor group between two subnormal subgroups of the given group is the identity automorphism.

22. Every infinite semi-group with one generator is isomorphic to the additive semi-group of the natural numbers. Each finite semi-group generated by one element is defined by one relation, viz., $a^{N+1} = a^{N+1-d}$, where N and d are natural numbers such that d is smaller than N. The semi-group defined by such a relation consists of the N elements $a, a^2, \ldots, a^N$ with the rule of multiplication $a^i a^j = a^{r(i+j)}$, where $1 \leq r(x) = x - q(x)d \leq N < r(x) + d$. There is just one idempotent. The multiples of this idempotent form a subgroup of order d.

23. Associate with each pair of sequences $a = (a_0, a_1, \ldots, a_r)$, $b = (b_0, b_1, \ldots, b_r)$ of integers for which $0 \leq a_0 < a_1 < \cdots < a_r$, $0 \leq b_0 < b_1 < \cdots < b_r$, a basis element $u(a, b)$ of a module $\mathfrak{M}$ over the rational integer ring $(r = 0, 1, 2, \ldots)$. Define

$$u(\pi(a), \pi'(b)) = \operatorname{sign} \pi \cdot \operatorname{sign} \pi' \cdot u(a, b)$$

for any permutations π of $a_0, a_1, \ldots, a_r$ and π' of $b_0, b_1, \ldots, b_r$ that map 0 onto 0 in case $a_0 = 0$ or $b_0 = 0$. For all other pairs of sequences of non-negative integers set $u(a, b) = 0$. Define the linear operators x_i, y_i of $\mathfrak{M}$ for $i = 1, 2, \ldots$ by setting

$$x_i u(a, b) = (1 - \operatorname{sign} a_0) u((i, a_1, \ldots, a_r), b)$$

$$y_i u(a, b) = (1 - \operatorname{sign} b_0) u(a, (i, b_1, \ldots, b_r)) - u((0, a_1, \ldots, a_r, a_0), (b_0, b_1, \ldots, b_r, i)),$$

where $u(a, b)$ runs over the basis of $\mathfrak{M}$. Extend the mapping.

Show that

a) $x_i x_k = y_i y_k = x_i c_{ik} = y_k c_{ik} = 0$, $x_j c_{ik} = c_{ik} x_j$, $y_j c_{ik} = c_{ik} y_j$, where $c_{ik} = x_i y_k - y_k x_i$;

b) $c_{11} c_{22} c_{33} \cdots c_{rr} \neq 0$ for $r = 1, 2, \ldots$;

c) The group $\mathfrak{A}$ generated by the automorphisms $1 + x_i = u_i$ and $1 + y_i = v_i$ $(i = 1, 2, \ldots)$ of $\mathfrak{M}$ is not subinvariant in the holomorph $\mathfrak{G}$ of $\mathfrak{A}$ over $\mathfrak{M}$, but the subgroups $\mathfrak{U} = \langle u_1, u_2, \ldots, \rangle$, $\mathfrak{B} = \langle v_1, v_2, \ldots \rangle$ are subinvariant in $\mathfrak{G}$ and generate $\mathfrak{A}$.

24. Two subinvariant subgroups $\mathfrak{A}$, $\mathfrak{B}$ of a group $\mathfrak{G}$ generate a subinvariant subgroup if $m(\mathfrak{A}, \mathfrak{G}) \leq 2$.

(Exercises 25—31 are due to Prof. J. Lambek and Dr. J. Riguet.)

25. Let Σ be a system of sets $\mathfrak{A}$, $\mathfrak{B}$, ... We consider the binary relations ${}_{\mathfrak{A}}R_{\mathfrak{B}}$ on Σ which are defined for any ordered pair of members $\mathfrak{A}$, $\mathfrak{B}$ of Σ as a subset $|{}_{\mathfrak{A}}R_{\mathfrak{B}}|$ of the set $\mathfrak{A} \times \mathfrak{B}$ of all ordered pairs $a \times b$ formed from an element a of $\mathfrak{A}$ and an element b of $\mathfrak{B}$. Thus, ${}_{\mathfrak{A}}R_{\mathfrak{B}}(a \times b)$ means that $a \times b$ belongs to $|{}_{\mathfrak{A}}R_{\mathfrak{B}}|$.

Show that

a) The binary relations on Σ form a poset if we define ${}_{\mathfrak{A}}R_{\mathfrak{B}} \leqq {}_{\mathfrak{C}}S_{\mathfrak{D}}$ to mean $|{}_{\mathfrak{A}}R_{\mathfrak{B}}| \leqq |{}_{\mathfrak{C}}S_{\mathfrak{D}}|$;

b) The multiplication that is defined by

$$ {}_{\mathfrak{A}}R_{\mathfrak{B}} \cdot {}_{\mathfrak{C}}S_{\mathfrak{D}} = {}_{\mathfrak{A}}T_{\mathfrak{D}} $$

where ${}_{\mathfrak{A}}T_{\mathfrak{D}}(a \times d)$ means the existence of an element b of $\mathfrak{B} \cap \mathfrak{C}$ such that ${}_{\mathfrak{A}}R_{\mathfrak{B}}(a \times b)$ and ${}_{\mathfrak{C}}S_{\mathfrak{D}}(b \times d)$ is uniquely defined and associative;

c) The identity ${}_{\mathfrak{A}}I_{\mathfrak{A}}$ that is defined by

$$ {}_{\mathfrak{A}}I_{\mathfrak{A}}(a \times b) $$

if and only if $a = b$ and a belongs to $\mathfrak{A}$ satisfies the rule

$$ {}_{\mathfrak{A}}I_{\mathfrak{A}} \cdot {}_{\mathfrak{A}}R_{\mathfrak{B}} = {}_{\mathfrak{A}}R_{\mathfrak{B}} \cdot {}_{\mathfrak{B}}I_{\mathfrak{B}}; $$

d) The *converse relation* ${}_{\mathfrak{B}}R_{\mathfrak{A}}^{-}$ of ${}_{\mathfrak{A}}R_{\mathfrak{B}}$ that is defined by

$$ {}_{\mathfrak{B}}R_{\mathfrak{A}}^{-}(b \times a) \text{ if and only if } {}_{\mathfrak{A}}R_{\mathfrak{B}}(a \times b) $$

is uniquely defined and satisfies the rules

$$ (R^{-})^{-} = R, \quad (RS)^{-} = S^{-}R^{-}, \quad {}_{\mathfrak{A}}I_{\mathfrak{A}}^{-} = {}_{\mathfrak{A}}I_{\mathfrak{A}}; $$

e) The *symmetry* of the relation ${}_{\mathfrak{A}}R_{\mathfrak{B}}$, i. e., the equation ${}_{\mathfrak{A}}R_{\mathfrak{B}} = {}_{\mathfrak{B}}R_{\mathfrak{A}}^{-}$ is already implied by the inequality ${}_{\mathfrak{B}}R_{\mathfrak{A}}^{-} \leq {}_{\mathfrak{A}}R_{\mathfrak{B}}$;

f) The *transitivity* of R, i. e., the inequality $R \cdot R \leqq R$, and the symmetry of R imply the equation $R \cdot R = R$.

26. If we think of ${}_{\mathfrak{A}}R_{\mathfrak{B}}$ as a many-valued mapping of a part of $\mathfrak{B}$ into $\mathfrak{A}$, i. e., if $R(a \times b)$ means that b is mapped onto a,

$R \cdot R^{-} \leqq {}_{\mathfrak{A}}I_{\mathfrak{A}}$ means that R is onto,

$R^{-} \cdot R \geq {}_{\mathfrak{B}}I_{\mathfrak{B}}$ means that R is universally defined,

$R \cdot R^{-} \leq {}_{\mathfrak{A}}I_{\mathfrak{A}}$ means that R is single-valued,

$R^{-} \cdot R \leq {}_{\mathfrak{B}}I_{\mathfrak{B}}$ means that R is one to one.

27. The congruence relation ${}_{\mathfrak{A}}R_{\mathfrak{A}}$ on $\mathfrak{A}$ is equal to a normal congruence relation on a subset of $\mathfrak{A}$ if and only if it is symmetric and transitive. The subset in question coincides with $\mathfrak{A}$ if and only if ${}_{\mathfrak{A}}I_{\mathfrak{A}} \leq R$ (*reflexivity* of R).

28. The binary relation ${}_{\mathfrak{A}}R_{\mathfrak{B}} = R$ is called *regular* if $RR^{-}R = R$. Show that

a) Regularity is already implied by the inequality $RR^{-}R \leq R$;

b) If R is regular, then $R \cdot R^{-}$ is equal to a normal congruence relation S on a subset $||S||$ of $\mathfrak{A}$ and $R^{-} \cdot R$ is equal to a normal congruence relation T on a subset $||T||$ of $\mathfrak{B}$ (use 27.);

c) There is, moreover, the natural one-to-one correspondence between $||S||/S$ and $||T||/T$ that maps the S-residue class represented by a onto the T-residue class represented by b, where ${}_{\mathfrak{A}}R_{\mathfrak{B}}(a \times b)$.

29. Let Σ be a system of groups having their unit element in common. The relation ${}_{\mathfrak{A}}R_{\mathfrak{B}}$ is called a *morphism* if the set $|{}_{\mathfrak{A}}R_{\mathfrak{B}}|$ is a subgroup of $\mathfrak{A} \times \mathfrak{B}$.

Show that

a) a relation $R = {}_{\mathfrak{A}}R_{\mathfrak{B}}$ is a morphism if and only if

$$R(1_{\mathfrak{A}} \times 1_{\mathfrak{B}}),$$
$$R(a \times b) \text{ implies } R(a^{-1} \times b^{-1}),$$
$$R(a \times b) \text{ and } R(c \times d) \text{ imply } R((ac) \times (bd));$$

b) The morphisms of Σ form a multiplicative semi-group containing all the identities ${}_{\mathfrak{A}}I_{\mathfrak{A}}$ ($\mathfrak{A}$ in Σ), and the morphism ${}_{\mathfrak{A}}R_{\mathfrak{A}}$ is a multiplicative normal congruence on a subgroup $||R||$ of $\mathfrak{A}$ if and only if ${}_{\mathfrak{A}}R_{\mathfrak{A}}$ is symmetric and transitive. In this case R is equal to the congruence relation of $||R||$ modulo the normal subgroup formed by the elements of $\mathfrak{A}$ that are R-congruent to the unit element of $\mathfrak{A}$. The corresponding subfactor group is uniquely determined by R and may be denoted by $\underline{R}$;

d) Every morphism is regular, and the one-to-one correspondence occurring in 28c) is an isomorphism.

30. (Goursat-Lambek.) For any subgroup $\mathfrak{U}$ of $\mathfrak{A} \times \mathfrak{B}$ the subgroup $\mathfrak{N}_{\mathfrak{A}} = \mathfrak{U} \frown \mathfrak{A}$ is normal in the $\mathfrak{A}$-component $\mathfrak{U}_{\mathfrak{A}} = \mathfrak{U}\,\mathfrak{B} \frown \mathfrak{A}$ and $\mathfrak{N}_{\mathfrak{B}} = \mathfrak{U} \frown \mathfrak{B}$ is normal in the $\mathfrak{B}$-component $\mathfrak{U}_{\mathfrak{B}} = \mathfrak{U}\,\mathfrak{B} \frown \mathfrak{B}$ of $\mathfrak{U}$ and are such that the mapping σ that maps the residue class $a\,\mathfrak{N}_{\mathfrak{A}}$ onto the residue class $b\,\mathfrak{N}_{\mathfrak{B}}$ (where $a \times b$ belongs to $\mathfrak{U}$) is an isomorphism between $\mathfrak{U}_{\mathfrak{A}}/\mathfrak{N}_{\mathfrak{A}}$ and $\mathfrak{U}_{\mathfrak{B}}/\mathfrak{N}_{\mathfrak{B}}$ (use 28. and 29.).

Conversely, if $\mathfrak{N}_{\mathfrak{A}} \lhd \mathfrak{U}_{\mathfrak{A}} \leq \mathfrak{A}$, $\mathfrak{N}_{\mathfrak{B}} \lhd \mathfrak{U}_{\mathfrak{B}} \leq \mathfrak{B}$ and if σ is a given isomorphism between $\mathfrak{U}_{\mathfrak{A}}/\mathfrak{N}_{\mathfrak{A}}$ and $\mathfrak{U}_{\mathfrak{B}}/\mathfrak{N}_{\mathfrak{B}}$, then the set $\mathfrak{U}$ of all the elements $a \times b$ of $\mathfrak{U}_{\mathfrak{A}} \times \mathfrak{U}_{\mathfrak{B}}$ for which $\sigma(a\,\mathfrak{N}_{\mathfrak{A}}) = b\,\mathfrak{N}_{\mathfrak{B}}$ is a subgroup of $\mathfrak{A} \times \mathfrak{B}$ such that $\mathfrak{U}_{\mathfrak{A}}$, $\mathfrak{N}_{\mathfrak{A}}$, $\mathfrak{U}_{\mathfrak{B}}$, $\mathfrak{N}_{\mathfrak{B}}$, σ are related to $\mathfrak{U}$ as was defined above (apply 29a)).

31. Prove the Lemma on Four Groups in Chap. II, § 5 by application of 28c) and 29a), d) to the morphism $R \cdot S$, where R and S are the binary relations on the group $\mathfrak{G}$ defined by the subfactorgroups $\mathfrak{U}/\mathfrak{u}$, $\mathfrak{B}/\mathfrak{v}$, respectively. Note that the isomorphism between $\underline{RSR} = \mathfrak{u}(\mathfrak{U} \frown \mathfrak{B})/\mathfrak{u}(\mathfrak{U} \frown \mathfrak{v})$ and $\underline{SRS} = \mathfrak{v}(\mathfrak{B} \frown \mathfrak{U})/\mathfrak{v}(\mathfrak{B} \frown \mathfrak{u})$ obtained according to 28c) is the same as the one obtained in the previous proof.

APPENDIX E

FURTHER EXERCISES FOR CHAP. IV, § 5

8. (Fitting.) Using (12) and (13) of Chap. II, § 6, prove that the inequality

$$\mathfrak{Z}_i(\mathfrak{A}\,\mathfrak{B}) \leq (\mathfrak{A} \cap \mathfrak{Z}_{i-1}(\mathfrak{B}))(\mathfrak{Z}_2(\mathfrak{A}) \cap \mathfrak{Z}_{i-2}(\mathfrak{B})) \cdots (\mathfrak{Z}_{i-1}(\mathfrak{A}) \cap \mathfrak{B})$$

holds for any two normal subgroups $\mathfrak{A}$ and $\mathfrak{B}$ of a group, and hence that the product of two nilpotent normal subgroups is nilpotent.

9. (Schenkman.) The element g of the group $\mathfrak{G}$ is called *weakly central* if, after a finite number of applications, the mapping of an arbitrary element x of $\mathfrak{G}$ onto the commutator $(g, x) = gxg^{-1}x^{-1}$ for any particular element x of $\mathfrak{G}$ always leads to the identity.

Show that

a) If g belongs to a nilpotent normal subgroup, then g is weakly central;

b) If for a prime p every element of p-power order of a finite group is weakly central, then the intersection of two different p-Sylow subgroups is always the identity. (Apply induction on the order of $\mathfrak{G}$, and use Theorem 7 of Chap. IV.)

c) If in a finite group the intersection of the p-Sylow subgroup S with any different p-Sylow subgroup is the identity, then the commutator of an element of S and an arbitrary element a of $\mathfrak{G}$ is in the normalizer of S if and only if a belongs to the normalizer of S;

d) If in a finite group the elements of p-power order are weakly central, then they form a normal subgroup;

e) A finite group is nilpotent if and only if every element is weakly central.

10. (Schenkman.) The *nilradical* of an arbitrary group $\mathfrak{G}$ consists of all elements g of $\mathfrak{G}$ with the property that for any weakly central element x of $\mathfrak{G}$, gx and $g^{-1}x$ are also weakly central. (For the definition of weakly central, see the preceding exercise.) Show that

a) The nilradical is a characteristic subgroup of $\mathfrak{G}$;

b) If $\mathfrak{A}$ is an abelian normal subgroup of the group $\mathfrak{G}$, g an element of the centralizer of $\mathfrak{A}$, x an element of $\mathfrak{G}$, and a an element of $\mathfrak{A}$, then $(gx, a) = (x, a)$;

c) Every nilpotent normal subgroup $\mathfrak{N}$ of $\mathfrak{G}$ belongs to the nilradical of $\mathfrak{G}$. (Apply induction on the class of $\mathfrak{N}$, and use b).)

d) The nilradical of a finite group is its maximal normal nilpotent subgroup. (Apply 9) and 10c).)

e) The nilradical of a finite group consists of all the elements that generate a subnormal subgroup of the full group.

11. Let $\mathfrak{N}$ be a normal subgroup of the group $\mathfrak{G}$, and let $\mathfrak{U}$, $\mathfrak{V}$ be two subgroups of $\mathfrak{G}$ satisfying

$$\mathfrak{U} \cap \mathfrak{N} = \mathfrak{V} \cap \mathfrak{N}, \quad \mathfrak{U}\mathfrak{N} = \mathfrak{V}\mathfrak{N}, \quad \mathfrak{U} \leq \mathfrak{V}.$$

Show that

$$\mathfrak{U} = \mathfrak{V}.$$

12. a) A group satisfies the maximal condition for subgroups if and only if every subgroup of the group is finitely generated.

b) If a group satisfies the maximal condition for subgroups, then so does every subgroup and every factor group.

c) If both a normal subgroup $\mathfrak{N}$ and the factor group of the group $\mathfrak{G}$ over $\mathfrak{N}$ satisfy the maximal condition for subgroups, then $\mathfrak{G}$ does so also. (Use 11.)

13. a) Every cyclic group and every finite group satisfies the maximal condition for subgroups.

b) If there exists a normal chain

$$\mathfrak{G} = \mathfrak{N}_0 \rhd \mathfrak{N}_1 \rhd \mathfrak{N}_2 \rhd \cdots \rhd \mathfrak{N}_s = 1$$

of a group $\mathfrak{G}$ such that each factor group $\mathfrak{N}_i/\mathfrak{N}_{i+1}$ either is finite or infinitely cyclic, then $\mathfrak{G}$ satisfies the maximal condition for subgroups. (Use 13a) and 12c).)

c) A finitely generated abelian group always satisfies the maximal condition for subgroups.

14. a) If the group $\mathfrak{G}$ is generated by the complex $\mathfrak{K}$ and if $\mathfrak{G}$ is nilpotent of class $c + 1$, then the subgroup $\mathfrak{U}_i$ generated by all higher commutators $(K_1, K_2, \ldots, K_i)$ of weight i in the components $K_1, K_2, \ldots, K_i$ each of which runs independently over $\mathfrak{K}$, coincides with $\mathfrak{Z}_i(\mathfrak{G})$. (Clearly $\mathfrak{U}_i \subseteq \mathfrak{Z}_i(\mathfrak{G})$, $(\mathfrak{K}, \mathfrak{U}_i) \subseteq \mathfrak{U}_{i+1}$. Since $\mathfrak{U}_c \subseteq \mathfrak{Z}_c(\mathfrak{G}) \subseteq \mathfrak{z}(\mathfrak{G})$ it follows that $(\mathfrak{G}, \mathfrak{U}_{c-1}) = (\langle\mathfrak{K}\rangle, \mathfrak{U}_{c-1}) \subseteq \mathfrak{U}_c$. By induction, $(\mathfrak{G}, \mathfrak{U}_i) \subseteq \mathfrak{U}_{i+1}$; hence $\mathfrak{U}_i \supseteq \mathfrak{Z}_i(\mathfrak{G})$.)

b) If the group $\mathfrak{G}$ is finitely generated and nilpotent, then all members of the descending central series are finitely generated. (Use a).)

c) (Jennings.) If the group $\mathfrak{G}$ is finitely generated and nilpotent, then it satisfies the maximal condition for subgroups. (Use 14b), 13c), and 12c).)

15. Let P be a group property. A group $\mathfrak{G}$ is said to be a *local P-group* if every finitely generated subgroup of $\mathfrak{G}$ has the property P. Show that

a) Every subgroup of a local P-group is a local P-group;

b) Every local P-subgroup of a group $\mathfrak{G}$ can be embedded into a maximal local P-subgroup of $\mathfrak{G}$.

16. a) Let $\mathfrak{K}$, $\mathfrak{L}$ be two non-empty complexes of a group $\mathfrak{G}$; let $\mathfrak{U} = \langle(\mathfrak{K}, \mathfrak{L}), \mathfrak{L}\rangle$; let $\mathfrak{V}$ be the smallest normal subgroup of $\mathfrak{U}$ containing $(\mathfrak{K}, \mathfrak{L})$; and let $\mathfrak{W}$ be the smallest normal subgroup of $\langle\mathfrak{K}, \mathfrak{L}\rangle$ that contains $(\mathfrak{K}, \mathfrak{L})$. Show that

$$\langle\mathfrak{V}, \mathfrak{K}\rangle \lhd \langle\mathfrak{V}, \mathfrak{K}, \mathfrak{L}\rangle$$

and hence that

$$\mathfrak{V} \subseteq \mathfrak{W} \subseteq \langle\mathfrak{V}, \mathfrak{K}\rangle.$$

Show that

$$\langle\mathfrak{W}, \mathfrak{K}\rangle \lhd \langle\mathfrak{K}, \mathfrak{L}\rangle, \ \langle\mathfrak{W}, \mathfrak{L}\rangle \lhd \langle\mathfrak{K}, \mathfrak{L}\rangle.$$

b) If $\mathfrak{K}$ is a finite subset of a locally nilpotent normal subgroup $\mathfrak{A}$ of $\mathfrak{G}$ and if $\mathfrak{L}$ is a finite subset of a locally nilpotent normal subgroup $\mathfrak{B}$ of $\mathfrak{G}$, then the subgroups $\mathfrak{U}$, $\mathfrak{V}$, $\mathfrak{W}$ defined under a) are finitely generated and nilpotent. (Use 14c).) Moreover $\langle\mathfrak{W}, \mathfrak{K}\rangle$, $\langle\mathfrak{W}, \mathfrak{L}\rangle$, and $\langle\mathfrak{K}, \mathfrak{L}\rangle$ are nilpotent. (Use Ex. 8.)

c) (Hirsch.) Any two locally nilpotent normal subgroups of a group generate a locally nilpotent normal subgroup. (Use 16b).)

d) Every group contains one and only one maximal normal locally nilpotent subgroup, which may be called its *Hirsch radical*.

e) Show that the Hirsch radical of a group is contained in the nilradical defined in Ex. 10.

f) The Hirsch radical of a finite group (more generally, of a group satisfying the maximal condition for subgroups) is its maximal nilpotent normal subgroup.

17. a) The normalizer $N_{\mathfrak{U}}$ of a maximal locally nilpotent subgroup $\mathfrak{U}$ of a group $\mathfrak{G}$ is its own normalizer. (Apply 16c).)

b) (Plotkin.) A group $\mathfrak{G}$ in which every proper subgroup is properly contained in its normalizer is locally nilpotent. (*Hint:* According to a) every maximal locally nilpotent subgroup of $\mathfrak{G}$ is contained in the Hirsch radical of $\mathfrak{G}$. Now apply 15b).)

18. For any set L and any ring $\mathfrak{o}$ the ring $M(\mathfrak{o}, L)$ of row-finite matrices over L with coefficients in $\mathfrak{o}$ is defined as the set of all formal sums $(\lambda_{ik}) = \Sigma\Sigma\lambda_{ik}e_{ik}$, where to each pair of elements i, k of L we have assigned a 'matrix unit' e_{ik} and where the coefficients λ_{ik} that belong to $\mathfrak{o}$ are such that for any fixed row index i all but a finite number of the coefficients λ_{ik} vanish. Define addition, multiplication, and multiplication by scalars according to the rules:

$$\Sigma\Sigma\lambda_{ik}e_{ik} + \Sigma\Sigma\mu_{ik}e_{ik} = \Sigma\Sigma(\lambda_{ik} + \mu_{ik})e_{ik}$$

$$\Sigma\Sigma\lambda_{ik}e_{ik} \cdot \Sigma\Sigma\mu_{ik}e_{ik} = \Sigma\Sigma(\sum_j \lambda_{ij}\mu_{jk})e_{ik}$$

$$\lambda\cdot \Sigma\Sigma\lambda_{ik}e_{ik} = \Sigma\Sigma\lambda\lambda_{ik}e_{ik}$$

$$\Sigma\Sigma\lambda_{ik}e_{ik} \cdot \lambda = \Sigma\Sigma\lambda_{ik}\lambda e_{ik}.$$

a) Prove that $M(\mathfrak{o}, L)$ is an $\mathfrak{o}$-ring. If $\mathfrak{o}$ contains a unit element, then the matrix $I = \Sigma\Sigma\delta_{ik}e_{ik}$ is the unit element of $M(\mathfrak{o}, L)$, and the mapping of λ onto λI provides an $\mathfrak{o}$-isomorphism of $\mathfrak{o}$ into $M(\mathfrak{o}, L)$.

b) For any subset S of L the subset $M(\mathfrak{o}, S, L)$ of all matrices (λ_{ik}) of $M(\mathfrak{o}, L)$ such that λ_{ik} vanishes whenever i or k does not belong to S is an $\mathfrak{o}$-subring of $M(\mathfrak{o}, L)$ that is $\mathfrak{o}$-isomorphic to $M(\mathfrak{o}, S)$.

c) If L is a poset, then the set $T(\mathfrak{o}, L)$ of all triangular matrices (λ_{ik}) which are characterized as row-finite matrices for which $\lambda_{ik} = 0$ if i is not contained in k, is an $\mathfrak{o}$-subring of $M(\mathfrak{o}, L)$.

d) Show that all matrices (λ_{ik}) of $M(\mathfrak{o}, L)$ with all but a finite number of columns vanishing and with the property that $\lambda_{ik} \neq 0$ implies that i is properly contained in k form a left ideal $N(\mathfrak{o}, L)$ of $T(\mathfrak{o}, L)$. Moreover, for every element A of $N(\mathfrak{o}, L)$ there is an exponent n for which $A^n = 0$.

e) If $\mathfrak{o}$ contains a unit element, then all the matrices $I + A$ ($A \in N(\mathfrak{o}, L)$) form a group $I + N(\mathfrak{o}, L)$. (Use the identity $(I - A)(I + A + A^2 + \cdots + A^{n-1}) = (I - A^n)$.)

f) The group $I + N(\mathfrak{o}, L)$ is locally nilpotent.

g) If there are arbitrary long chains

$$i_0 < i_1 < i_2 \cdots < i_l, \quad i_{j+1} \nsubseteq i_j \text{ for } j = 0, 1, \ldots, l-1$$

in the poset L, then the group $I + N(\mathfrak{o}, L)$ is not nilpotent. (Let L be for example, the ordered set of all natural numbers.) If the length of properly increasing chains is bounded, then $I + N(\mathfrak{o}, L)$ is nilpotent.

h) For a subposet S of L, the set of matrices $I + N(\mathfrak{o}, L) \cap M(\mathfrak{o}, S, L)$ is a subgroup of $I + N(\mathfrak{o}, L)$ isomorphic to $I + N(\mathfrak{o}, S)$. This subgroup is its own normalizer in $I + N(\mathfrak{o}, L)$ if any element of L is properly contained in an element of S and if all elements of L not belonging to S are equivalent. (For example, let L be the ordered set of all natural numbers, and let S be the subset of all elements $\neq 1$.)

19. A group every element of which is of finite order is called *periodic*. Show that

a) Every subgroup and every factor group of a periodic group is periodic;

b) If a normal subgroup $\mathfrak{N}$ of a group $\mathfrak{G}$ as well as the factor group $\mathfrak{G}/\mathfrak{N}$ is periodic, then $\mathfrak{G}$ is periodic;

c) If $\mathfrak{N}$ is a normal periodic subgroup and if $\mathfrak{U}$ is a periodic subgroup of the group $\mathfrak{G}$, then $\mathfrak{U}\mathfrak{N}$ is a periodic subgroup of $\mathfrak{G}$;

d) In every group $\mathfrak{G}$ there is precisely one maximal normal periodic subgroup $\mathfrak{T}(\mathfrak{G})$ called the *torsion subgroup* of $\mathfrak{G}$;

e) If the normal subgroup $\mathfrak{N}$ of a group is periodic, then $\mathfrak{T}(\mathfrak{G}/\mathfrak{N}) = \mathfrak{T}(\mathfrak{G})/\mathfrak{N}$.

20. A group is called *torsion-free* if it contains no element of finite order other than 1. Show that

a) Every free group is torsion-free;

b) Every subgroup of a torsion-free group is torsion-free;

c) If the factor group of the group $\mathfrak{G}$ over the normal subgroup $\mathfrak{N}$ is torsion-free, then $\mathfrak{N}$ contains the torsion subgroup $\mathfrak{T}(\mathfrak{G})$ defined in 18d);

d) If both $\mathfrak{N}$ and $\mathfrak{G}/\mathfrak{N}$ are torsion-free, then $\mathfrak{G}$ is torsion-free;

e) In a locally nilpotent group any two periodic subgroups generate a periodic subgroup (use Ex. 6);

f) Every periodic subgroup of a locally nilpotent group is contained in its torsion subgroup (use 19e));

g) The factor group of a locally nilpotent group over its torsion subgroup is torsion-free (use 19a), 18b));

h) If the center of a group is torsion-free, then its second center is also torsion-free (use (9), Chap. II, § 6); hence it follows by induction that all the members of the ascending central series are torsion-free;

i) A finitely generated abelian group is torsion-free if and only if it is a free abelian group;

j) (Jennings.) A finitely generated nilpotent group $\mathfrak{G}$ is torsion-free if and only if there is a finite chain of normal subgroups of $\mathfrak{G}$

$$\mathfrak{G} = \mathfrak{N}_0 \supset \mathfrak{N}_1 \supset \mathfrak{N}_2 \supset \cdots \supset \mathfrak{N}_r = 1$$

such that the factor groups $\mathfrak{N}_i/\mathfrak{N}_{i+1}$ are infinite cyclic for $i = 0, 1, \ldots, r-1$. (Use 14c, 20h, 20i.)

21. A group $\mathfrak{G}$ is called a group of *finite rank* if there is a finite normal chain

$$\mathfrak{G} = \mathfrak{N}_0 \geq \mathfrak{N}_1 \geq \mathfrak{N}_2 \geq \cdots \geq \mathfrak{N}_s = 1,$$

for which each factor group $\mathfrak{N}_i/\mathfrak{N}_{i+1}$ either is periodic or infinite cyclic. Show that

a) The number $r = r(\mathfrak{G})$ of infinite cyclic factor groups $\mathfrak{N}_i/\mathfrak{N}_{i+1}$ is independent of the choice of the normal chain by which it is determined (apply Schreier's Refinement Theorem); hence we have obtained a group invariant which may be called the *rank of the group*;

b) Every subgroup and every factor group of a group of rank r is of finite rank not greater than r;

c) If the normal subgroup $\mathfrak{N}$ of a group $\mathfrak{G}$ is of rank r and if $\mathfrak{G}/\mathfrak{N}$ is of rank r', then $\mathfrak{G}$ is of rank $r + r'$;

d) For a group $\mathfrak{G}$ of finite rank, $r(\mathfrak{G}) = r(\mathfrak{G}/\mathfrak{T}(\mathfrak{G}))$;

e) The rank of a group of finite rank is characterized as the maximum number of independent elements, where we call the elements $A_1, A_2, \ldots, A_\pi$ of a group independent if A_i $(i = 1, 2, \ldots, \pi)$ is of infinite order modulo the smallest normal subgroup $\mathfrak{A}_{i-1}$ of $\langle A_1, A_2, \ldots, A_i\rangle$ containing the elements $A_1, A_2, \ldots, A_{i-1}$.

22. (MacLain.) Prove that

a) For any two subgroups $\mathfrak{A}$, $\mathfrak{B}$ of a group, $(\mathfrak{A}, \mathfrak{B})$ is always normal in $\langle\mathfrak{A}, \mathfrak{B}\rangle$;

b) If $\mathfrak{B}$ is contained in $(\mathfrak{A}, \mathfrak{B})$ and if $\langle\mathfrak{A}, \mathfrak{B}\rangle$ is nilpotent, then $\mathfrak{B} = 1$;

c) If $\mathfrak{B}$ is finitely generated and contained in $(\mathfrak{A}, \mathfrak{B})$ and is not 1, then $\langle\mathfrak{A}, \mathfrak{B}\rangle$ contains a finitely generated subgroup that is not nilpotent;

d) A minimal normal subgroup of a locally nilpotent group always belongs to the center of the group.

APPENDIX F

FURTHER EXERCISES FOR CHAP. IV, § 1

4. (Gaschütz.) Let us consider the relation between an extension $\mathfrak{G}$ of the abelian group $\mathfrak{N}$ with the factor group $\mathfrak{E}$ and the extension $\mathfrak{H}$, contained in $\mathfrak{G}$, of $\mathfrak{N}$ with a subgroup $\mathfrak{U}$ of finite index n in $\mathfrak{E}$. We assume that $\mathfrak{H}$ splits over $\mathfrak{N}$. Using a right representative system $\mathfrak{F}$ of $\mathfrak{E}$ over $\mathfrak{U}$ we can decompose $\mathfrak{G}$ into the cosets $S_{\varrho u}\mathfrak{N}$, where ϱ runs over $\mathfrak{F}$ and u runs over $\mathfrak{U}$, such that $S_{\varrho u} = S_\varrho S_u$, $S_{uu'} = S_u S_{u'}$, $S_x S_y = C_{x,y} S_{xy}$, $S_x N S_x^{-1} = N^x$ for x, y in $\mathfrak{E}$, N in $\mathfrak{N}$; moreover, the $C_{x,y}$ are in $\mathfrak{N}$.

Show that

a) $S_{xu} = S_x S_u$ for x in $\mathfrak{E}$, u in $\mathfrak{U}$;

b) $C_{x,y} = C_{x,yu}$ for x, y in $\mathfrak{E}$, u in $\mathfrak{U}$;

c) The element a_σ obtained by forming the product over all $C_{\varrho,\sigma}$ with ϱ running over $\mathfrak{F}$ is independent of the choice of the representative system $\mathfrak{F}$ (where σ is a fixed element of $\mathfrak{E}$);

d) $C^n_{\sigma,\tau} = a^\sigma_\tau a_\sigma a^{-1}_{\sigma\tau}$ for σ, τ in $\mathfrak{E}$;

e) If the mapping of N onto N^n is an automorphism of $\mathfrak{N}$, then an extension $\mathfrak{G}$ of $\mathfrak{N}$ with the factor group $\mathfrak{E}$ splits if and only if the extension of $\mathfrak{N}$ with the factor group $\mathfrak{U}$ contained in $\mathfrak{G}$ splits over $\mathfrak{N}$;

f) An extension $\mathfrak{G}$ of a finite abelian group $\mathfrak{N}$ with a finite factor group $\mathfrak{E}$ splits over $\mathfrak{N}$ if and only if the extension of $\mathfrak{N}$ contained in $\mathfrak{G}$, with each Sylow subgroup of $\mathfrak{E}$ as factor group, splits over $\mathfrak{N}$;

g) Assuming $C_{\sigma,\tau} = 1$ for σ, τ in $\mathfrak{E}$ and T_σ being a second representative system of $\mathfrak{G}$ over $\mathfrak{N}$ that satisfies the condition $T_\sigma T_\tau = T_{\sigma\tau}$, $T_u = S_u$ for σ, τ in $\mathfrak{E}$, u in $\mathfrak{U}$, we have

$$T_\sigma = b_\sigma S_\sigma,\ b_{\sigma\tau} = b_\sigma b_{\tau u},\ b_u = 1,\ b_{\sigma u} = b_\sigma,\ b^n_\sigma = \delta^{1-\sigma},$$

where b_σ denotes a certain element of $\mathfrak{N}$ depending on σ, and δ is the product of the elements b_σ with ϱ running over $\mathfrak{F}$;

h) Under the assumption that the extension $\mathfrak{G}$ over $\mathfrak{N}$ splits and that the mapping of N onto N^n is an automorphism of $\mathfrak{N}$, two representative subgroups $\mathfrak{E}_1$, $\mathfrak{E}_2$ of $\mathfrak{G}$ over $\mathfrak{N}$ are conjugate under $\mathfrak{G}$ if and only if the representative subgroups $\mathfrak{E}_1 \cap \mathfrak{H}$, $\mathfrak{E}_2 \cap \mathfrak{H}$ of $\mathfrak{H}$ over $\mathfrak{N}$ obtained by restricting $\mathfrak{E}$ to $\mathfrak{U}$ are conjugate under $\mathfrak{G}$;

i) Two representative subgroups of a splitting extension $\mathfrak{G}$ of a finite abelian group $\mathfrak{N}$ with a finite factor group $\mathfrak{E}$ are conjugate over $\mathfrak{G}$ if and only if the representative subgroups obtained by restricting $\mathfrak{E}$ to any of the Sylow subgroups are conjugate under $\mathfrak{G}$. The condition that every Sylow subgroup of $\mathfrak{G}$ splits over its intersection with $\mathfrak{N}$ is equivalent to this condition.

5. (Example showing that Gaschütz' Theorem does not hold if the normal subgroup is not abelian.) Let $\mathfrak{G}_{48}$ be the multiplicative group formed by the 2×2-matrices of determinant 1 with its entries in $GF(3^2)$ for which a suitable non-vanishing scalar multiple with its entries in $GF(3)$ exists. Let $\mathfrak{G}_8$ be the normal subgroup consisting of $\pm I_2$ and of the six matrices that have their entries in $GF(3)$ and that have sum

zero over the main diagonal. Let $\mathfrak{G}_{16}$ be the 2-Sylow subgroup of $\mathfrak{G}_{48}$ which is generated by $\mathfrak{G}_8$ and by the matrix $\begin{pmatrix} i & 0 \\ 0 & -i \end{pmatrix}$, where $i^2 = -1$; let $\mathfrak{G}_3$ be the 3-Sylow subgroup generated by the matrix $\begin{pmatrix} 1 & 1 \\ 0 & 1 \end{pmatrix}$; let $\mathfrak{G}_{48.48}$, $\mathfrak{G}_{16.16}$, and $\mathfrak{G}_{8.8}$, $\mathfrak{G}_{3.3}$ be the groups of all 4×4-matrices of the form $\begin{pmatrix} A & 0 \\ 0 & B \end{pmatrix}$, with A, B contained in $\mathfrak{G}_{48}$, $\mathfrak{G}_{16}$, $\mathfrak{G}_8$, and $\mathfrak{G}_3$, respectively; and let $\mathfrak{G}_2$ be the normal subgroup of $\mathfrak{G}_{48.48}$ consisting of I_4 and $-I_4$. Then the group $\mathfrak{G} = \mathfrak{G}_{48.48}/\mathfrak{G}_2$ does not split over its normal subgroup $\mathfrak{N} = \mathfrak{G}_{8.8}/\mathfrak{G}_2$, whereas both a 2-Sylow subgroup and a 3-Sylow subgroup of $\mathfrak{G}/\mathfrak{N}$ correspond to splitting extensions of $\mathfrak{N}$, viz., $\mathfrak{G}_{16.16}/\mathfrak{G}_2$ and $\mathfrak{G}_{3.3}\,\mathfrak{G}_{8.8}/\mathfrak{G}_2$ respectively.

6. The Φ-subgroup of a finite group $\mathfrak{G}$ does not contain a Sylow subgroup $\neq 1$ of $\mathfrak{G}$. (*Hint*: Apply Ex. 7 of Chap. IV § 3, and Theorem 25.)

APPENDIX G

A THEOREM OF WIELANDT

An addendum to Chap. IV

In Chap. II, § 4 we saw that a finite group $\mathfrak{G}$ having 1 as center is isomorphic to the group of inner automorphisms of $\mathfrak{G}$. Also, for an element a of the automorphism group $\mathfrak{G}_1$ of $\mathfrak{G}$ we have $a\underline{x}a^{-1} = \underline{x^a}$, where $\underline{x}$ denotes the inner automorphism corresponding to the element x of $\mathfrak{G}$. Hence an equation $a\underline{x}a^{-1} = \underline{x}$ implies that $x^a x^{-1}$ is contained in the center of $\mathfrak{G}$ and hence that x^a is equal to x. By identifying each element x of $\mathfrak{G}$ with the corresponding inner automorphism, we may consider $\mathfrak{G} = \mathfrak{G}_0$ as a normal subgroup of $\mathfrak{G}_1$ such that the centralizer of $\mathfrak{G}_0$ in $\mathfrak{G}_1$ is 1.

Since $\mathfrak{G}_1$ is again finite and has center 1, we may continue with the construction by extending $\mathfrak{G}_1$ to the automorphism group $\mathfrak{G}_2$ of $\mathfrak{G}_1$, etc. We obtain a sequence of finite groups

$$\mathfrak{G} = \mathfrak{G}_0 \lhd \mathfrak{G}_1 \lhd \mathfrak{G}_2 \lhd \cdots, \tag{1}$$

called the *automorphism tower* of $\mathfrak{G}$, each member of which is contained as a normal subgroup in the succeeding subgroup and has centralizer 1. We may raise the question, Is the sequence of the groups in (1) strictly increasing? Or is one of the groups $\mathfrak{G}_i$ isomorphic with its group of automorphisms? In the second case we say that the automorphism tower is of finite height. In this case $\mathfrak{G}_i = \mathfrak{G}_{i+1} = \cdots$ is a complete group, and we will have succeeded in embedding the given group with center 1 subnormally in a complete finite group with center 1.

The solution to this problem was found by *Wielandt* and gave rise to interesting theorems of a general nature concerning the subnormal subgroups of a group, which we will now consider.

Definition: A property P of groups is called *normally persistent* if

1. Every group isomorphic to a group with property P also has property P;

2. In a given group the subgroup generated by a system of normal subgroups having property P also has property P.

Examples of normally persistent properties are

1. the property of being a p-group,

2. the property of being semi-simple,

3. *Local finiteness:* Finitely many elements of $\mathfrak{G}$ always generate a finite subgroup.[1]

4. Local nilpotency (see Exx. 15, 16c), Appendix E),

5. Periodicity (see Ex. 19c), Appendix E).

If, in the statement of condition 2., the word 'normal' is replaced by 'subnormal', a property P satisfying 1. and 2. is called *subnormally persistent.*

It is immediate that subnormal persistence implies normal persistence. Conversely,

THEOREM 28: *Normal persistence implies subnormal persistence.*

Proof: For brevity we call a group having the normally persistent property P a *P-group*. Let B be a set of subnormal P-subgroups of the group $\mathfrak{G}$. We want to prove that the subgroup $\mathfrak{H}$ generated by all the subgroups belonging to B is a P-subgroup of $\mathfrak{G}$.

First of all, let us assume that the function $m(\mathfrak{X}, \mathfrak{G})$ defined on B and indicating the minimal length of a normal chain from $\mathfrak{X}$ to $\mathfrak{G}$, is bounded by a number n. We apply induction on n. If $n = 0$, then every member of B coincides with $\mathfrak{G}$, and hence $\mathfrak{G} = \mathfrak{H}$ is a P-group. If $n = 1$, then every member of B is a normal P-subgroup of $\mathfrak{G}$. Because of the normal persistence the subgroup $\mathfrak{H}$ generated by its own normal P-subgroups is also a P-subgroup of $\mathfrak{G}$. Now let $n > 1$, and assume that the statement is true in the case that $m(\mathfrak{X}, \mathfrak{G}) < n$ for all members $\mathfrak{X}$ of B. For any member $\mathfrak{X}$ of B there exists a normal chain $\mathfrak{X} = \mathfrak{X}_0 \lhd \mathfrak{X}_1 \lhd \mathfrak{X}_2 \lhd \cdots \lhd \mathfrak{X}_n = \mathfrak{G}$ from $\mathfrak{X}$ to $\mathfrak{G}$ of length n. Taking any conjugate $\mathfrak{X}^t$ of $\mathfrak{X}$ (t in $\mathfrak{G}$) we have the normal chain $\mathfrak{X}^t = \mathfrak{X}_0^t \lhd \mathfrak{X}_1^t \lhd \mathfrak{X}_2^t \lhd \cdots \lhd \mathfrak{X}_{n-1}^t = \mathfrak{X}_{n-1}$ of length $n - 1$ from $\mathfrak{X}^t$ to $\mathfrak{X}_{n-1}$. All conjugates of $\mathfrak{X}$ under $\mathfrak{H}$ are isomorphic to $\mathfrak{X}$ and hence are P-subgroups, and furthermore we have $m(\mathfrak{X}^t, \mathfrak{X}_{n-1}) < n$. By the induction hypothesis, the subgroup $\overline{\mathfrak{X}}$ generated by the conjugates of $\mathfrak{X}$ under $\mathfrak{H}$ is a normal P-subgroup of $\mathfrak{H}$. All normal subgroups $\overline{\mathfrak{X}}$ generate a P-subgroup $\overline{\mathfrak{H}}$ of $\mathfrak{H}$. Since $\mathfrak{H}$ is a subgroup of $\overline{\mathfrak{H}}$, it follows that $\mathfrak{H} = \overline{\mathfrak{H}}$.

---

[1] It has to be proved that two locally finite normal subgroups $\mathfrak{N}_1$, $\mathfrak{N}_2$ of a group $\mathfrak{G}$ generate a locally finite normal subgroup. Let $x_1, x_2, \ldots, x_r$ be finitely many elements of $\mathfrak{N}_1\mathfrak{N}_2$. Then $x_i = y_i z_i$, where y_i is an element of $\mathfrak{N}_1$, z_i an element of $\mathfrak{N}_2$, and the subgroup $\mathfrak{U}$ generated by $z_1, z_2, \ldots, z_r$ is finite. The finitely many conjugates of the elements y_i under $\mathfrak{U}$ are in $\mathfrak{N}_1$ and hence all together generate a finite subgroup $\mathfrak{V}$ of $\mathfrak{G}$ which is normal in the subgroup $\mathfrak{W}$ of $\mathfrak{G}$ generated by $\mathfrak{V}$ and $\mathfrak{U}$; hence $\mathfrak{W} = \mathfrak{V}\mathfrak{U}$ is generated by $y_1, y_2, \ldots, y_r, z_1, z_2, \ldots, z_r$, so that the index of $\mathfrak{W}$ over $\mathfrak{V}$ is equal to the index of $\mathfrak{U}$ over the intersection of $\mathfrak{U}$ and $\mathfrak{W}$. Thus $\mathfrak{W}$ is finite. The subgroup generated by $x_1, x_2, \ldots, x_r$, being a subgroup of $\mathfrak{W}$, is finite.

Since for any subnormal P-subgroup $\mathfrak{X}$ of $\mathfrak{G}$ the minimal length of a normal chain from any conjugate of $\mathfrak{X}$ to $\mathfrak{G}$ is the same, it follows from the preceding that any subgroup generated by some conjugates of $\mathfrak{X}$ is a P-subgroup.

For any system B of subnormal P-subgroups of $\mathfrak{G}$ the normal subgroup $\overline{\mathfrak{X}}$ of $\mathfrak{H}$ generated by the conjugates under $\mathfrak{H}$ of each member $\mathfrak{X}$ of B is a P-subgroup; hence we see as above that the members of B generate a P-subgroup $\mathfrak{H}$ of $\mathfrak{G}$, Q. E. D.

The smallest normal subgroup of a group $\mathfrak{G}$ containing a given subgroup $\mathfrak{U}$ is the subgroup generated by all the conjugates of $\mathfrak{U}$. Applying Theorem 28, we obtain Corollary 1.

COROLLARY 1: *The normal subgroup generated by the conjugates of a subnormal subgroup having the normally persistent property P is a subgroup with property P.*

Theorem 28, applied to Examples 1. and 2., gives Corollaries 2 and 3, respectively.

COROLLARY 2: *Any number of subnormal p-subgroups of a group generate a p-subgroup. In particular, the normal subgroup generated by the conjugates of a subnormal p-subgroup is a p-subgroup.*

COROLLARY 3: *Any number of subnormal semi-simple subgroups of a group generate a semi-simple subgroup. In particular, the normal subgroup generated by the conjugates of a subnormal semi-simple subgroup is semi-simple.*

Theorem 28 can be extended in the case of a group $\mathfrak{G}$ subnormally embedded in another finite group $\mathfrak{H}$ if there is known an ascending chain

$$1 = \mathfrak{B}_0 \lhd \mathfrak{B}_1 \lhd \mathfrak{B}_2 \lhd \cdots \lhd \mathfrak{B}_r = \mathfrak{G} \tag{2}$$

of normal subgroups extending from 1 to $\mathfrak{G}$ such that $\mathfrak{B}_{i+1}/\mathfrak{B}_i$ is either a p_{i+1}-group or a semi-simple group.

THEOREM 29: *With proper choice of the chain* (2) *the subgroups $\mathfrak{H}_i$ of $\mathfrak{H}$ generated by the subgroups conjugate to $\mathfrak{B}_i$ under $\mathfrak{H}$ together with $\mathfrak{H}$ form an ascending chain $1 = \mathfrak{H}_0 \lhd \mathfrak{H}_1 \lhd \mathfrak{H}_2 \lhd \cdots \lhd \mathfrak{H}_r \lhd \mathfrak{H}$ of normal subgroups of $\mathfrak{H}$ such that*

1. *$\mathfrak{B}_{i+1}/\mathfrak{B}_i$ and $\mathfrak{H}_{i+1}/\mathfrak{H}_i$ are either p_{i+1}-groups and there is a normal subgroup $\mathfrak{U}_{i+1}$ of $\mathfrak{G}$ with a p_{i+1}-factor group; or $\mathfrak{B}_{i+1}/\mathfrak{B}_i$ and $\mathfrak{H}_{i+1}/\mathfrak{H}_i$ are semi-simple groups, in which case we set $\mathfrak{U}_{i+1} = \mathfrak{G}$; in any case*

$$(\mathfrak{U}_{i+1}, \mathfrak{H}_{i+1}) \subseteq \mathfrak{B}_{i+1}\mathfrak{H}_i; \tag{3}$$

2. *The intersection of* $\mathfrak{H}_i$ *and* $\mathfrak{G}$ *is* $\mathfrak{B}_i$ *for* $i = 0, 1, \ldots, r$.

Proof: If $\mathfrak{G} = 1$, then set $\mathfrak{H}_0 = 1$. If $\mathfrak{G} \neq 1$, then $\mathfrak{G}$ contains a smallest normal subgroup different from 1 which is either an abelian group of prime exponent p_1 or a semi-simple group. In the former case, let $\mathfrak{B}_1$ be the maximal normal p_1-subgroup of $\mathfrak{G}$. In the latter, let $\mathfrak{B}_1$ be the maximal semi-simple normal subgroup of $\mathfrak{G}$. In any case, in view of Corollaries 2 and 3 to Theorem 28, $\mathfrak{B}_1$ is uniquely determined. Similarly, form $\mathfrak{B}_2/\mathfrak{B}_1$ in $\mathfrak{G}/\mathfrak{B}_1$, etc. Continuing this construction, we obtain a chain (2). For the corresponding $\mathfrak{H}_i$ the factor group $\mathfrak{H}_{i+1}/\mathfrak{H}_i$ is the normal subgroup of $\mathfrak{H}/\mathfrak{H}_i$ generated by the subnormal subgroup $\mathfrak{B}_{i+1}\mathfrak{H}_i/\mathfrak{H}_i$.

Since the factor group $\mathfrak{B}_{i+1}\mathfrak{H}_i/\mathfrak{H}_i$ is isomorphic to the factor group of $\mathfrak{B}_{i+1}$ over the intersection of $\mathfrak{B}_{i+1}$ and $\mathfrak{H}_i$, it follows from Corollaries 2 and 3 of Theorem 28 that either $\mathfrak{B}_{i+1}/\mathfrak{B}_i$ and $\mathfrak{H}_{i+1}/\mathfrak{H}_i$ are both p_{i+1}-groups or they are both semi-simple. We claim that $\mathfrak{B}_i = \mathfrak{G} \cap \mathfrak{H}_i$. This is obvious for $i = 0$. Assume that $\mathfrak{H}_i \cap \mathfrak{G} = \mathfrak{B}_i$. We observe that $\mathfrak{H}_{i+1} \cap \mathfrak{G}/\mathfrak{B}_i$ is isomorphic to a factor group of $\mathfrak{H}_{i+1}/\mathfrak{H}_i$ and hence is a p_{i+1}-group in the first case and is semi-simple in the second case. Since in both cases $\mathfrak{B}_{i+1}$ is contained in the intersection of $\mathfrak{H}_{i+1}$ and $\mathfrak{G}$, we conclude from the maximal property of $\mathfrak{B}_{i+1}$ that $\mathfrak{H}_{i+1} \cap \mathfrak{G} = \mathfrak{B}_{i+1}$.

In order to prove (3), let us observe that there is a normal chain $\mathfrak{G} = \mathfrak{G}_0 \lhd \mathfrak{G}_1 \lhd \mathfrak{G}_2 \lhd \cdots \lhd \mathfrak{G}_s = \mathfrak{H}$. Let $\mathfrak{G}_{ij} = \mathfrak{H}_i(\mathfrak{G}_j \cap \mathfrak{H}_{i+1})$, so that $\mathfrak{H}_i$ is properly contained in $\mathfrak{G}_{i0} = \mathfrak{H}_i\mathfrak{B}_{i+1}$ and $\mathfrak{G}_{i0} \lhd \mathfrak{G}_{i1} \lhd \cdots \lhd \mathfrak{G}_{is} = \mathfrak{H}_{i+1}$. Let $\mathfrak{g}_{ij}$ be the set of all elements x of $\mathfrak{G}$ satisfying the condition $(x, \mathfrak{H}_{i+1}) \subseteq \mathfrak{G}_{ij}$. For an element x of $\mathfrak{g}_{ij}$ and an element h of $\mathfrak{H}_{i+1}$ we find that $(x, h^{-1}) = w$ is contained in $\mathfrak{G}_{ij}$. By induction on n, we find that $(x^n, h^{-1}) = w\,w^x \cdots w^{x^{n-1}}$. Since x, as an element of $\mathfrak{G}$, is also an element of $\mathfrak{G}_{j-1}$, we find that $x^m w x^{-m} w^{-1}$ already belongs to $\mathfrak{H}_i(\mathfrak{G}_{j-1} \cap \mathfrak{H}_{i+1})$, which is the same as $\mathfrak{G}_{i,\,j-1}$; hence the two elements w and w^{x^m} are congruent modulo $\mathfrak{G}_{i,\,j-1}$, and thus the element $(x^n, h^{-1}) = ww^x \cdots w^{x^{n-1}}$ is congruent to w^n. It follows that $(x^{\mathfrak{G}_{ij}:\mathfrak{G}_{i,j-1}}, h^{-1})$ is contained in $\mathfrak{G}_{i,\,j-1}$ for h^{-1} an arbitrary element of $\mathfrak{H}_{i+1}$; therefore $x^{\mathfrak{G}_{ij}:\mathfrak{G}_{i,j-1}}$ belongs to $\mathfrak{g}_{i,\,j-1}$. Hence $x^{\mathfrak{G}_{ij}:\mathfrak{G}_{i,j-2}}$, being equal to $(x^{\mathfrak{G}_{ij}:\mathfrak{G}_{i,j-1}})^{\mathfrak{G}_{i,j-1}:\mathfrak{G}_{i,j-2}}$, belongs to $\mathfrak{g}_{i,\,j-2}$, etc. Lastly $x^{\mathfrak{G}_{ij}:\mathfrak{G}_{i0}}$ belongs to $\mathfrak{g}_{i0}$. Since, as follows from the preceding, $\mathfrak{g}_{i0}$ contains all the powers of each of its elements, and since $\mathfrak{G}_{ij} : \mathfrak{G}_{i0}$ is a divisor of $\mathfrak{H}_{i+1} : \mathfrak{H}_i$, it follows that $x^{\mathfrak{H}_{i+1}:\mathfrak{H}_i}$ belongs to $\mathfrak{g}_{i0}$ for all x contained in the set $\mathfrak{g}_{is}$, $\mathfrak{g}_{is}$ being the same as $\mathfrak{G}$.

When $\mathfrak{B}_{i+1}/\mathfrak{B}_i$ is a p_{i+1}-group, let $\mathfrak{G}/\mathfrak{U}_{i+1}$ be the maximal p_{i+1}-factor group of $\mathfrak{G}$; then $\mathfrak{U}_{i+1}$ is generated by all the p_{i+1}^n-th powers of the elements of $\mathfrak{G}$, with n large. Since in addition $\mathfrak{H}_{i+1}/\mathfrak{H}_i$ is a p_{i+1}-group, it follows that $\mathfrak{U}_{i+1}$ is contained in $\mathfrak{g}_{i0}$, and hence (3) holds.

If, on the other hand, $\mathfrak{H}_{i+1}/\mathfrak{H}_i$ is semi-simple, then by the results of Appendixes B and C,[1] the group $\mathfrak{G}\mathfrak{H}_i/\mathfrak{H}_i$, being subnormal in $\mathfrak{H}/\mathfrak{H}_i$, is normal in $\mathfrak{G}\mathfrak{H}_{i+1}/\mathfrak{H}_i$. Hence $(\mathfrak{G}, \mathfrak{H}_{i+1})$ is contained in $\mathfrak{H}_{i+1} \cap \mathfrak{G}\mathfrak{H}_i$. But by the modular law, $\mathfrak{H}_{i+1} \cap (\mathfrak{G}\mathfrak{H}_i) = (\mathfrak{H}_{i+1} \cap \mathfrak{G})\mathfrak{H}_i = \mathfrak{B}_{i+1}\mathfrak{H}_i$, and thus $(\mathfrak{G}, \mathfrak{H}_{i+1})$ is contained in $\mathfrak{B}_{i+1}\mathfrak{H}_i$, Q. E. D.

THEOREM 30: *If together with the assumptions of Theorem* 29 *we have the centralizer of* $\mathfrak{G}$ *in* $\mathfrak{H}$ *equal to* 1, *then the order of* $\mathfrak{H}$ *is bounded by a constant depending only on* $\mathfrak{G}$.

Proof: In the case that $\mathfrak{B}_{i+1}/\mathfrak{B}_i$ is a p_{i+1}-group, we have for any element x of $\mathfrak{G}$

$$(Z(\mathfrak{u}_{i+1}) \cap \mathfrak{H}_{i+1})^x = Z(\mathfrak{u}_{i+1}^x) \cap \mathfrak{H}_{i+1}^x = Z(\mathfrak{u}_{i+1}) \cap \mathfrak{H}_{i+1};$$

hence

$$(Z(\mathfrak{u}_{i+1}) \cap \mathfrak{H}_{i+1}) \lhd (Z(\mathfrak{u}_{i+1}) \cap \mathfrak{H}_{i+1})\mathfrak{G}. \tag{4}$$

Let $\mathfrak{p}$ be a p_{i+1}-Sylow subgroup of $\mathfrak{G}$. It is contained in a p_{i+1}-Sylow subgroup $\mathfrak{P}$ of $(Z(\mathfrak{u}_{i+1}) \cap \mathfrak{H}_{i+1})\mathfrak{G}$. The intersection of $\mathfrak{z}(\mathfrak{P})$ with $Z(\mathfrak{u}_{i+1}) \cap \mathfrak{H}_{i+1}$ belongs to $Z(\mathfrak{u}_{i+1}) \cap Z(\mathfrak{P})$ and hence to $Z(\mathfrak{u}_{i+1}) \cap Z(\mathfrak{p}) = Z(\mathfrak{u}_{i+1}\mathfrak{p}) = Z(\mathfrak{G})$. But by assumption, the centralizer of $\mathfrak{G}$ in $\mathfrak{H}$ is 1; hence

$$\mathfrak{z}(\mathfrak{P}) \cap Z(\mathfrak{u}_{i+1}) \cap \mathfrak{H}_{i+1} = 1. \tag{5}$$

Because of (4), we find that

$$\mathfrak{P} \cap Z(\mathfrak{u}_{i+1}) \cap \mathfrak{H}_{i+1} \text{ is normal in } \mathfrak{P}. \tag{6}$$

From (5), (6) and from the theorems on p-groups proved in Chap. IV, § 2, we conclude that

$$\mathfrak{P} \cap Z(\mathfrak{u}_{i+1}) \cap \mathfrak{H}_{i+1} = 1.$$

Hence p_{i+1} does not divide the order of $Z(\mathfrak{u}_{i+1}) \cap \mathfrak{H}_{i+1}$. A fortiori, p_{i+1} does not divide the index of $Z(\mathfrak{u}_{i+1}) \cap \mathfrak{H}_{i+1}$ over $Z(\mathfrak{u}_{i+1}) \cap \mathfrak{H}_i$; this index,

---

[1] We have to show that a group $\mathfrak{C}$ generated by a subnormal semi-simple group $\mathfrak{A}$ and a subnormal group $\mathfrak{B}$ contains $\mathfrak{B}$ as a normal subgroup. By Corollary 3 of Theorem 28, the normal subgroup $\overline{\mathfrak{A}}$ of $\mathfrak{C}$ generated by the conjugates of $\mathfrak{A}$ under $\mathfrak{C}$ is semi-simple. Let $\mathfrak{B} = \mathfrak{B}_0 \lhd \mathfrak{B}_1 \lhd \cdots \lhd \mathfrak{B}_s = \mathfrak{C}$ be a normal chain of minimal length s extending from $\mathfrak{B}$ to $\mathfrak{C}$. Since $\mathfrak{B}_{s-1} \cap \overline{\mathfrak{A}}$ is normal in $\overline{\mathfrak{A}}$ and $\overline{\mathfrak{A}}$ is semi-simple, it follows that there is one and only one subgroup $\mathfrak{A}_1$ of $\overline{\mathfrak{A}}$ for which $\overline{\mathfrak{A}}$ is the direct product of $\mathfrak{A}_1$ and $\mathfrak{B}_{s-1} \cap \overline{\mathfrak{A}}$. Transformation of this decomposition with an element x of $\mathfrak{B}_{s-1}$ yields a direct decomposition of $\overline{\mathfrak{A}}$ into the direct product of $\mathfrak{A}_1^x$ and $\mathfrak{B}_{s-1} \cap \overline{\mathfrak{A}}$; hence $\mathfrak{A}_1^x = \mathfrak{A}_1$; and hence $\mathfrak{A}_1$ is normal in $\mathfrak{C}$, and $\mathfrak{C}$ is the direct product of $\mathfrak{A}_1$ and $\mathfrak{B}_{s-1}$. If $s > 1$, then we conclude in a similar way that $\mathfrak{B}_{s-1} \cap \overline{\mathfrak{A}}$ is the direct product of $\mathfrak{B}_{s-2} \cap \overline{\mathfrak{A}}$ and some subgroup $\mathfrak{A}_2$, so that $\mathfrak{B}_{s-1}$ is the direct product of $\mathfrak{A}_2$ and $\mathfrak{B}_{s-2}$, $\mathfrak{C}$ is the direct product of $\mathfrak{A}_1$, $\mathfrak{A}_2$ and $\mathfrak{B}_{s-2}$, and hence $\mathfrak{B}_{s-2}$ is normal in $\mathfrak{C}$, which contradicts the minimal property of s. Hence $s \leq 1$, and $\mathfrak{B}$ is normal in $\mathfrak{C}$.

because of the Second Isomorphism Theorem, is equal to the index of $(Z(\mathfrak{U}_{i+1}) \frown \mathfrak{H}_{i+1})\, \mathfrak{H}_i$ over $\mathfrak{H}_i$, which is a divisor of $\mathfrak{H}_{i+1} : \mathfrak{H}_i$, a power of p_{i+1}. Therefore $Z(\mathfrak{U}_{i+1}) \frown \mathfrak{H}_{i+1}$ is equal to $Z(\mathfrak{U}_{i+1}) \frown \mathfrak{H}_i$, so that

(7) $$Z(\mathfrak{U}_{i+1}) \frown \mathfrak{H}_{i+1} \text{ is contained in } \mathfrak{H}_i.$$

According to (3), each element x of $\mathfrak{H}_{i+1}$ determines the function (x, y) of y, an element of $\mathfrak{U}_{i+1}$, with values in $\mathfrak{H}_i \mathfrak{B}_{i+1}$. There are at most $(\mathfrak{H}_i \mathfrak{B}_{i+1} : 1)^{\mathfrak{U}_{i+1}:1}$ such functions. We have $(x, y) = (x', y)$ for all y of $\mathfrak{U}_{i+1}$ if and only if $xyx^{-1} = x'yx'^{-1}$, i. e., x is right congruent to x' modulo $Z(\mathfrak{U}_{i+1})$; hence

$$\mathfrak{H}_{i+1} : (Z(\mathfrak{U}_{i+1}) \frown \mathfrak{H}_{i+1}) \leq (\mathfrak{H}_i \mathfrak{B}_{i+1} : 1)^{\mathfrak{U}_{i+1}:1}.$$

From (7) it follows that

$$\mathfrak{H}_{i+1} : \mathfrak{H}_i \leq \mathfrak{H}_{i+1} : (Z(\mathfrak{U}_{i+1}) \frown \mathfrak{H}_{i+1}).$$

Noting that $\mathfrak{H}_i \mathfrak{B}_{i+1} : \mathfrak{H}_i = \mathfrak{B}_{i+1} : (\mathfrak{H}_i \frown \mathfrak{B}_{i+1})$ and that one has $\mathfrak{H}_i \mathfrak{B}_{i+1} : 1 = (\mathfrak{H}_i \mathfrak{B}_{i+1} : \mathfrak{H}_i)(\mathfrak{H}_i : 1)$, we find that

(8) $$\mathfrak{H}_{i+1} : 1 \leq (\mathfrak{B}_{i+1} : \mathfrak{B}_i)^{\mathfrak{U}_{i+1}:1} (\mathfrak{H}_i : 1)^{(\mathfrak{U}_{i+1}:1)+1}.$$

On the other hand, when $\mathfrak{B}_{i+1}/\mathfrak{B}_i$ is semi-simple, we set $\mathfrak{U}_{i+1} = \mathfrak{G}$. Thus (3) is again satisfied. According to our assumption concerning $\mathfrak{G}$, the centralizer of $\mathfrak{U}_{i+1}$ is 1; hence (7) is also satisfied. From (3) and (7) we conclude, as before, that the inequality (8) holds. Finally, we have to note that $Z(\mathfrak{G}) = 1$ implies $Z(\mathfrak{H}_r) = 1$, and hence $\mathfrak{H} : 1$ is a divisor of the order of the automorphism group of $\mathfrak{H}_r$. Consequently, a very rough estimate is

(9) $$\mathfrak{H} : 1 \leq (\mathfrak{H}_r : 1)!$$

Set

$$M_0 = 1,$$
$$M_{i+1} = (\mathfrak{B}_{i+1} : \mathfrak{B}_i)^{\mathfrak{U}_{i+1}:1} M_i^{(\mathfrak{U}_{i+1}:1)+1} \quad \text{for } i = 0, \ldots, r-1,$$
$$M = (M_r)!.$$

Then (8) and (9) together yield

$$\mathfrak{H} : 1 \leq M, \quad \text{Q. E. D.}$$

Theorem 31 (Wielandt): *The automorphism tower of a finite group $\mathfrak{G}$ with center 1 is of finite height.*

Proof: Let $\mathfrak{G} = \mathfrak{G}_0 \lhd \mathfrak{G}_1 \lhd \mathfrak{G}_2 \lhd \cdots$ be the automorphism tower of $\mathfrak{G}$. The center of each member is 1. In order to apply Theorem 30, we must prove that the centralizer of $\mathfrak{G}$ in $\mathfrak{G}_j$ is 1.

In fact, we can show that the centralizer of $\mathfrak{G}_i$ in $\mathfrak{G}_j$ is 1 if $j > i$.[1] This has already been shown in the case $j - i = 1$. We complete the proof by induction on $k = j - i$. Assume that the assertion is true for $j - i = k \geq 1$, and let $j - i = k + 1$. Then the centralizer of $\mathfrak{G}_i$ in $\mathfrak{G}_{j-1}$ is 1. The normalizer of $\mathfrak{G}_i$ in $\mathfrak{G}_{j-1}$ is $\mathfrak{G}_{i+1}$. For, $\mathfrak{G}_{i+1}$ belongs to the normalizer of $\mathfrak{G}_i$ in $\mathfrak{G}_{j-1}$, and if a is one of the elements of this normalizer, then the transformation by a defines an automorphism $\bar{a}$ of $\mathfrak{G}_i$; hence $\bar{a}$ is obtained by transformation of $\mathfrak{G}_i$ with an element a' of the automorphism group $\mathfrak{G}_{i+1}$, and $a^{-1}a'$ belongs to the centralizer of $\mathfrak{G}_i$ in $\mathfrak{G}_{j-1}$; i. e., $a^{-1}a' = 1$, $a = a'$, and a is contained in $\mathfrak{G}_{i+1}$. An element b of the centralizer of $\mathfrak{G}_i$ in $\mathfrak{G}_j$ transforms the normalizer of $\mathfrak{G}_i$ in $\mathfrak{G}_{j-1}$ into the normalizer of $\mathfrak{G}_i^b = \mathfrak{G}_i$ in $\mathfrak{G}_{j-1}^b = \mathfrak{G}_{j-1}$; i. e., $\mathfrak{G}_{i+1}^b = \mathfrak{G}_{i+1}$. Consequently, b belongs to the normalizer of $\mathfrak{G}_{i+1}$ in $\mathfrak{G}_j$ which, by the induction hypothesis and by the subsequent statement concerning the normalizer, is equal to $\mathfrak{G}_{i+2}$. For an element x of $\mathfrak{G}_{i+1}$ and an element y of $\mathfrak{G}_i$ we have: $byb^{-1} = y$, xyx^{-1} is contained in $\mathfrak{G}_i$, and bxb^{-1} is contained in $\mathfrak{G}_{i+1}$; therefore

$$xyx^{-1} = y^x = by^xb^{-1} = bxb^{-1}byb^{-1}bx^{-1}b^{-1} = y^{(x^b)};$$

and therefore $x = x^b$. The element b belongs to the centralizer of $\mathfrak{G}_{i+1}$ in $\mathfrak{G}_{i+2}$, and hence $b = 1$.

[1] This short demonstration is in accordance with a communication from D. G. Higman.

APPENDIX H

FURTHER EXERCISES FOR CHAP. V, § 1

Exercise: For an odd prime p let $\left(\frac{a}{p}\right) = 0,\ 1,\ -1$ when $a \equiv 0$, when $a \equiv x^2 \not\equiv 0$ is solvable, and when $a \equiv x^2$ is not solvable modulo p, respectively. Show that

a) $\left(\frac{a}{p}\right) \equiv a^{\frac{p-1}{2}} \pmod{p}$;

b) If $a \not\equiv 0 \pmod{p}$, then $\left(\frac{a}{p}\right) = (-1)^{\mu(a)}$, where $\mu(a)$ denotes the number of solutions of the congruence $ax \equiv -y\,(p)$ satisfying $0 < x,\ y \leqq \frac{p-1}{2}$;

c) $\left(\frac{ab}{p}\right) = \left(\frac{a}{p}\right)\left(\frac{b}{p}\right)$.

(*Hint:* Compute the transfer from the multiplicative group $\mathfrak{G}$ of the prime residue classes modulo p to the subgroup $\mathfrak{U}$ of $\pm 1 \pmod{p}$ in a) according to (19) and in b) by taking $1, 2, \ldots, \frac{p-1}{2} \pmod{p}$ as representative system.)

FREQUENTLY USED SYMBOLS

$\mathfrak{G}$	GROUP (p. 1)
$\mathfrak{U}$	SUBGROUP (p. 10)
$\mathfrak{G}:\mathfrak{U}$	INDEX of $\mathfrak{G}$ with respect to $\mathfrak{U}$ = number of left (right) cosets (p. 10)
$\mathfrak{K}$	COMPLEX = subset of a group (p. 19)
$\mathfrak{K}^x$	Complex transformed by x = set of all xKx^{-1} (p. 25)
$N_{\mathfrak{K}}$	NORMALIZER of $\mathfrak{K}$ = group of all x which transform $\mathfrak{K}$ into itself (p. 26)
$Z_{\mathfrak{K}}$	CENTRALIZER of $\mathfrak{K}$ = group of all x which are permutable with every element of $\mathfrak{K}$ (p. 50)
$\mathfrak{N}$	NORMAL SUBGROUP = subgroup which is transformed into itself by all elements (p. 23)
$\mathfrak{G}/\mathfrak{N}$	FACTOR GROUP of $\mathfrak{G}$ over $\mathfrak{N}$ = group of cosets of $\mathfrak{G}$ by $\mathfrak{N}$ (p. 38)
$\mathfrak{z}$	CENTER OF $\mathfrak{G}$ = group of all elements commuting with every element of $\mathfrak{G}$ (p. 27)
$J_{\mathfrak{G}}$	Group of all INNER AUTOMORPHISMS (transformations) of $\mathfrak{G}$ (p. 48)
$A_{\mathfrak{G}}$	Group of all AUTOMORPHISMS of $\mathfrak{G}$ (p. 48)
A/J	Group of OUTER AUTOMORPHISMS of $\mathfrak{G}$ (p. 48)
Φ	A subgroup of $\mathfrak{G}$ = intersection of $\mathfrak{G}$ with its maximal subgroups (p. 49)
$(a,b)=aba^{-1}b^{-1}$	COMMUTATOR of a with b (p. 18)
$(a,b,c)=(a,(b,c))$	(p. 81)
$(\mathfrak{U},\mathfrak{V})$	mutual COMMUTATOR GROUP = group of all (U, V) (p. 81)
$\mathfrak{G}'=D\mathfrak{G}=(\mathfrak{G},\mathfrak{G})$	COMMUTATOR GROUP of $\mathfrak{G}$ (p. 67)
$\mathfrak{G}/\mathfrak{G}'$	FACTOR COMMUTATOR GROUP (p. 67)
$D^i\mathfrak{G}=D(D^{i-1}\mathfrak{G})$	i-TH DERIVATIVE of $\mathfrak{G}$ (p. 79)
k	degree of METABELIAN group $\mathfrak{G}$, so that $D^{k-1}\mathfrak{G}\neq D^k\mathfrak{G}=e$ (p. 79)
$\mathfrak{G}=\mathfrak{Z}_1\supseteq\mathfrak{Z}_2\supseteq\mathfrak{Z}_3\cdots$	DESCENDING CENTRAL SERIES (p. 155) so that
$\mathfrak{Z}_i=(\mathfrak{G},\mathfrak{Z}_{i-1})$	is the i-th Reidemeister commutator group
$e=\mathfrak{z}_0\subseteq\mathfrak{z}_1\subseteq\mathfrak{z}_2\cdots$	ascending central series (p. 50) so that
$\mathfrak{z}_i$	is the i-th center of $\mathfrak{G}$, hence $\mathfrak{z}_i/\mathfrak{z}_{i-1}$ is the center of $\mathfrak{G}/\mathfrak{z}_{i-1}$
c	Class of the nilpotent group $\mathfrak{G}$, hence $\mathfrak{z}_{c-1}\neq\mathfrak{z}_c=\mathfrak{G}$ and $\mathfrak{Z}_c\neq\mathfrak{Z}_{c+1}=e$
S_p	is a SYLOW p-GROUP of $\mathfrak{G}$ (p. 135)
N_p	Normalizer of S_p (p. 135)
z_p	Center of S_p (p. 135)
$d(\mathfrak{G})$	The minimal number of independent generators of $\mathfrak{G}$ (p. 141)
$k_d=(p^d-1)\,(p^{d-1}-1)\ldots(p-1)$ (p. 142)	
$\mathfrak{G}'(p)$	p-commutator group = intersection of all normal subgroups with abelian p-factor group (p. 158)

$\mathfrak{G}/\mathfrak{G}'(p)$	p-factor commutator group = maximal abelian p-factor group (p. 158)
$\mathfrak{D}_p$	Intersections of all normal subgroups with index a power of p (p. 159)
$\mathfrak{G}/\mathfrak{D}_p$	Maximal p-factor group (p. 159)
$\in$	$x \in \mathfrak{G}$ means: x is an element of $\mathfrak{G}$
$\subset$	$\mathfrak{U} \subset \mathfrak{G}$ means: $\mathfrak{U}$ is a proper subgroup of $\mathfrak{G}$
$\cup$	$\mathfrak{U} \cup \mathfrak{V}$ is the sum of the sets $\mathfrak{U}$ and $\mathfrak{V}$
$\cap$	$\mathfrak{U} \cap \mathfrak{V}$ is the intersections of $\mathfrak{U}$ and $\mathfrak{V}$
$\subseteq$	$\mathfrak{U} \subseteq \mathfrak{G}$ ($\mathfrak{G} \supseteq \mathfrak{U}$) means: $\mathfrak{U}$ is a subgroup of $\mathfrak{G}$
$\triangleleft$	$\mathfrak{U} \triangleleft \mathfrak{G}$ ($\mathfrak{G} \triangleright \mathfrak{U}$) means: $\mathfrak{U}$ is a normal subgroup of $\mathfrak{G}$
$\triangleleft\triangleleft$	$\mathfrak{U} \triangleleft \triangleleft \mathfrak{G}$ ($\mathfrak{G} \triangleright \triangleright \mathfrak{U}$) means: $\mathfrak{U}$ is a subnormal subgroup of $\mathfrak{G}$
$\{\mathfrak{U}, \mathfrak{V}\}$	or $\langle \mathfrak{U}, \mathfrak{V} \rangle$ denotes the subgroup generated by the two subgroups $\mathfrak{U}, \mathfrak{V}$ of a group
$\times$	$\mathfrak{A} \times \mathfrak{B}$ denotes the direct product of the groups $\mathfrak{A}, \mathfrak{B}$
$*$	$\mathfrak{A} * \mathfrak{B}$ denotes the free product of the groups $\mathfrak{A}, \mathfrak{B}$
$\leq$	$a \leq b$ ($b \geq a$ or $a\mathsf{C}b$) means: the poset element a is contained in the poset element b
$<$	$a < b$ means: $a \leq b$, but not $b \geq a$; also denoted by: $a > b$
K	$a\mathsf{K}b$ means: the lattice element a is Kurosh-invariant in the lattice element b
N	$a\mathsf{N}b$ means: a is normal in b
NN	$a\mathsf{NN}b$ means: a is subnormal in b
J	$a\mathsf{J}b$ denotes the join of a and b
M	$a\mathsf{M}b$ denotes the meet of a and b
a/b	denotes the factor lattice of a over b

BIBLIOGRAPHY

Any one who is interested in obtaining a view of the entire field of group theory should consult the article by Wilhelm Magnus "Allgemeine Gruppentheorie" which appears in the *Enzyklopädie der Mathematischen Wissenschaften,* Band I, 1. Heft 4, Teil 1, No. 9, pp. 1-51, 1939. The article written for the earlier edition of the *Enzyklopädie* by Burkhardt (IA. 6, pp. 208-226) is also of interest. Both of these articles contain generous bibliographies.

Texts on Group Theory

English

BIRKHOFF and MACLANE. *A Survey of Modern Algebra,* New York, 1941.

BURNSIDE, W. *Theory of Groups of Finite Order,* Cambridge, 1897. Second edition, 1911.

CARMICHAEL, R. D. *Introduction to the Theory of Groups of Finite Order,* Boston, 1937.

HILTON, H. *An Introduction to the Theory of Groups of Finite Order,* Oxford, 1908.

KUROSH, A. G. *The Theory of Groups,* I *and* II, New York, 1956. (English translation of *Teoriya Grupp,* 2nd ed.)

LEDERMAN, W. *Introduction to the Theory of Finite Groups,* New York, 1949.

MATHEWSON, L. C. *Elementary Theory of Finite Groups,* Boston, 1930.

MILLER, BLICHFELDT and DICKSON. *Theory and Application of Finite Groups,* New York 1916. Second edition, 1938.

VAN DER WAERDEN, B. L. *Modern Algebra* I, New York, 1949. (English translation of *Moderne Algebra* I.)

German

BAUMGARTNER, L. *Gruppentheorie,* Berlin, 1921 (Sammlung Göschen No. 837).

SPEISER, A. *Die Theorie der Gruppen von endlicher Ordnung,* Berlin, 1937, also New York, 1945.

VAN DER WAERDEN, B. L. *Moderne Algebra* I. 3rd ed. Berlin, 1937.

French

DUBREIL, P. *Algèbre,* Paris, 1946.

GALOIS, E. *Oeuvres mathématiques,* Paris, 1897, pp. 25-61.

JORDAN, C. *Traité des substitutions,* Paris, 1870.

DE SEGUIER, J. A. *Théorie des groupes finis.* Vol. I, Paris, 1904. Vol. II, Paris, 1912.

SERRET, J. A. *Cours d'algébre supérieure.*Vol. II, Paris, 7th ed. 1928.

Russian

GRAVE, P. A. *Theory of Finite Groups,* Kiew 1908.

SCHMIDT, O. J. *Abstract Group Theory,* Kiew 1916. 2nd ed. 1933.

Japanese

SONO, S. *Group Theory,* Tokyo, 1928.

Another general reference should be mentioned: the two volumes of G. A. MILLER *Collected Works,* University of Illinois Press, 1934. The remaining references are not general but are either direct sources or amplifications of the topics taken up in this book. They are listed according to chapter and section.

CHAPTER I

§ 1. P. LORENZEN. *Ein Beitrag zur Gruppenaxiomatik.* Math. Zeit. 49 (1944) p. 313-327.

K. PRACHAR. *Zur Axiomatik der Gruppen.* Akad. Wiss. Wien. S-B IIa. 155 (1947) p. 97-102.

§ 10. G. FROBENIUS. *Über einen Fundamentalsätz der Gruppentheorie*. Berl. Sitz. 1903 p. 987, and 1907 p. 428.

P. DUBUQUE. *Une généralisation des théorèmes de Frobenius et Weisner*. Rec. Math. (Moscow) 5(47) (1939) p. 189-196.

C. S. FU. *On Frobenius' theorem*. Quart. J. Math. Oxford. Ser. 1. 17 (1946) p. 253-256.

B. A. KRUTIK. *Über einige Eigenschaften der endlichen Gruppen*. Rec. Math. (Math. Sbornik) N. S. 0152 (1942) p. 239-247.

S. LUBELSKI. *Verallgemeinerung eines Frobeniusschen gruppentheoretischen Satzes*. Act. Acad. Lima 6 (1945) p. 133-137.

CHAPTER II

§ 4. Definition and discussion of the principal properties the subgroup from Frattini. See: Miller, Blichfeldt, Dickson, *op. cit.* p. 52. Also see: B. NEUMANN, *Some remarks on infinite groups*. Lond. Math. Soc. 12 (1937), and O. ORE, *Contributions to the theory of groups of finite order*. Duke Math. J. 5 (1934) p. 431-460.

Paper on the holomorph of a group: GARRETT BIRKHOFF. A theorem on transitive groups. Proc. Camb. Phil. Soc. 29 (1933) p. 257-259.

§ 5. The Jordan-Hölder Theorem: C. JORDAN. Journal de math. (2) 14 (1869) p. 139, shows that the orders of the factors in different composition series are the same.

O. HÖLDER, Math. Ann. 34 (1889) shows that the composition factors of a finite group are unique up to order in the sense of isomorphism.

SCHREIER. *Über den Jordan-Hölderschen Satz*. Hamb. Abh. 6 p. 300, gives the definition of refinement and the refinement theorem.

H. ZASSENHAUS. *Zum Sätz von Jordan-Hölder-Schreier*. Hamb. Abh. 10:106. gives the explicit procedure for obtaining refinements with the theorem on four groups.

O. ORE. *Structures and group theory*. Duke Math. Journal, Part I, Vol. 3, p. 149-174, and Part II, Vol. 4, p. 247-269, give the general structure theory.

A. KUROSH. Composition systems in infinite groups. Rec. Math. (Mat. Sbornik) N. S. 16 (58), p. 59-72 (1945).

§5. For lattice theory in general, see: GARRETT BIRKHOFF, *Lattice Theory*, New York, 1940, 2nd ed., 1948.

Concerning Kurosh invariance, see: A. G. KUROSH, *The Jordan-Hölder Theorem in arbitrary lattices*, Sbornik Pamyati Grave (1940) pp. 110-116.

§ 6. The definition of commutator form and the theorems come from P. HALL. *A contribution to the theory of groups of prime-power orders*. Proc. Lond. Math. Soc. II 36 (1933) p. 29-95.

R. BAER. The higher commutator subgroups of a group. Bull. Amer. Math. Soc. 50 (1944) p. 143-160.

For the extension of the notion of solvability to the infinite case see: R. BAER. *Nilpotent groups and their generalizations*. Trans. Amer. Math. Soc. 47 (1940) p. 393-434.

H. FITTING. In the Jahr. Deutsch. Math. Verein. 48 (1938) p. 77-141.

K. A. HIRSH. *On infinite solvable groups*. Proc. Lond. Math. Soc. (2), Part I, 44 (1938) p. 53-60; Part II, 44 (1938) p. 336-344; Part III, 49 (1946) p. 184-194.

P. HALL. *The construction of soluble groups*. J. Reine Angew. Math. 182, p. 206-214 (1940).

W. MAGNUS. *Neuere Ergebnisse über auflösbare Gruppen*. Jahr. D. Math. Verein. 47 (1937), p. 69-78.

O. SCHMIDT. *Infinite solvable groups*. Rec. Math. (Mat. Sb.) N. S. 17 (59) (1945), p. 363-375.

§ 7. Theorem 16: E. WITT, *Über die Kommutativität endlicher Schiefkörper*. Hamb. Abh. 8 (1931), p. 413.

Almost fields: ZASSENHAUS, *Über endliche Fastkörper*. Hamb. Abh. 11, p. 187-220.

Rings of matrices: N. JACOBSON, *The Theory of Rings*. Math. Surveys 2, Amer. Math. Soc. (1943). See especially the bibliography.

CHAPTER III

§ 2. See the paper of Fitting cited in the footnote.

F. KIOKEMEISTER. *A note on the Schmidt-Remak Theorem.* Bull. Amer. Math. Soc. 53 (1947) p.957-958.

O. ORE. *A remark on the normal decompositions of groups.* Duke Math. J. 5 (1939) p. 172-173.

§§ 3-5. IYANAGA. *Zum Beweis des Hauptidealsatzes.* Hamb. Abh. 10 (1934).

§§ 6-8. SCHREIER. *Über die Erweiterung von Gruppen.* Teil I: Monatshefte für Math. u. Phys., Bd. 34. Teil II: Hamb. Abh. 4, p. 321.

R. BAER. *Erweiterung von Gruppen und ihren Isomorphismen.* Math. Zeit. 38 (1934) p. 375-416. *Groups with abelian central quotient group.*

Trans. Amer. Math. Soc. 44 (1938) p. 357-386.

Y. A. GOL'FAND. *On an isomorphism between extensions of groups.* Doklady Ak. Nauk. S.S.S.R. (N. S.) 60, 1123-1125 (1948).

S. EILENBERG and S. MACLANE. *Group extensions and homology.* Ann. of Math. (2) 43 (1942) p. 757-831. *Cohomology theory in abstract groups.* I. Ann. of Math. (2) 48 (1947) p. 51-78. II. (Group extensions with a non-abelian kernel) Ann. of Math. (2) 48 (1947) p. 326-341.

B. H. NEUMANN. *Adjunction of elements to groups.* J. Lond. Math. Soc. 18 (1943) p. 4-11.

K. SHODA. *Über die Schreiersche Erweiterungstheorie.* Proc. Imp. Acad. Tokyo 19 (1943) p. 518-519.

§ 9. See IYANAGA. *op. cit.*

CHAPTER IV

Theorem 13: WIELAND. Math. Zeit. 41 (1936).

§ 1. R. BAER. *Sylow theorems for infinite groups.* Duke Math. J. 6 (1940) p. 598-614.

A. P. DIETZMAN (Dicman). *On an extensions of Sylow's theorem.* Ann. of Math. (2) 48 (1947) p. 137-146. And *On Sylow's theorem*: Doklady Akad. Nauk S.S.S.R. (N. S.) 59 (1948), p. 1235-1236.

P. GOLDBERG. The Silov p-groups of locally normal-groups. Rec. Math. (Math Sbornik) N. S. 19 (61) (1946) p. 451-460.

H. WIELANDT. *p-Sylowgruppen und p-Faktorgruppen.* J. Reine Angew. Math. 182 (1940), p. 180-193.

§ 3. See Baer reference under II, 6. Also Fitting and P. Hall.

O. SCHMIDT. Über unendliche spezielle Gruppen. Rec. Math. (Mat. Sbornik) N. S. 8 (59) (1940) p. 363-375.

§§ 4-5. See first cited paper of P. Hall.

CHAPTER V

§ 1. G. HANNINK. *Verlagerung und Nichteinfachheit von Gruppen.* Monatsh. Math. Phys. 50 (1942) p. 207-233.

§ 2. O. GRÜN. *Beiträge zur Gruppentheorie.* I. Crelle (1935) 1.

§ 3. ZASSENHAUS. *Über endliche Fastkörper, op. cit.*

§ 4. IYANAGA. *op. cit.*

E. WITT. *Bemerkungen zum Beweis des Hauptidealsatzes von S. Iyanaga.* Hamb. Abh. 11 p. 221.

APPENDIX B AND G

H. WIELANDT, *Eine Verallgemeinerung der invarianten Untergruppe,* Math. Zeit. 45 (1939) pp. 209-244.

APPENDIX D

Exercises 25-31. RIGUET, *Relations binaires,* Bull. Soc. Math. France 76 (1948), pp. 114-155; RIGUET, *Quelques proprietes des relations difonctionelles,* C. R. Acad. Sci. Paris 230 (1950), pp. 1,999-2,000; also, a forthcoming paper of J. LAMBEK (Canadian Journal of Math.)

AUTHOR INDEX

INDEX